CAMBRIDGE LIBRARY COLLECTION

*Books of enduring scholarly value*

## History

The books reissued in this series include accounts of historical events and movements by eye-witnesses and contemporaries, as well as landmark studies that assembled significant source materials or developed new historiographical methods. The series includes work in social, political and military history on a wide range of periods and regions, giving modern scholars ready access to influential publications of the past.

## A History of the Royal Society

The Royal Society has been dedicated to scientific inquiry since the seventeenth century and has seen a long line of illustrious scientists and thinkers among its fellowship. The society's Assistant Secretary and Librarian, Charles Richard Weld (1813–1869), spent four years writing this two-volume *History of the Royal Society*, published in 1848, which also includes illustrations by his wife, Anne. Weld's aim was to document the 'rise, progress, and constitution' of the society. He charts how the informal meetings of like-minded men engaged in scientific pursuits in the mid-1600s developed into a prestigious society that by 1830 counted as one of the world's most influential scientific institutions. Volume 1 covers the period to 1755, describing the society's origins and key moments in its growth, with a focus on its governance, benefactors and organisation. It also contains biographies of presidents including Samuel Pepys and Isaac Newton.

Cambridge University Press has long been a pioneer in the reissuing of out-of-print titles from its own backlist, producing digital reprints of books that are still sought after by scholars and students but could not be reprinted economically using traditional technology. The Cambridge Library Collection extends this activity to a wider range of books which are still of importance to researchers and professionals, either for the source material they contain, or as landmarks in the history of their academic discipline.

Drawing from the world-renowned collections in the Cambridge University Library, and guided by the advice of experts in each subject area, Cambridge University Press is using state-of-the-art scanning machines in its own Printing House to capture the content of each book selected for inclusion. The files are processed to give a consistently clear, crisp image, and the books finished to the high quality standard for which the Press is recognised around the world. The latest print-on-demand technology ensures that the books will remain available indefinitely, and that orders for single or multiple copies can quickly be supplied.

The Cambridge Library Collection will bring back to life books of enduring scholarly value (including out-of-copyright works originally issued by other publishers) across a wide range of disciplines in the humanities and social sciences and in science and technology.

# A History of the Royal Society

## *With Memoirs of the Presidents*

VOLUME 1

CHARLES RICHARD WELD

CAMBRIDGE UNIVERSITY PRESS

Cambridge, New York, Melbourne, Madrid, Cape Town, Singapore,
São Paolo, Delhi, Dubai, Tokyo, Mexico City

Published in the United States of America by Cambridge University Press, New York

www.cambridge.org
Information on this title: www.cambridge.org/9781108028172



This edition first published 1848
This digitally printed version 2011

ISBN 978-1-108-02817-2 Paperback

P. Stopford del. J. W. Lowry fc.

MEETING ROOM OF THE ROYAL SOCIETY, CRANE COURT.

# A HISTORY OF THE ROYAL SOCIETY,

WITH

MEMOIRS OF THE PRESIDENTS.

COMPILED FROM AUTHENTIC DOCUMENTS

BY

CHARLES RICHARD WELD, ESQ.,

*BARRISTER-AT-LAW;*

ASSISTANT-SECRETARY AND LIBRARIAN TO THE ROYAL SOCIETY.

IN TWO VOLUMES.

VOLUME THE FIRST.

LONDON:
JOHN W. PARKER, WEST STRAND.

M.DCCC.XLVIII.

Cambridge:
Printed at the University Press.

TO

THE MOST HONOURABLE

SPENCER JOSHUA ALWYNE COMPTON,

MARQUIS OF NORTHAMPTON,

*PRESIDENT;*

AND TO

THE COUNCIL AND FELLOWS

OF

THE ROYAL SOCIETY:

THIS HISTORY

IS MOST RESPECTFULLY DEDICATED.

# PREFACE.

IT will probably be expected, and not unreasonably, that some account should be given of the motives and circumstances which have called this work into existence;—and the Author readily avails himself of the privilege accorded by long custom to introduce here a few explanatory observations.

This is the more necessary as some readers will recall to mind three works already existing, each entitled, *History of the Royal Society;* and it might be considered that a fourth book on the subject was scarcely, if at all, required. But the most superficial examination of these *Histories*, will show how very deficient they are in information relating to the rise and progress of the Society. Bishop Sprat's work, published in 1667, manifestly cannot be regarded as a history of the Society, seeing that the Institution had only just been organized when the book was written. It even fails to give us a satisfactory account of the origin, or the early proceedings of the Society; the author having laboured much more diligently to defend the Fellows from the attacks and criticism of Aristotelian philosophers, than with any other object: he tells us, indeed, that "the objections and cavils of the detractors of so noble an Institution, did make it necessary for him to write of it, not in the way of a plain history, but as *an apology*."

The next *History of the Society*, is that by Dr. Birch, in four quarto volumes, published in 1756. Here again, the work fails to redeem the promise of its Title, for although occupying so large a space, it breaks off at the year 1687, and treats only of the scientific proceedings of the Society, with a reproduction of many papers which were read at the Meetings, and printed in the *Philosophical Transactions.*

The third publication, which was written by Dr. Thomson, appeared in 1812; and this again, although styled *History of the Society*, is, as the author states, "an attempt to elucidate the *Philosophical Transactions:*" in fact, the entire volume is filled with rapid sketches of the progress of science, and an analysis of the papers in the *Transactions.*

It is thus evident that no work, marking the Society's progress from a period antecedent to its incorporation until our own time, exists; for it would be vain to seek in the above *Histories* for information respecting the endowments of the Society, or indeed any other fact connected with its civil history.

The want of such a publication suggested itself very soon after my being appointed to the offices which I now hold, from the difficulty of replying to questions relative to the early events and constitution of the Society, without first making a laborious search through the Archives. It consequently appeared to me, that a thorough acquaintance with the Society's civil history would increase my efficiency as Assistant-Secretary; and this could only be arrived at by a diligent examination of the voluminous records under my charge.

The circumstances, however, which more immediately gave rise to this work, were briefly these:—Having received instructions, immediately after my appointment, to visit the Society's estate at Acton, respecting some legal difficulty of tenure, the details of which I was ignorant of, and had not leisure at the time to ascertain; it occurred to me, while riding down to the property, that some account of the Society, containing at least every fact of importance relating to it, would be useful to the Fellows, and might at the same time prove interesting to a considerable portion of the scientific world. The idea thus conceived, soon assumed a more definite character, and but little time elapsed before it was acted on. But I felt it to be my duty, as it was my inclination, to consult Dr. Roget, Senior-Secretary to the Society, on the subject; and meeting with the warmest encouragement from him, and other officers of the Society, I commenced the undertaking, which, it may be as well to state, has occupied the principal portion of my leisure hours for nearly four years.

The examination of the Archives, the Journal, Register, and Council-books, comprising some hundreds of volumes, with several thousand letters, was a formidable task; but I soon found that the work could not be compiled from these documents alone. It was the custom, in the early days of the Society, for the Secretaries to have the custody of the books and papers, many of which, on their decease, were not returned to the Society by their executors, and have since been presented to the British Museum; a locality, it may be observed, far less appropriate for their preservation than the Royal Society's library, to which, indeed, they in justice belong. Thus several volumes of Hooke's papers are in the National Library, besides letters and other documents written by Oldenburg, Wallis, Wren, Sloane, &c. To these it became, of course, important to refer; and such use has been made of them as was necessary for the purposes of this work. The State-Paper Office, the Archives of the Lord Chamberlain's Office, and the Bodleian Library, which I also visited, have furnished me with valuable matter.

From this statement it will be understood that my main object has been to render a faithful account of the rise, progress, and constitution of the Royal Society, and to record its most important proceedings. I need scarcely observe that the work partakes more of a civil, than of a scientific character; and indeed it is my earnest wish that it should be regarded only as a contribution towards some future philosophical history of the Society, which, proceeding from an abler pen than mine, shall at once embrace the entire subject.

Scientific matters, it is true, are occasionally treated of in the present work, but only in an historical light;—for example, when the Dean and Chapter of St. Paul's requested the Society to protect their Cathedral from lightning, the manner in which the protection was effected is recorded; but no attempt is made to enter into the argument, whether or not the means employed were based on the soundest philosophical principles:—this manifestly appertains to another branch of inquiry, and if carried out in cases of a similar nature, would require far more space than two volumes can afford. Thus, all the great scientific labours, which originated either in the Society as a body, or from its Fellows individually, will be found historically narrated and elucidated in every case, as far as possible, by original and authentic documents from the Society's Archives.

A sketch of the revival of literature and science in Italy, and the development of scientific institutions in that country was deemed a fitting introduction to that of the Royal Society; and especially as it might assist in forming a correct estimate of the labours of the small band of truth-seeking philosophers, who founded an association which has acquired world-wide renown, and whose members have probably done more than any other body of men to benefit the community by rendering science available for the practical purposes of life.

A few words require to be said with respect to the biographies of some of the Presidents, which may not be thought so full as could be desired; but having devoted much time in search of information respecting them, with generally very indifferent success, I could only arrive at the conclusion, that the subjects of my research were persons who had done little in practical science, and therefore did not enjoy an extended reputation. Happily this remark applies to but two or three of the number; with the rest, the difficulty consisted more in condensing than in collecting the materials at command.

I am well aware that omissions, though not I trust of any great consequence, will be detected; but without attempting apologies for the sake of conciliating the criticism of the reader, which is his privilege as it is his right, I must remind him, that it was considered more judicious to bring these volumes into a compass of reasonable size, than to extend them to dimensions, which, in these days, could hardly be expected to meet with approval.

It has been deemed advisable to close the history of the Society, with the election of the Duke of Sussex, in 1830. But this year, considered with reference to a period of very nearly two centuries, may be said to belong to our own times; and the contents of the concluding Chapter, with the publication of the recently amended Statutes, will enable the reader to form a correct idea of the present state and constitution of the Society.

SOMERSET HOUSE,
*June*, 1848.

---

# CONTENTS

OF

# THE FIRST VOLUME.

---

## CHAPTER I.



## CHAPTER II.

## CHAPTER III.



## CHAPTER IV.



## CHAPTER V.

## CHAPTER VI.



## CHAPTER VII.



## CHAPTER VIII.

## CHAPTER IX.



## CHAPTER X.

## CHAPTER XI.



## CHAPTER XII.



## CHAPTER XIII.

## CHAPTER XIV.



## CHAPTER XV.

## CHAPTER XVI.



## CHAPTER XVII.

# ILLUSTRATIONS.

# HISTORY

OF

# THE ROYAL SOCIETY.

## CHAPTER I.

State of Literature and Science in Europe during the Twelfth and Thirteenth Centuries—Prevalence of Ecclesiastical Authority—Its baneful effect on Learning—Labours of Barlaam, Petrarch, and Boccaccio—Their zeal for Classical Literature—Restoration of Classical Learning in Italy—Discovery of Ancient Manuscripts—Patronage of learning by the Medici family—Philosophy not advanced—Roger Bacon—Other English Philosophers—Leonardo da Vinci—His great philosophical acquirements—Establishes an Academy of Arts at Milan—Theoretical Reformers of Science in the Sixteenth Century—Cæsalpinus—Campanella—Ramus—Bruno—Galileo—Francis Bacon—Establishment of Scientific Societies—Institution of the Academia Secretorum Naturæ—Academy at Venice—Accademia dei Lyncei—Della Crusca—Del Cimento—Great number of Academies and Scientific Institutions founded in Italy—Their quaint Titles—Establishment of a Society of Antiquaries in England in 1572—Its dissolution by James I.—Proposition for a Royal Academy in England—Curious Scheme for its Incorporation—Charles I. grants a Special License for establishing a Scientific Institution styled 'Minerva's Museum'—Objects of the Institution—Duties of the Professors—Scientific and Literary Societies in Germany—Their brief existence—Number of Books printed in the principal Cities of Europe from the Invention of Printing to the beginning of the Sixteenth Century—Establishment of the French Academy—Its subsequent Incorporation with the Academy of Sciences—First Institution for the Investigation of Science out of Italy, established in England—Laplace's Opinion of Scientific Societies.

ON the fair land of Italy rose the intellectual sun, whose vivifying rays penetrated, and eventually dispersed, the gloomy clouds of deep ignorance and superstition, that hung over Europe from the decline of the Roman Empire.

The long interval, from the overthrow of Roman civilization to the fourteenth century, is mainly characterized by the absolute reign of ecclesiastical authority. In common with all earthly powers, Philosophy was compelled to bow before the altar of Church despotism, and was solely employed in the defence of a subtile and mystical theology.

"It was an offence" says Dr. Whewell, "against religion as well as reason, to reject the truth; and the truth could be but one. In this manner arose that claim, which the doctors of the Church put forth, to control men's opinions upon all subjects[1]." Tennemann, in his *Geschichte der Philosophie*, mentions a papal rescript, admonishing the members of an university to "be content with the landmarks of science already fixed by their fathers, to have due fear of the curse pronounced against him who removeth his neighbour's landmark, and not to incur the blame of innovation or presumption[2]."

"The Italians in the thirteenth century," remarks Professor Ranke, "manifested a grand tendency towards searching investigation, intrepid pursuit of truth, noble aspirations, and high prophetic visions of discovery. Who shall say whither this tendency would have led? But the Church marked out a line which they were not to overstep,—woe to him who ventured to pass it[3]!"

Barlaam, a Calabrian monk, and afterwards Bishop of Locri, had considerable influence in reviving a

---

[1] *Phil. Ind. Sc.*, Vol. II. p. 151. [2] Vol. VIII. p. 461.
[3] *Hist. of the Popes*, Vol. I. p. 502.

taste for classical literature in Italy. The Calabrian Churches were long attached to the throne of Constantinople, and several of the monks of St. Basil pursued their studies on mount Athos, and the schools of the East. Barlaam visited Constantinople in the early part of the fourteenth century; and, when sent, in 1339, as ambassador to Pope Benedict the Twelfth, had the good fortune to be the first to revive beyond the Alps the memory, or rather the writings, of Homer[4]. According to Petrarch and Boccaccio, he was a man far in advance of the age in which he lived, and gifted with profound learning and genius[5].

At the court of Avignon he formed an intimate acquaintance with Petrarch, and had considerable influence on the studies of this immortal poet[6], who stands prominently forward as one of the great regenerators of literature; the pioneer and handmaid to science. At Petrarch's recommendation, the Republic of Florence founded a chair of Grecian literature in 1360[7]; and we are informed that the highest honour was attached to classical learning, and that Petrarch, Dante, and Boccaccio, obtained a degree of celebrity for their erudition unequalled by the schoolmen of the middle ages.

These men, says Tennemann, *waren die ersten Dichter von ausgezeichneter ôriginalitat, die zugleich mit hoher Achtung, gegen die Classiker erfüllt waren, und den Enthusiasmus verbreiteten, mit welchem die griechische Literatur aufgenommen wurde*[8].

---

[4] Gibbon, *Dec. and Fall of the Rom. Em.*, Vol. XII. p. 116.

[5] Boccaccio, *de Genealog. Deorum*, l. xv. c. 6.

[6] *Mém. sur la Vie de Petrarque*, Tom. I. p. 406.

[7] Sismondi, *Ital. Rep.*

[8] *Gesch.*, Vol. IX. p. 15.

The Italians may, indeed, be said to have discovered as it were anew the ancient world of literature; and a passion for erudition spread throughout the land, with an ardour and rapidity highly remarkable, when contrasted with the ignorance of the preceding centuries[9].

The wealthy inhabitants of the principal Italian cities became ardent cultivators of literature and philosophy, in the true original meaning of this word, and all Europe was ransacked to discover the manuscripts of classical authors. The success of the search, however, was greatly impeded by the monks, whose interest it was to keep the literary treasures to themselves, as the transcribing of them was a source of considerable emolument. Poggio, a contemporary of Petrarch, has given us a melancholy account of the barbarous ignorance of the monastic possessors of some of the ancient manuscripts. When attending the Council at Constance, he visited the Convent of Saint Gallo, situated about twenty miles from that town, where, he was informed, certain manuscripts were kept, and this, on examination, proved to be the fact. His words record graphically the condition in which they were found: *Tra una grandissima copia di libri dic' egli che lungo sarebbe l'annoverare, trovammo un Quintiliano ancor sano e salvo, ma pien di polvere e d'immondezza; perciocchè eran que' libri nella biblioteca, non com' il loro onor richiedeva, ma sepolti in una oscura e tetra prigione, cioè nel fondo di una torre, in cui non si getterebbon nemmeno i*

[9] Hallam says, "Petrarch was more proud of his Latin poem called *Africa*, than of the sonnets and odes which have made him immortal."

*dannati a morte*[10]. In this dungeon of a turret, which formed part of the convent, Poggio discovered, besides the Quintilian, several rare manuscripts, including a portion of the Argonautics of Valerius Flaccus[11].

There is a story told by Chapelain, the poet, that the tutor of a Marquis di Rouville, having sent to Saumur for some rackets, found upon the parchment composing them, the titles of the 8th, 10th, and 11th Decads of Livy: on applying to the racket-maker, the latter stated that a pile of parchment volumes, some of which contained the history of Livy, had been procured from the abbess of Fontevrault, and that out of these he had made *une multitude très grande de battoirs.* Exaggerated as this story may be, it certainly derives some degree of credibility from the well-known fact, that Sir Robert Cotton rescued the original Magna Charta from the hands of a tailor, who was on the point of cutting it up for measures. Brighter days were, however, at hand: literature and the arts and sciences found a congenial abode at Florence, and patrons in the Medici family. Cosmo, the *Pater Patriæ*, born in 1389, and created Prior in 1416, was the head of a commercial establishment, which had counting-houses in all the great cities of Europe, and in the Levant. He cultivated literature with ardour; and his palace, one of the most sumptuous in Florence, was the habitual resort of artists, poets, and learned men. At the suggestion of Germistus Pletho, a native of the Morea, Cosmo founded a Platonic Academy in Florence, at first, for

---

[10] Tiraboschi, *Stor. del Lit. Ital.*, Vol. VI. p. 121.

[11] Aurispa was also very successful in his search for ancient MSS. In 1423 he brought home to Venice 238 volumes. Tiraboschi, Vol. VI. p. 102.

the purpose of studying the works of Plato; though, subsequently, Italian literature, and particularly the study and explanation of the poetry of Dante were added. This Academy was greatly patronized by Lorenzo de' Medici, who was in the habit of giving parties at his villa at Fiesole, when certain portions of the works of Plato were discussed[12]. "Nor was mere philology, the sole or leading pursuit to which so truly noble a mind accorded its encouragement. He sought in ancient learning something more elevated than the narrow, though necessary, researches of criticism. In a villa overhanging the towers of Florence, on the steep slope of that lofty hill crowned by the mother-city, the ancient Fiesole, in gardens which Tully might have envied, with Ficino, Landino, and Politian at his side, he delighted his hours of leisure with the beautiful visions of Platonic philosophy, for which the summer stillness of an Italian sky appears the most congenial accompaniment[13]."

Philosophy, however, during the thirteenth, fourteenth, and fifteenth centuries, was, as far as progress is concerned, literally at a stand still. As Dr. Whewell remarks: "The same knots were tied and untied; the same clouds were formed and dissipated[14]."

Our countryman, Roger Bacon, a giant of profound understanding in his generation, must be cited as an exception to the general class of schoolmen—to this general "opake of nature and of soul." Born as early as 1214, his learning was such as to gain him the title of *Doctor Mirabilis* from his brother Franciscan monks; and his *Opus Majus*, addressed

---

[12] Roscoe, *Life of Lorenzo de' Medici.*

[13] Hallam, *Int. to the Lit. of Eur.*, Vol. I. p. 243.

[14] *Hist. Ind. Sci.*, Vol. I. p. 340.

to Pope Clement IV., displays a mode of philosophizing far in advance of the age in which he lived. This is particularly evident in the 6th part, entitled, *De scientia experimentali*[15].

But the labours of Bacon brought forth no fruit. "Some change, disastrous to the fortunes of science, must have taken place about 1230, soon after the foundation of the Dominican and Franciscan orders. Nor can we doubt that the adoption of the Aristotelian philosophy by these two orders, was one of those events which most tended to defer for three centuries, the reform which Roger Bacon urged, as a matter of crying necessity, in his own time[16]."

Other English names might be adduced to prove that this country was not deficient during the 13th, 14th, and 15th centuries, in men whose abilities were of a high philosophical order, though their writings do not render them so well known to posterity as Roger Bacon. Professor De Morgan, in showing that this country held a conspicuous rank in the philosophy of the middle ages, gives a list of no fewer than

[15] Roger Bacon's philosophy is well illustrated by the following passage from the above work:—*Duo sunt modi cognoscendi; scilicet per argumentum et experimentum. Argumentum concludit et facit nos concludere questionem; sed non certificat neque removet dubitationem, ut quiescat animus in intuitu veritatis, nisi eam inveniat via experientiæ; quia multi habent argumenta ad scibilia sed quia non habent experientiam negligunt ea neque vitant nociva nec persequuntur bona. Si enim aliquis homo qui nunquam vidit ignem probavit per argumenta sufficientia quod ignis comburit et lædit res et destruit, nunquam propter hoc quiesceret animus audientis nec ignem vitaret antequam poneret manum vel rem combustibilem ad ignem, ut per experientiam probaret quod argumentum edocebat; sed assumptâ experientiâ combustionis certificatur animus et quiescit in fulgore veritatis quo argumentum non sufficit, sed experientia.* p. 446.

[16] *Phil. Ind. Sci.*, Vol. II. p. 173.

ninety six English mathematical and astronomical writers between 1068 and 1599, and remarks, that he had no doubt but that it might still be enlarged; and Captain Smyth, in his *Celestial Cycle*, says, alluding to this period, that "England has contributed her full quota to the series of philosophical and zealous enquirers who have so largely opened the human intellect."

Leonardo da Vinci claims notice, as combining in a most remarkable manner, the professions of mathematician and engineer, as well as painter, sculptor, and architect. He was born in 1452, and, during a considerable portion of his life, devoted himself to science, looking to nature as the true source of knowledge. He maintained that "in the study of the sciences which depend on mathematics, those who do not consult nature, but authors, are not the children of nature, but only her grandchildren. She is the true teacher of men and genius." "But," he adds, "see the absurdity of men. They turn up their noses at a man who prefers to learn from nature herself, rather than from authors who are only her clerks[17]." His Manuscripts, which remain unpublished at Milan, and in the library of the Institute at Paris[18], abundantly attest his diligence and application, particularly to mechanics, which he styled *il paradiso delle scienze matematiche perchè con quella si viene al frutto delle scienze matematiche*[19]. Dr. Whewell states that "he is no inconsiderable figure in the prelude

[17] Venturi, *Essai sur les Ouvrages de Leonard da Vinci.*

[18] Two Volumes are at Milan, one in Paris, which was abstracted from Milan when Napoleon's army was in possession of that city.

[19] Libri, *Hist. des. Sci. Mat. en Italie*, in which many extracts from his MSS. are given.

to the great discoveries in both these sciences;" and Humboldt pronounces him "the greatest physicist of the fifteenth century, combining distinguished mathematical knowledge with the most admirable and profound insight into nature. If the physical views of Leonardo da Vinci had not remained buried in his manuscripts, the field of observation, which the new world offered, would have been already cultivated scientifically in many of its parts before the great epoch of Galileo, Pascal, and Huygens. Like Francis Bacon, and a full century before him, he regarded induction as the only sure method in natural science; *dobbiamo comminciare dall' esperienza, e per mezzo di questa scoprine la ragione*[20].

Whilst residing at Milan, under the patronage of Ludovico il Moro, he established an Academy of Arts, in which the study of Geometry was much cultivated, under the auspices of himself and of his friend Pacioli[21].

Passing on to the sixteenth century, we find the names of individuals, whose philosophic doctrines are stamped with the elements of change. Bernardinus Telesius, born at Cosenza, in the kingdom of Naples, in 1508; Andrew Cæsalpinus, born at Arezzo in 1520; Thomas Campanella, born at Stilo in Calabria, in 1568; Peter Ramus[22], born in Picardy in 1515; Giordano Bruno[23], born at Nola about 1550;

---

[20] *Cosmos,* Vol. II. p. 285.

[21] Pacioli, *Divina proportione. Amorette Memorie,* p. 40.

[22] He is celebrated as having successfully defended for an entire day the thesis;—"All that Aristotle said is not true;" in those days a dangerous experiment. Tennemann, IX. p. 420.

[23] The name of Bruno is dear to the English philosopher; at the cost of his life he maintained his enlightened scientific views, being

Galileo[24], born at Pisa in 1564, and others of less note, were all men devoted to science, whose minds were possessed with grand and vigorous ideas, and who unshrinkingly supported their philosophy at the price of imprisonment and martyrdom.

Turning to our own country, we have the proud boast of claiming Francis Bacon, whose labours probably did more to clear away the wretched systems which trammelled and shut out the light of scientific truth, than any other philosopher of his age. Of him more will be said presently.

A natural consequence of a similarity of pursuits was association; and, accordingly, we find societies springing up under the fostering hand of the new reformers.

The first Society instituted for the investigation of physical science, was that established at Naples, in the year 1560, with the name of *Academia Secretorum Naturæ*. The Members of this Academy appear to have laboured diligently under the presidency of G. Porta, but their studies were prematurely brought to a close, the Academy having been dissolved by the ecclesiastical authorities[25].

---

burnt at Rome, in February 1600, declaring to his unjust judges, as the flames were rising around him, "Your sentence strikes more fear into *your* souls than *I now* feel." Tiraboschi, XI. p. 437. Bruno had taken refuge from his persecutors in England about the year 1583, where, according to Libri, he published his *Spaccio della bestia trionfante.*

[24] It is painful to pass this great name with the mere mention of it. In an introductory chapter of this kind, space would be wanting to give even an outline of Galileo's labours. Their effect in advancing science is, happily, so well known, as to render this unnecessary.

[25] Tiraboschi, Vol. VII. p. 153.

The pontificate of Leo X. was chiefly distinguished by the encouragement given to the arts and literature. The Roman Academy flourished under his protection, and in 1514, above one hundred professors attached to it received salaries.—Independently of this Academy, Leo founded another for the sole study of the Greek language, where, according to Tiraboschi, a Greek press was established, and the scholiasts on Homer printed. "Those were happy days," says Sadolet, writing in 1529 to Angelo Colocci, "when, in your suburban gardens, or mine on the Quirinal, or in the circus, or by the banks of the Tiber, we held those meetings of learned men, all recommended by their own virtues, and by public reputation. Then it was, after a repast, which the wit of the guests rendered exquisite, we heard poems or orations recited to our great delight, productions of the ingenious Casanuova, the sublime Vida, the elegant and correct Biroaldo, and many others, still living, or now no more[26]."

We must not omit to notice the Academy established at Venice by Aldo Manuzio, for the ostensible purpose of promoting literature generally, but more immediately, as Zeno states in his *Notizie de Manuzi*, to superintend the production of the works issuing from the celebrated press of the founder, and of rendering them as perfect and elegant as possible[27]. That

---

[26] *Epist.*, p. 225, edit. 1554.

[27] This recalls to mind the method adopted by Stephens of Paris who was in the habit of hanging up his proofs of the classical authors in public places, that they might receive the corrections of the learned. It is said that his New Testament, called *O Mirificam*, (because the Preface begins with these words), has not a single fault.

this Academy laboured efficiently, and was of signal use to literature, and incidentally to science, there is no doubt.—The fame of Aldus, and his press, extended in a very short time to England. The author of a quaint old book published in the sixteenth century, entitled *Janva Lingvarum*, says, in the section treating of the means to be employed in order to write elegantly, *Primo colligendæ sunt ex autoribus qui ex professo de his tractant phrases similiter et elegantiæ, quales ad linguam Latinam habet Aldus Manutius*. The principal members of the above Academy were, Mesuro, Peter Bembo, afterwards created a Cardinal; Gabrielli, Navagero, Rinieri, Sanuto, Ramberti, and Egnazio[28]. Sicily also boasted of her Academies. Several were established in Palermo, between the years 1549 and 1588, under, as Captain Smyth informs us, the "ostensible names of *Gli Ebbri*, or drunken; *Riaccesi*, or re-ignited; *Addolorati*, or grieved; *Geniali*, or sympathetic; *Animosi*, or intrepid, and others[29]." That of the *Riaccesi* was one of the most famous, and another entitled the *Sregolati*, cultivated science.

The example thus set by the principal cities in Italy was speedily imitated by other towns in that country; and we find an extraordinary number of Academies and Societies springing up, the majority of which enjoyed but a very ephemeral existence. The members of these institutions were sometimes ladies, who formed societies, where they recited their poetical compositions. Tiraboschi, in his elaborate

---

[28] Hallam states that the members of this Academy may be considered as literary partners of the noble-minded printer.

[29] *Sicily*, p. 41.

work, has given a list of no fewer than 171 Academies, instituted for the cultivation of literature and science, independent of the Universities. The titles of some of these Societies are extremely curious, and in many instances ludicrous. Thus we have, The Inflammable; the Pensive; the Intrepid; the Humourists; the Unripe; the Drowsy; the Rough; the Dispirited; the Solitary; the Fiery; the Lyncean (of which Galileo was a member); and the Della Crusca, literally, of the bran or chaff, in allusion to its great object, which was to sift the flour of language from the bran. This celebrated Academy, founded at Florence in 1582, for the purpose of purifying the national tongue, and which published the first edition of its well-known Dictionary in 1612, adopted for its device a sieve, with the motto, *Il più bel fior ne coglia*[30], and the Lyncean used as their symbol rain dropping from a cloud, with the motto, *Redit agmine dulci*[31]. The strange desire that was manifested to give many of these Institutions, avowedly established for noble purposes, absurd names, was not long in meeting with appropriate ridicule, of which a proof will be found in the work of Menkenius, *De Charlatanerie Eruditorum*.

---

[30] "The Academy still assembles in the Palazzo Ricardi for the formalities of holding meetings, and granting diplomas. The backs of their arm-chairs are in the shape of a winnowing shovel; the seats represent sacks; every member takes a name allusive to the miller's calling, and receives a grant of an estate, properly described by metes and bounds, in Arcadia." Murray's *Hand-Book, Northern Italy*, p. 549.

[31] See, for some curious particulars relative to this Society, Mr. D. Bethune's *Life of Galileo*. After slumbering for many years, it was reorganized about forty years ago. Some of our philosopher are members of this Society.

The Accademia del Cimento demands particular mention, as being the first Institution whose members earnestly and successfully devoted themselves to the investigation of physical science. It was established at Florence, on the 19th June, 1657, under the patronage of the Grand Duke Ferdinand II., and by desire of his brother Leopold, who acted upon the advice of Viviani, the great geometrician. The Academicians assembled in the Ducal Palace, when, according to Tiraboschi, the royal host and his family mingled as equals with the most humble members of the Academy. "The name this Society assumed gave promise of their fundamental rule, the investigation of truth by experiment alone. The number of Academicians was unlimited, and all that was required as an article of faith, was the abjuration of all faith, and a resolution to inquire into truth, without regard to any previous sect of philosophy. This Academy lasted, unfortunately, but ten years in vigour. It is a great misfortune for any literary institution to depend on one man, and especially on a prince, who, shedding a factitious, as well as sometimes a genuine lustre round it, is not easily replaced without a diminution of the world's regard. Leopold, in 1667, became a Cardinal, and was thus withdrawn from Florence; other members of the Accademia del Cimento died or went away, and it rapidly sunk into insignificance[32]." This Society, however, did not cease to exist without leaving a record of its labours. A volume containing reports of the Experiments made by the Academy, was printed in 1666, including, with many others, those on the supposed incompressibility

---

[32] Hallam, *Int. to Lit. of Europe*, Vol. IV. p. 562.

of water, the universal gravity of bodies, and the property of electrical matters. The work testifies to the diligence and inquiring spirit of the members[33]. Amongst these, Castellio and Torricelli, disciples of Galileo, were the most illustrious; to them are due many great discoveries in the science of hydraulics; whilst the invention of the barometer alone, renders the name of Torricelli immortal. He made this discovery in 1643; and in 1648, Pascal, by his celebrated experiment on the Puy de Dome, established the theory of atmospheric pressure beyond dispute.

The rapid rise of the numerous Academies and Societies, together with the establishment of Universities and Libraries, abundantly attest the extraordinary ardour that prevailed throughout Italy for the revival of literature and the advancement of science. Up to this period, however, there is no record of the establishment of any Academy or Institution in France or Germany similar to those in Italy. "But it is deserving of remark," says Mr. Hallam, "that one sprung up in England, not indeed of the classical and polite character that belonged to the *Inflammati* of Padua, or the *Della Crusca* of Florence, yet useful in its objects, and honourable alike to its members and to the country. This was the Society of Antiquaries, founded by Archbishop Parker, in 1572. Their object was the preservation of ancient documents, illustrative of history, which the recent dissolution of religious houses, and the shameful devastation attending it, had exposed to great peril. They intended also, by the reading of papers at their meetings, to keep alive the love and knowledge of English Antiquity. In the second of these objects this Society

---

[33] Tiraboschi, Vol. VIII. p. 242.

was more successful than in the first; several short dissertations, chiefly by Arthur Agard, their most active member, having been afterwards published. The Society comprised very reputable names, especially of lawyers, and continued to meet till early in the reign of James, who, from some jealousy, thought fit to dissolve it[34]."

The introduction to the first Volume of the *Archæologia* contains an interesting account of this Society. They intended, it appears, to apply to Queen Elizabeth for a Charter of Incorporation, as "An Academy for the Studye of Antiquity and History, under a President, two Librarians, and a number of Fellows, with a body of Statutes; the Library to be called 'The Library of Queen Elizabeth,' and to be well furnished with scarce books, original charters, muniments, and other MSS.: the Members to take the oath of supremacy, and another, to preserve the Library; the Archbishop and the great officers of state for the time being to visit the Society every five years; the place of meeting to be in the Savoy, or the dissolved priory of Saint John of Jerusalem, or elsewhere[35]."

It does not appear that a Charter was granted; and although the Society comprised some very eminent men, and met weekly in the apartments of Sir William Dethike, in the Herald's Office, it was eventually dissolved by James I. about the year 1604[36].

---

[34] *Int. to Lit.*, Vol. II. p. 262.

[35] The present Society of Antiquaries, established in Feb. 1717, and incorporated in 1751, is the representative of this Elizabethan progenitor. It will be remembered that Leland was appointed 'Royal Antiquary' by Henry VIII.

[36] This sovereign, by letters patent, bearing date Aug. 13, 1609,

In 1616, or 1617, a scheme for founding a Royal Academy in England was started by Edmund Bolton, an eminent scholar and antiquary of that period. There is some account of this design in the introduction to the *Archæologia*; but we are indebted to the Rev. Joseph Hunter, F.S.A., for a more ample sketch, in a paper contained in the thirty-second volume of that work. Without adducing the various authorities which have enabled him to furnish us with his interesting paper, we pass to the following extract, premising that Bolton had gained the patronage of influential noblemen.

"The inception of the design is to be referred to the year 1616 or 1617. This was in the second year of Villiers' introduction at Court; and there can hardly be a doubt that Bolton saw, in the rising influence of his countryman and distant kinsman, a circumstance favourable to the success of his design It must also be mentioned, to the honour of Villiers, that he was a lover and encourager of the arts and literature by natural inclination. The subject was first moved to him, having then become Marquis of Buckingham; and Bolton was introduced by him to the King when at Newmarket, in 1617. There and then the first outline of the project was presented to his Majesty. The Marquis of Buckingham also spoke of it in parliament, where the design was favourably received by many of the lords. This was probably in March 1621, when the Marquis opened a design for a

granted the Temple to the Benchers of the two Societies, called the Inner and Middle Temple. The Temple is spoken of as being "of the number of those four most famous Colleges of Europe."

college for the education of the young nobility, as we find on the journals."

Bolton's petition proposes that the Title of the Institution shall be "KING JAMES HIS ACADEME OR COLLEGE OF HONOUR." The King was very well disposed towards it. "He grew," says Bolton, "so favorable to it, that besides approbation of the whole, because it was purely for the public; it finally pleased him, after some years had passed from the time of the first overture thereof, to enlarge the institution itself with more grants and faculties than were desired."

It was ultimately settled that the Academy was "to consist of three classes of persons, who were to be called *Tutelaries*, *Auxiliaries*, and *Essentials*. The *Tutelaries* were to be Knights of the Garter, with the Lord Chancellor, and the Chancellors of the two Universities. The *Auxiliaries* were to be lords and others, selected out of the flower of the nobility, and councils of war, and of the new plantations. The *Essentials*, upon whom the weight of the work was to lie, were to be persons called out from the most able and most famous lay gentlemen of England, masters of families, or being otherwise men of themselves, and either living in the light of things, or without any title of profession, or art of life, for lucre: such persons being already of other bodies."

The paper proceeds to specify the duties of the members, who were to wear a green ribbon, with the device of the Society composed of the four capital letters J. R. F. C. (*Jacobus Rex, Fundator Collegii*), intertwined with a thread under a crown imperial: and who were to honour, love, and serve one another, according to St. John: *Non tantum verbo et lingua, sed opere et veritate.*

A list of the members is contained in Bolton's original manuscript, amongst which it is very interesting to find the name of Sir Kenelm Digby, one of the original members of the Royal Society, by whom the latter is connected with the proposed Academy. "He was," says Mr. Hunter, "almost the only natural philosopher and experimenter who was a member of the proposed Academy, and placed there rather, it may be presumed, as a philologer than a philosopher, for he was both."

The death of the King, which occurred in March 1625, seems to have been fatal to the completion of the undertaking; for, although his successor was favourably disposed towards it, political events interfered, and led to its final abandonment.

Ten years subsequent to the dissolution of the foregoing Academy, another attempt was made to establish a scientific Institution under the patronage of Charles I., who, in the eleventh year of his reign, granted a special licence under the privy seal, to found a College, or Academy, with the title of *Minerva's Museum*, for the instruction of the young nobility in the liberal arts and sciences. Sir Francis Kynaston, who was one of the four Esquires of the King's body, was appointed Governor, or Regent; and the following extract from the licence sets forth the objects of the proposed Institution:

*Rex, omnibus ad quos &c. Salutem. Cum Nos ex speciali gratiâ nostrâ et ex benevolo erga universum Populum Regni nostri* Magne Britanie *affectu publico commodo et utilitati prospicientes comperimus et intelleximus, quod erectio et institutio alicujus Domus, Societatis aut Hospicii admodum Collegii seu Cenobii, infra civitatem regiam nostram augustam* London, *aut prope illam foret per quam commoda et necessaria ad eruditionem filiorum Nobilium nostrorum et aliorum ingenue natorum, in omnibus Literis humanioribus artibus, liberabilibus et scientiis licitis et laudabilibus quarum ope et adminiculo prefati subditi nostri adolescentes artibus et scientiis predictis ornati et instructi, magis docti et idonei et evaderent tam ad publica Reipublice munera exequenda quam ad officia Magistratuum gerenda.*

*Cum autem fidelis et dilectus famulus noster,* Franciscus Kinaston, *Miles unus è quatuor corporis nostri armigeris, ex singulari ejus erga Nos fide et observantia et haud vulgari erga Patriam suam amore, studioque non minori promovendi Literas et excolendi animos nobilioris hujus seculi juventutis, artibus ingenuis et scientiis generoso homine dignis, ejusque predicte Domus erectioni et institutioni suppecias attulit, ut propria sua industria invenit et proprias et propriis suis sumptibus emit et comparavit edificium sive domum satis aptam et commodam, qui fuit tale Hospitium sive Collegium in quo viri imprimis docti et literati convenirent, docerent, erudirent et instituerent discipulos suos, in omnium artium facultatibus.*

*Cumque prefatus* Franciscus Kinaston *plurima paravit Instrumenta Mathematica et Musica, Libros, Codices, Manuscripta, Picturas, Imagines cum aliis rebus antiquis, raris et exoticis plurimum estimandis et haud mediocris valoris, ad pleniorem prefati Hospitii seu Societatis usum et ornatum spectantibus.*

The Licence goes on to declare that the College shall be erected in Covent Garden, in the parish of St. Martin-in-the-Fields, in the county of Middlesex; that it shall be designated the *Musæum Minervæ,* and that the following six professors shall be attached to it: *Edward May*, Doctor of Philosophy and Medicine; *Thomas Hunt*, Bachelor of Music; *Nicholas Phiske*, Professor of Astronomy; *John Spidell*, Professor of Geometry; *Walter Salter*, Professor of Languages; and *Michael Mason*, Professor of Fencing. The Licence is dated Canbury, June 26, 1635[37].

A very curious and scarce tract, entitled *The Constitutions of the Mvsævm Minervæ*, published in 1636, gives some further information of this institution. It opens thus:—

"To the noble and generous well-wishers to vertuous actions and learning: The Regent and Professours of the *Musæum Minervæ* wish all honour and happinesse.

"Howbeit publick actions and undertakings doe usually receive no preface, it being needlesse to divulge that which of itself will be exposed to all men's censures; nevertheless, new enterprises (how good or just soever they be) are commonly subject at least to suspicion, if not unto oblique interpretation; which frequent experience as well as in other things hath manifested in this our new institution of an Academy in England; which, though already it hath been justified and approved by the wisdomes of the King's most sacred Majestie, confirmed by his Majestie's Letters Patents, and ratified under the hands and seales of the Right Honorable the Lord Keeper of the Great Seale of England, and the two Lord Chief Justices. Yet for a further and more full satisfaction of all men, as well ignorant detractours, as vertuous favorers of this designe, some remonstrance may not seem impertinent, but

---

[37] The Licence is printed entire in Rymer's *Fœdera*, Vol. XIX. p. 638.

rather necessary to be printed and published, for the better understanding of what hath been undertaken."

The objects of the institution are then given, by which it appears, as stated in the licence, that the instruction of 'wealthy gentrie,' as they are styled, was the principal design of the founders. They say: "One great end is to give language and instruction, with other ornaments of travell, unto our gentlemen before their undertaking any long journeys into forreigne parts. For it is found, by lamentable experience, that noblemen and gentlemen, for want of an Academy here, are, as it were, necessitated to send their sonnes beyond the seas for education: where, through change of climate and dyet, they become subject to sicknesses and immature death, than otherwise they might have been; we leave it to carefull, prudent parents to consider how necessary the institution of such an Academy in London is, in which special order may be taken for the bringing up of young gentlemen; untill both for yeares and learning they may be fit, as well to travell and make benefit of their tyme abroad, as to gain some knowledge how to prevent the dangers both of forreigne air and dyet."

The aristocratic tendency of the institution may be judged by the first rule:

"Every man that shall be admitted into the said Musæum shall bring a testimoniall of his arms and gentry, and his coate armour tricked on a table, to be conserved in the Musæum[38]."

[38] The arms of the Academy are annexed to the title-page of the above tract. They consist of a shield argent, with two swords crossed surmounting an open book, with a satyr and mermaid for supporters.

The duties of the Professors are thus allotted:

| | |
|---|---|
| The Doctour of Philosophie and Physick shall reade and professe these:— | *Physioligie, Anatomie, or any other parts of Physick.* |
| The Professour of Astronomie shall teach these:— | *Astronomie, Opticks, Navigation, Cosmographie.* |
| The Professour of Geometrie shall teach these:— | *Arithmetique, Analyticall Algebra, Geometrie, Fortification, Architecture.* |
| The Professour of Musick shall teach these:— | *Skill in Singing, and musick to play upon—Organ, Lute, Violl, &c.* |
| The Professour of Languages shall teach these:— | *Hebrew, Greek, Latine, Italian, French, Spanish, High Dutch.* |
| The Professour of Defence shall teach these:— | *Skill at all weapons and wrestling, also Riding, Dancing, and Behaviour.* |

Our limits forbid enlarging our extracts from this curious publication: enough has been given to show that an Academy of Sciences was imagined, care being taken that the fashionable arts of fencing and dancing should go hand in hand with the severer studies of astronomy and mathematics. But it must be borne in mind, that the College was destined for noble youths, to whom, at that period, a knowledge of fencing was of more vital importance than the acquisition of the problems of Euclid. The time was too unsettled to allow so fair a project to ripen, and it is almost needless to state, that Minerva's Museum never attained its contemplated greatness.

About the same period that the above plans were brought forward, an academy, designated *Die frucht-*

*bringende Gesellschaft*, or, 'The Fruitful Society,' was established at Weimar[39].

It was royally patronized, and gave promise of advancing literature and science; the members, however, accomplished but little, and the institution ceased to exist, with others which subsequently sprung up, imitating those of Italy only as regarded the adoption of their fanciful titles. "They are gone," exclaims Bouterwek, "and have left no clear vestige of their existence." With the foregoing exceptions, Italy seems to have been the principal labourer in the great work of mental culture; for, as Mr. Hallam observes, "Neither England, nor France, nor Germany, seemed aware of the approaching change[40]." Yet, in the latter country, an engine had been invented, to which literature, art, and science, are more indebted than to all the patronage they have received from monarchs or princes. Petit-Radel, in his *Recherches sur les Bibliothèques Anciennes et Modernes*, gives an enumeration of the number of books, or editions, published in different parts of Europe, from the date of the invention of printing to the beginning of the sixteenth century. These numbers are: at Venice, 2789; Rome, 972; Paris, 789; Strasburgh, 298; Westminster, 99; London, 31; Oxford, 7; and Spain and Portugal, 126. Thus, although printing was invented in Germany, we see that Italy, above other European nations, was the first country to avail herself of its inestimable advantages[41].

---

[39] Bouterwek gives 1617 as the the date of its foundation.

[40] *Europe during the Middle Ages*, Vol. II. p. 527.

[41] Mr. Hallam, in his *Int. to the Lit. of Europe*, gives the titles dates, and places of publication of the first editions of the prin-

France was at length moved to follow the stirring example of Italy, and the French Academy was established. This Institution sprang from a "private Society of men of letters at Paris, who, about the year 1629, agreed to meet once a week, to converse on all subjects, and especially on literature. Such among them as were authors communicated their works, and had the advantage of free and fair criticism. This continued for three or four years with such harmony and mutual satisfaction, that the old men who remembered this period, says their historian, Pelisson, 'looked back upon it as a golden age.' They were but nine in number, of whom Gombauld and Chapelain are the only names by any means famous, and their meetings were at first very private. More, by degrees, were added: among others, Boisrobert, a favourite of Richelieu. The Cardinal, pleased with an account of this Society, suggested their incorporation. This, it is said, was unpleasing to every one of them, and some proposed to refuse it; but the consideration that the offers of such a man were not to be slighted, overpowered their modesty, and they consented to become a royal institution. They now enlarged their numbers, created officers, and began to keep registers of their proceedings. These records commence March 13, 1634. The name of '*French Academy*' was chosen after some deliberation. They were established by letters patent in January 1635; which the parliament of Paris enregistered with great reluctance, requiring, not only a letter from Richelieu, but an express order from the King; and when this was completed, in

---

cipal Greek and Latin authors. These amount to seventy-nine; no fewer than sixty-one of which were printed in Italy.

July 1637, it was with a singular proviso, that the Academy should meddle with nothing but the embellishment and improvement of the French language, and such books as might be written by themselves, or by others who should desire their interference[42]. The professed object of the Academy was to purify the language from vulgar, technical, or ignorant usages, and to establish a fixed standard. The Academicians undertook to guard scrupulously the correctness of their own works; examining the arguments, the method, the style and the structure, of each particular word. It was proposed by one that they should swear not to use any word which had been rejected by a plurality of votes. They soon began to labour in their vocation, always bringing words to the test of good usage, and deciding accordingly. These decisions are recorded in their registers. Their numbers were fixed by the letters patent at forty; having a director, chancellor, and secretary; the two former changed every two, afterwards every three, months; the last was chosen for life. They read discourses weekly, which, by the titles of some that Pelisson has given us, seem rather trifling, and in the style of the Italian Academies: but this practice was soon disused. Their more important and ambitious occupations were to compile a dictionary and a grammar. Chapelain drew up the scheme of the former, in which it was determined, for the sake of brevity, to give no quotations, but to form it from about twenty-six good authors in prose, and twenty in verse[43]."

---

[42] It will be remembered that the *Sorbonne* College (founded in 1255), included the study of the French language, as part of the foundation of the school of theology. The first printing presses in Paris were established at this college.

[43] Hallam, *Int. to Lit. of Europe*, Vol. III. pp. 643—5.

Pelisson relates that soon after the formation of the Academy, its ability and impartiality were tried by Richelieu, who called on the body to pronounce an opinion upon the *Cid* of Corneille, to which work he had a strong dislike. At first, the members were most unwilling to give any opinion, but, as the Cardinal was not a man to take excuses, a committee was eventually appointed, who prepared a report, which was, on the whole, favourable to Corneille. Mr. Hallam states that the "*sentimens de l'Académie* were drawn up with great good sense and dignity." This Institution was subsequently incorporated with the 'Academy of Sciences,' and that of 'Inscriptions and Belles Lettres[44].'

Although France thus early founded a Society for the cultivation of literature, yet to England belongs the high honour of being the first country, after Italy, to establish a Society for the investigation and advancement of physical science. "The period was arrived when experimental philosophy, to which Bacon had held the torch, and which had already made considerable progress, especially in Italy, was finally established on the ruins of arbitrary figments and partial inductions."

The development and advancement of science are signally indebted to three Associations: the Academy del Cimento at Florence, which, as we have seen, endured but for a short time; the Royal Society of London; and the Academy of Sciences at Paris. Laplace has well said: *Le principal avantage des Académies, est l'esprit philosophique qui doit s'y introduire, et de là se répandre dans toute une nation et sur tous les objets. Le savant isolé, peut se livrer sans*

[44] Fontenelle, Vol. v. p. 23.

*crainte à l'esprit de système: il n'entend que de loin la contradiction qu'il éprouve. Mais dans une société savante, le choc des opinions systématiques finit bientot par les détruire; et le désir de se convaincre mutuellement, établit nécessairement entre les membres, la convention de n'admettre que les résultats de l'observation et du calcul. Aussi l'experience a-t-elle montré que depuis l'origine des Académies, la vraie philosophie s'est généralement répandue. En donnant l'exemple de tout soumettre à l'examen d'une raison sévère, elles ont fait disparaître les préjugés qui trop long-temps avaient régné dans les sciences, et que les meilleurs esprits des siècles précédens avaient partagés. Leur utile influence sur l'opinion, a dissipé des erreurs accueillies de nos jours, avec un enthousiasme qui, dans d'autres temps, les aurait perpétuées. Egalement éloignées de la crédulité qui fait tout admettre, et de la prévention qui porte à rejeter tout ce qui s'écarte des idées reçues; elles ont toujours sur les questions difficiles et sur les phénomènes extraordinaires, sagement attendu les réponses de l'observation et de l'expérience, en les provoquant par des prix et par leurs propres travaux. Mesurant leur estime autant à la grandeur et à la difficulté d'une découverte, qu'à son utilité immédiate, et persuadées par beaucoup d'exemples, que la plus stérile en apparence, peut avoir, un jour, des suites importantes; elles ont encouragé la recherche de la vérité sur tous les objets, n'excluant que ceux qui, par les bornes de l'entendement humain, lui seront à jamais inaccessibles. Enfin c'est de leur sein que se sont élevées ces grandes théories que leur généralité met au-dessus de la portée du vulgaire, et qui se répandant par de nombreuses applications, sur la nature et sur les arts, sont devenues d'inépuisables sources de lumières et de jouis-*

*sances. Les gouvernemens sages convaincus de l'utilité des sociétés savantes, et les envisageant comme l'un des principaux fondemens de la gloire, et de la prospérité des empires, les ont instituées et placées près d'eux, pour s'éclairer de leurs lumières dont souvent ils ont retiré de grands avantages*[45]."

It must appear singular, that early as our English Universities were founded, some philosophical association should not have grown out of them prior to those of which we have any record. Oxford, it is true, did cultivate mathematics to an extent as remarkable for the period, as the present fostering of that science by the sister University; and this in an age when the fifth proposition of Euclid was often the halting-place of the philosopher. But when we remember that all learning was overlaid and encumbered by scholastic theology, with its heavy net-work of mystical dogmatism, we shall cease to feel surprised that so little was done. Had ecclesiastical authority exercised less sway, we should probably have to go farther for the history of our scientific Institution.

To the Royal Society attaches the renown of having, from its foundation, applied itself with untiring zeal and energy to the great objects of its institution; and we now behold it, venerable in years, yet shewing no symptom of decay, and regarded with admiration and esteem by the civilized world.

Having now rapidly sketched the rise and progress of literary and scientific societies; we shall, in the next chapter, proceed to the immediate object before us; the origin and foundation of the Royal Society.

---

[45] *Précis de l'Histoire de l'Astronomie*, p. 99.

# CHAPTER II.

Origin of Royal Society involved in some obscurity—Wallis's Account—Oxford Philosophical Society—Their Regulations—Influential in promoting Establishment of Royal Society—Hooke's Answer to Cassini's Statement respecting the Society—Boyle's Account of the Invisible College—Sprat's description of Meetings at Gresham College—Interruption occasioned by Civil Wars—Evelyn's design for a Philosophical College—Cowley's Proposition for the establishment of a College—Sir William Petty's Scheme for a Gymnasium Mechanicum, or College of Tradesmen—Plan of building a Philosophical Institution at Vauxhall—Domestic troubles postpone the establishment of any Scientific or Literary Society.

---

1645—55.

THE origin of the Royal Society, in common with that of many other illustrious institutions, is involved in some obscurity; for though the year 1660 may be regarded as the date of its establishment, yet there is no doubt that a society of learned men were in the habit of assembling together, to discuss scientific subjects, for many years previously to the above time.

In the Publisher's Appendix to his Preface of Thomas Hearne's edition of Peter Langtoft's Chronicle, we find, under the head of "Dr. Wallis's account of some passages of his own life," written in January 1696, 7, the following interesting extract:—

"About the year 1645, while I lived in London, (at a time when, by our civil wars, academical studies were much interrupted in both our Universities) beside the conversation of divers eminent divines, as to matters theological, I had the opportunity of being acquainted with divers worthy persons, inquisitive into natural philosophy, and other parts of human learning; and particularly of what hath been called the *New Philosophy*, or *Experimental Philosophy*. We did by agreements, divers of us, meet weekly in London on a certain day, to treat and discourse of such affairs; of which number were *Dr. John Wilkins* (afterward *Bishop of Chester*), *Dr. Jonathan Goddard*, *Dr. George Ent*, *Dr. Glisson*, *Dr. Merret* (Drs. in Physick), *Mr. Samuel Foster*, then Professor of Astronomy at Gresham College, *Mr. Theodore Hank*[1], (a German of the Palatinate, and then resident in London, who, I think, gave the first occasion, and first suggested those meetings), and many others.

"These meetings we held sometimes at *Dr. Goddard's* lodgings in *Wood Street* (or some convenient place near), on occasion of his keeping an operator in his house for grinding glasses for telescopes and microscopes; sometimes at a convenient place in *Cheapside*[2], and sometimes at *Gresham College*, or some place near adjoyning.

"Our business was (precluding matters of theology and state-affairs), to discourse and consider of *Philosophical Enquiries*, and such as related thereunto: as *Physick*, *Anatomy*, *Geometry*, *Astronomy*, *Navigation*, *Staticks*, *Magneticks*, *Chymicks*, *Mechanicks*, and natural *Experiments;* with the state of these studies, as then cultivated at home and abroad. We then discoursed of the *circulation of the blood*, *the valves in the veins*, *the venæ lacteæ*, *the lymphatick vessels*, *the Copernican hypothesis*, *the nature of comets and new stars*, *the satellites of Jupiter*, *the oval*

---

[1] Doubtless *Haak*.

[2] The convenient place to which Dr. Wallis refers was the Bull-Head Tavern, in Cheapside.

*shape* (as it then appeared) *of Saturn, the spots in the sun, and its turning on its own axis, the inequalities and selenography of the Moon, the several phases of Venus and Mercury, the improvement of telescopes, and grinding of glasses for that purpose, the weight of air, the possibility, or impossibility of vacuities, and nature's abhorrence thereof, the Torricellian experiment in quicksilver, the descent of heavy bodies, and the degrees of acceleration therein*; and divers other things of like nature. Some of which were then but new discoveries, and others not so generally known and imbraced, as now they are, with other things appertaining to what hath been called *The New Philosophy*, which from the times of *Galileo* at *Florence*, and *Sir Francis Bacon (Lord Verulam)* in England, hath been much cultivated in *Italy, France, Germany*, and other parts abroad, as well as with us in *England*.

"About the year 1648, 1649, some of our company being removed to Oxford (first *Dr. Wilkins*[3], then I, and soon after Dr. Goddard), our company divided. Those in London continued to meet there as before (and we with them, when we had occasion to be there), and those of us at Oxford; with

[3] Aubrey in his lives of eminent men, states that Wilkins "was the principal reviver of experimental philosophy (*secundem mentem Domini Baconi*) at Oxford, where he had weekely an experimental philosophicall clubbe, which began 1649, and was the incunabile of the Royall Society. When he came to London, they met at the Bull's-Head Tavern, in Cheapside; *e.g.* 1659, 1660, and after, till it grew too big for a clubbe, and so they came to Gresham Colledge parlour." Vol. III. p. 583. And in his Life of *Seth Ward, Bishop of Salisbury*, he says, "The beginning of philosophicall experiments was at Oxon (1649), by Dr. Wilkins, Seth Ward, Ralph Bathurst, &c. &c."

*Dr. Ward* (since *Bishop of Salisbury*), *Dr. Ralph Bathurst* (now *President of Trinity College in Oxford*), *Dr. Petty* (since Sir William Petty), *Dr. Willis* (then an eminent physician in *Oxford*), and divers others, continued such meetings in Oxford, and brought those studies into fashion there; meeting first at Dr. Petty's lodgings (in an apothecarie's house), because of the convenience of inspecting drugs, and the like, as there was occasion; and after his remove to Ireland (though not so constantly), at the lodgings of *Dr. Wilkins*, then Warden of Wadham College, and after his removal to *Trinity College in Cambridge*, at the lodgings of the *Honourable Mr. Robert Boyle*, then resident for divers years in Oxford."

The original Minutes of the Philosophical Society of Oxford are preserved in the Ashmolean Museum. At the commencement of the first volume the regulations are inserted, which will probably be perused with considerable interest[4].

"OCTOBER 23, 1651. ORDERED,

"1. That no man be admitted, but with the consent of the major part of the company.

"2. That the votes for admission (to the intent they may be free, and without prejudice,) be given in secret; affirmatives by blanks, negatives by printed papers put into the box.

"3. That every man's admission be concluded the next day after it is proposed; so as at the passing of it there be at the least eleven present.

---

[4] The Council of the Royal Society, on my proposition, ordered a copy of these Minutes to be made, for preservation in the archives of the Society.

"4. That every one pay for his admission an equal share to the money in stock, and two-third parts of it for the instruments in stock, answerable to the number of the Company.

"5. If any of the Company (being resident in the University) do willingly absent himself from the weekly meeting, without speciall occasion, by the space of six weeks together, he shall be reputed to have left the Company, his name from thenceforth to be left out of the catalogue.

"6. That if any man doe not duly upon the day appoynted perform such exercise, or bring in such experiment as shall be appoynted for that day, or in case of necessity provide that the course be supplyed by another, he shall forfeit to the use of the company for his default 2*s.* 6*d.*, and shall perform his task notwithstanding, within such reasonable time as the company shall appoint.

"7. That one man's fault shall not (as formerly) be any excuse for him that was to succeed the next day, but the course shall goe on.

"8. That the time of meeting be every Thursday, before two of the clock[5]."

The Oxford Society was a powerful auxiliary to the Royal Society. When occupied at the Ashmolean Museum making researches for this work, I examined the Minute-books of the former Society, and found that frequent mention is made of the Royal Society. It was the custom mutually to communicate the principal labours of the respective Fellows to each society, by

[5] Ashmolean MSS. No. 1810.

which means science was materially advanced[6]. Much credit is due to the Oxford Society for the pains taken to enlist learned bodies in their good cause. In a letter to the heads of Universities, they say, "We would by no means be thought to slight or undervalue the philosophy of Aristotle, which hath for many ages obtained in the schools. But have (as we ought), a great esteem for him, and judge him to have been a very great man, and think those who do most slight him, to be such as are less acquainted with him. He was a great enquirer into the history of nature, but we do not think (nor did he think), that he had so exhausted the stock of knowledge of that kind as that there would be nothing left for the enquiry of after-times, as neither can we of this age hope to find out so much, but that there will be much left for those that come after us[7]." The letter proceeds earnestly to request the assistance of members of colleges, &c., towards the great work of advancing scientific knowledge. It would lead us out of our proper path to follow the Oxford Society further; it will be sufficient to state, that they met, at irregular intervals, until 1690, in which year their meetings terminated, much to the regret of philosophers[8].

Reverting to the Royal Society, we have some further particulars regarding it from Dr. Wallis, in a curious and scarce tract, entitled *A Defence of the*

[6] By a letter of Dr. Wallis's, in the archives of the Royal Society, it appears that such experiments were made at Oxford as could not be conveniently carried on in London.

[7] Ashmolean MSS.

[8] Dr. Plot, in a letter to the Master of University College, written in 1694, and preserved in the Bodleian Library, laments the cessation of the meetings of the Oxford Philosophical Society.

*Royal Society, an Answer to the Cavils of Dr. William Holder*, published in 1678. Dr. Holder affirmed that "divers ingenious persons in Oxford laid the first foundation of the Royal Society." In answer to this, Dr. Wallis says, "I take its first ground and foundation to have been in London, about the year 1645, if not sooner, when Dr. Wilkins, (then chaplain to the Prince Elector Palatine, in London) and others, met weekly at a certain day and hour, under a certain penalty, and a weekly contribution for the charge of experiments, with certain rules agreed upon amongst us. When (to avoid diversion to other discourses, and for some other reasons) we barred all discourses of divinity, of state-affairs, and of news, other than what concerned our business of Philosophy. These meetings we removed soon after to the Bull Head in Cheapside, and in term-time to Gresham College, where we met weekly, at Mr. Foster's lecture (then Astronomy Professor there), and, after the lecture ended, repaired, sometimes to Mr. Foster's lodgings, sometimes to some other place not far distant, where we continued such enquiries, and our numbers increased.

"About the years 1648, 9 some of our company were removed to Oxford; first, Dr. Wilkins, then I, and soon after, Dr. Goddard, whereupon our company divided. Those at London (and we, when we had occasion to be there,) met as before. Those of us at Oxford, with Dr. Ward, Dr. Petty, and many others of the most inquisitive persons in Oxford, met weekly (for some years) at Dr. Petty's lodgings, on the like account, to wit, so long as Dr. Petty continued in Oxford, and for some while after, because of the conveniences we had there (being the house of an apothecary) to view, and make use of, drugs and other like matters, as there was occasion.

"Our meetings there were very numerous and very considerable. For, beside the diligence of persons studiously inquisitive, the novelty of the design made many to resort thither; who, when it ceased to be new, began to grow more remiss, or did pursue such inquiries at home. We did afterwards (Dr. Petty being gone for Ireland, and our numbers growing less) remove thence; and (some years before His Majesty's return) did meet at Dr. Wilkins's lodgings in Wadham College. In the meanwhile, our company at Gresham College being much again increased, by the accession of divers eminent and noble persons, upon His Majesty's return, we were (about the beginning of the year 1662) by his Majesty's grace and favour, incorporated by the name of the Royal Society[9]," &c.

Hooke, in his *Answer to some Particular Claims of M. Cassini*, states that he, "M. Cassini, is in error concerning the beginning and original of the Royal Society. Concerning which he might have been much better informed if he had taken notice of what has been said concerning it, but that, it seems, did not suit so well to his design of making the French to be the first. He makes Mr. Oldenburg to have been the instrument who inspired the English with a desire to imitate the French, in having philosophical clubs, or meetings, and that this was the occasion of founding the Royal Society, and making the French the first. I will not say that Mr. Oldenburg did rather inspire the French to follow the English, or, at least, did help them, and hinder us. But it is well known who were

[9] p. 8.

the principal men that began and promoted that design, both in London and Oxford; and that a long while before Mr. Oldenburg came into England. And not only these philosophical meetings were before Mr. Oldenburg came from Paris; but the Society itself was begun before he came hither; and those who then knew Mr. Oldenburg, understood well enough how little he himself knew of philosophick matters[10]."

In the life of the Hon. Robert Boyle, prefixed to his Works, are some letters in which that eminent philosopher[11] alludes to the Royal Society before its incorporation, under the title of the *Invisible College.* In a communication to Mr. Marcombes, dated London, October 22, 1646, he says:

"The other humane studies I apply myself to are natural philosophy, the mechanics, and husbandry, according to the principles of our new philosophical college, that values no knowledge but as it has a tendency to use. And therefore I shall make it one of my suits to you, that you would take the pains to inquire a little more thoroughly into the ways of hus-

---

[10] *Phil. Exper.* p. 388.

[11] It is worthy of record, that Boyle entertained a plan of giving 12,000*l.* to purchase confiscated lands in Ireland; the profits from which were to be devoted to the promotion of knowledge. It appears that there was some prospect of this scheme being carried into execution, as Oldenburg thus alludes to it in a letter to Boyle, written at Saumur in 1657: "I am hugely pleased that the Council has granted your desires for the promotion of knowledge, which I suppose to be those that were couched in a certain petition you were pleased to impart to me at Oxford; wherein, if I remember well, a matter of twelve thousand pounds sterling was offered to purchase confiscated lands and houses within Ireland, to employ the profits in the entertainment of an agent, secretary, translators, for keeping intelligence, distributing rewards, &c., in order to the end aforesaid."

bandry, &c. practised in your parts; and when you intend for England, to bring along with you what good receipts or choice books of any of these subjects you can procure; which will make you extremely welcome to our *Invisible College.*" And in a letter to Mr. Francis Tallents, Fellow of Magdalen College, Cambridge, dated London, February 1646-7, he says, "The best on't is, that the corner-stones of the *invisible*, (or, as they term themselves, the philosophical college,) do now and then honour me with their company, which makes me as sorry for those pressing occasions that urge my departure, as I am at other times angry with that solicitous idleness, that I am necessitated to, during my stay: men of so capacious and searching spirits, that the school-philosophy is but the lowest region of their knowledge; and yet, though ambitious to lead the way to any generous design, of so humble and teachable a genius, as they disdain not to be directed to the meanest, so he can plead reason for his opinion; persons that endeavour to put narrow-mindedness out of countenance, by the practice of so extensive a charity that it reaches unto every thing called man, and nothing less than an universal goodwill can content it. And, indeed, they are so apprehensive of the want of good employment, that they take the whole body of mankind for their care. But, lest my seeming hyperbolical expressions should more prejudice my reputation than it is in any way able to advantage theirs, and I be thought a liar for telling so much truth, I will conclude their praises with the recital of their chiefest fault, which is very incident to almost all good things; and that is, that there is not enough of them."

In May, 1647, Boyle again alludes to the *Invisible*

*College*, in a letter to Hartlib, which leaves little doubt that he meant by this title that assembly of learned and high-minded men, who sought, by a diligent examination of natural science, which was then called the *New Philosophy*, an alleviation from the harrowing scenes incidental to the civil wars.

"For such a candid and impassionate Company as that was," says Dr. Sprat, in his *History of the Royal Society*, "and for such a gloomy season, what could have been a fitter subject to pitch upon than *Natural Philosophy?* To have been always tossing about some theological question, would have been to have made that their private diversion, the excess of which they themselves disliked in the publick; to have been eternally musing on *civil business*, and the distresses of their country, was too melancholy a reflection: it was nature alone which could pleasantly entertain them in that estate.—Their *meetings* were as frequent as their affairs permitted: their proceedings rather by action than discourse, chiefly attending some particular trials in *Chymistry* or *Mechanicks:* they had no rules nor method fixed: their intention was more to communicate to each other their discoveries, which they could make in so narrow a compass, than an united, constant, or regular inquisition. Thus they continued, without any great intermissions, till about the year 1658. But those being called away to several parts of the nation, and the greatest number of them coming to London, they usually met at Gresham College, at the Wednesday's and Thursday's lectures of Dr. Wren and Mr. Rooke[12]; where there joyn'd with

---

[12] Ward, in his *Life of Rooke*, says, that "he was very zealous and serviceable in promoting the great and useful institution of the

them several eminent persons of their common acquaintance: The *Lord Viscount Brouncker*, the now *Lord Brereton, Sir Paul Neil, Mr. John Evelyn, Mr. Henshaw, Mr. Slingsby, Dr. Timothy Clark, Dr. Ent, Mr. Ball, Mr. Hill, Dr. Crone*, and divers other gentlemen, whose inclinations lay the same way. This custom was observed once, if not twice a-week, in term-time; till they were scattered by the miserable distractions of that fatal year; till the continuance of their meetings there might have made them run the hazard of the fate of *Archimedes:* for then the place of their meeting was made a *quarter for soldiers*."

"This day," says the author of the foregoing extract, in a letter to Mr. Wren, written in 1658, and published in the *Parentalia, or Memoirs of the Times of Bishop Wren*, 1658, "I went to visit Gresham College, but found the place in such a nasty condition, so defiled, and the smells so infernal, that if you should now come to make use of your tube, it would be like Dives looking out of hell into heaven. Dr. Goddard, of all your colleagues, keeps possession, which he could never be able to do, had he not before prepared his nose for camp-perfumes by his voyage into Scotland, and had he not such excellent restoratives in his cellar. The soldiers, by their violence which they

---

Royal Society." Dr. Ward, Bishop of Exeter, was much attached to him; he gave the Society, in memory of his friend, a large pendulum clock, made by Fromantel, then esteemed a great rarity, which was set up in the room where they met in Gresham College, and afterwards placed in the outer hall of their house at Crane Court. An inscription to the memory of Rooke was engraved on the dial-plate. Rooke was elected Gresham Professor of Astronomy in 1652, which he resigned in 1657 for the Chair of Geometry.

put on the Muses' seats, have made themselves odious to all the ingenious world; and if we pass by their having undone the nation, this crime we shall be never able to forgive them."

And Matthew Wren, in a letter to Christopher Wren, dated Oct. 25, 1658, also published in the *Parentalia*, says, "Yesterday, being the first day of term, I resolv'd to make an experiment whether Dr. Horton entertained the new auditory of Gresham with any lecture; for I took it for granted, that if his divinity could be spar'd, your mathematicks would not be expected. But at the gate I was stop'd by a man with a gun, who told me there was no admission upon that account, the College being reform'd into a garrison[13]."

Although the distracted state of the times prevented a continuance of the scientific meetings at Gresham College, the philosophers did not abandon their cause in despair.

In 1659, the philanthropic Evelyn addressed a letter to the Hon. Robert Boyle, containing his plan for the institution of a Scientific College. In his diary, under the date Sept. 1, 1659, he writes, "I commu-

---

[13] Dr. Birch states that Gresham College could not have been occupied by soldiers until 1659, and adduces in proof the letter of Matthew Wren, quoted above, written in 1659, and not, as its date affirms, on Oct. 25, 1658. Although 1659 may justly be denominated the *year of distractions*, yet there is no reason to suppose that the evidence of Dr. Sprat and Mr. Wren, as to the occupancy of Gresham College by troops in 1658, and the consequent dispossession of the professors and other scientific men, is erroneous; indeed, it is distinctly stated in the MS. of the *Parentalia*, which is in the Library of the Royal Society, that "in 1658 the College was garrisoned by the rebels, and the professors driven out."

nicated to Mr. Robert Boyle my proposal for erecting a Philosophic-mathematic College."

The letter, which is published in Boyle's Works, is so interesting as characteristic of the times, and so strikingly illustrative of the desire on the part of the learned men of that day to establish an institution for the investigation and cultivation of Science, that an account of the rise of the Royal Society would be incomplete without it. It is dated Say's Court, Sept. 3, 1659.

"Noble Sir,

"Together with these testimonies of my cheerful obedience to your commands, and a faithful promise of transmitting the rest, if yet there remain anything worthy your acceptance amongst my unpolished and scattered collections, I do hereby make bold to trouble you with a more minute discovery of the design, which I casually mentioned to you, concerning my great inclination to redeem the remainder of my time, considering *quam parum mihi supersit ad metas;* so as may best improve it to the glory of God Almighty, and the benefit of others. And since it has proved impossible for me to attain to it hitherto (though in this my private and mean station), by reason of that fond morigeration to the mistaken customs of the age, which not only rob men of their time, but extremely of their virtue and best advantages; I have established with myself, that it is not to be hoped for without some resolution of quitting these incumbrances, and instituting such a manner of life for the future as may best conduce to a design so much breathed after and, I think, so advantageous. In order to this, I propound that since we are not to hope for a mathematical college, much less a *Solomon's* house, hardly a friend in this sad *catalysis,* and *inter hos armorum strepitus*, a period so uncharitable and perverse, why might not some gentlemen, whose geniuses are greatly suitable, and who desire nothing more than to give a good example, preserve science, and cultivate themselves, join together in society, and resolve upon some orders and economy to be mutually observed, such as shall best become the end of their union; if I cannot say without a kind of singularity, because the thing is new; yet such, at least, as shall be free from pedantry, and all affectation? The possibility, Sir, of this is so obvious, that I confess were I not an aggregate person, and so obliged, as well by my own nature, as the laws of decency and their merits, to provide for my dependents, I would cheerfully devote my small fortune towards it, by which I might hope to assemble some small number together, who might resign themselves to live profitably and sweetly together. But since I am unworthy so great a happiness, and that it is not now in my power, I propose that if any one worthy person, and *queis meliore luto*, so qualified as Mr. Boyle, will join in the design (for not with every one, rich, and learned, there are very few disposed, and it is the greatest difficulty to find the man), we would not doubt in a short time (by God's assistance) to be possessed of the most blessed life that virtuous persons could wish or aspire to in this miserable and uncertain pilgrimage, whether considered as to the present revolutions, or what may happen for the future in all human probability. Now, Sir, in what instances, and how far this is practicable, permit me to give you an account of, by the calculations which I have deduced for our little foundation.

"I propose the purchasing of thirty or forty acres of land, in some healthy place, not above twenty-five miles from London, of which a good part should be tall wood, and the rest upland pastures, or downs, sweetly irrigated. If there were not already an house, which might be converted, &c., we would erect upon the most convenient site of this, near the wood, our building, viz. one handsome pavilion, containing a refectory, library, withdrawing-room, and a closet; this the first story; for we suppose the kitchen, larders, cellars and offices, to be contrived in the half story under ground. In the second should be a fair lodging-chamber, a pallet-room, gallery, and a closet; all which should be well and very nobly furnished, for any worthy person that might desire to stay any time, and for the reputation of the College. The half story above for servants, wardrobes, and like conveniences. To the entry forefront of this a court, and at the other backfront a plot walled in of a competent square, for the common seraglio, disposed into a garden; or it might be only carpet, kept curiously, and to serve for bowls, walking, or other recreations, &c.; if the company please. Opposite to the house, towards the wood, should be erected a pretty chapel; and at equal distances (even with the flanking walls of the square), six apartments or cells, for the members of the Society, and not contiguous to the pavilion, each whereof should contain a small bed-chamber, an outward room, a closet, and a private garden, somewhat after the manner of the Carthusians. There should likewise be an elaboratory with a repository for rarities and things of nature, aviary, dove-house, physick-garden, kitchen-garden, and a plantation of orchard-fruit, &c.; all uniform buildings, but of single stories, or a little elevated. At convenient distance towards the olitory-garden should be a stable for two or three horses, and a lodging for a servant or two. Lastly, a garden-house and conservatory for tender plants.

"The estimate amounts thus. The pavilion 400*l.*, chapel 150*l.*, apartments, walls, and out-housing 600*l.* The purchase of the fee for thirty acres, at 15*l.* per acre, eighteen years' purchase, 400*l.* The total 1550*l.*; 1600*l.* will be the utmost.

"Three of the cells or apartments, that is one moiety, with the appurtenances, shall be at the disposal of one of the founders, and the other half at the others.

"If I and my wife take up two apartments (for we are to be decently asunder; however I stipulate, and her inclination will greatly suit with it, that shall be no impediment to the Society, but a considerable advantage to the œconomick part), a third shall be for some worthy person; and to facilitate the rest, I offer to furnish the whole pavilion completely, to the value of 500*l.* in goods and moveables, if need be for seven years, till there be a publick stock, &c.

"There shall be maintained at the publick charge, only a Chaplain, well qualified, an ancient woman to dress the meat, wash, and do all such offices; a man to buy provisions, keep the garden, horses, &c.; a boy to assist him and serve within.

"At one meal a-day of two dishes only, (unless some little extraordinary upon particular days, or occasions, then never exceeding three) of plain and wholesome meat; a small refection at night: wine, beer, sugar, spice, bread, fish, fowl, candle, soap, oats, hay, fuel, &c., at 4*l.* per week, 200*l.* per annum; wages 15*l.*; keeping the gardens 20*l.*; the chaplain 20*l.* per annum. Laid up in the treasury yearly 145*l.*, to be employed for books, instruments, drugs, trials, &c. The total 400*l.* a-year, comprehending the keeping of two horses for the chariot or the saddle, and two kine; so that 200*l.* per annum will be the utmost that the founders shall be at to maintain the whole Society, consisting of nine persons (the servants included), though there should no others join capable to alleviate the expense: but if any of those who desire to be of the Society be so qualified as to support their own particulars, and allow for their proportion, it will yet much diminish the charge; and of such there cannot want some at all times, as the apartments are empty.

"If either of the founders thinks expedient to alter his condition, or that any thing do *humanitus contingere*, he may resign to another, or sell to his colleague, and dispose of it as he pleases; yet so as it still continue the institution.

"ORDERS.

"At six in summer, prayers in the chapel. To study till half an hour after eleven. Dinner in the refectory till one. Retire till four. Then called to conversation (if the weather invite) abroad, else in the refectory. This never omitted, but in case of sickness.

"Prayers at seven. To bed at nine. In the winter the same, with some abatements for the hours, because the nights are tedious and the evenings' conversation more agreeable. This in the refectory. All play interdicted, sans bowls, chess, &c. Every one to cultivate his own garden. One month in spring a course in the elaboratory on vegetables, &c. In the winter a month on other experiments. Every man to have a key of the elaboratory, pavilion, library, repository, &c. Weekly fast; Communion once every fortnight, or month at least. No stranger easily admitted to visit any of the Society, but upon certain days weekly, and that only after dinner. Any of the Society may have his commons to his apartment, if he will not meet in the refectory, so it be not above twice a-week. Every Thursday shall be a musick meeting at conversation hours. Every person of the Society shall render some publick account of his studies weekly, if thought fit; and especially shall be recommended the promotion of experimental knowledge, as the principal end of the institution. There shall be a decent habit and uniform used in the college. One month in the year may be spent in London, or any of the Universities, or in a perambulation for the publick benefit, &c.; with what other orders shall be thought convenient, &c.

"Thus, Sir, I have in haste (but, to your loss, not in a laconic stile) presumed to communicate to you (and truly, in my life, never to any but yourself) that project which for some time has traversed my thoughts; and therefore, far from being the effect either of an impertinent or trifling spirit, but the result of mature and frequent reasonings. And, Sir, is not this the same that many noble personages did, at the confusion of the empire by the barbarous Goths, when St. Hierome, Eustochium, and others, retired from the pertinencies of the world to the sweet recesses and societies in the East, till it came to be burthened with the vows and superstitions, which can give no scandal to our design, that provides against all such snares?

"Now to assure you, Sir, how pure and unmixed the design is from any other than the publick interest, propounded by me, and to redeem the time to the noblest purposes, I am thankfully to acknowledge, that as to the common forms of living in the world, I have little reason to be displeased at my present condition, in which, I bless God, I want nothing conducing either to health or honest diversion extremely beyond my merit; and therefore would I be somewhat choice and scrupulous in my collegue; because he is to be the most dear person to me in the world. But oh! how I should think it designed from heaven, and *tanquam numen διοπετὴς*, did such a person as Mr. Boyle, who is alone a society of all that were desirable to a consummate felicity, esteem it a design worthy his embracing! Upon such an occàsion how would I prostitute all my other concernments! how would I exult! and, as I am, continue upon infinite accumulations and regards.

"Sir, his most humble, and
most obedient servant,
EVELYN.

"If my health permit me the honour to pay my respects to you, before you leave the town, it will bring you a rude plot of the building, which will better fix the idea, and shew what symmetry it holds with this description."

Liberal as were the proposals contained in this letter from the noble-minded Evelyn, they were not responded to, though in all probability, they had some effect in hastening the establishment of the Royal Society. The times were however so distracted, that it could hardly be expected that men whose lives were in daily peril, would bestow much time or thought on science; and thus, Evelyn's scheme met with no other encouragement than that of being carefully treasured amongst Boyle's papers.

Cowley's *Proposition for the Advancement of Experimental Philosophy*, was first published at about the same date as Evelyn's letter. This is a most curious document, but, unfortunately, of such a length as to preclude its entire insertion here. Its principal points are:

1. "That the Philosophical Colledge be situated within one, two, or (at farthest) three miles of London, and, if it be possible, to find that convenience upon the side of the river, or very near it.

2. "That the revenue amount to £4000 a year.

3. "That the company consist of;—twenty philosophers or professors; sixteen young scholars, servants to the professors; a chaplain; a baily for the revenue; a manciple for the provisions of the house; two gardeners; a master-cook; an under cook; a butler; an under-butler; a chirurgeon; two lungs, or chymical servants; a library-keeper, who is likewise to be apothecary, druggist, and keeper of instruments, engines, &c.; an officer to feed and take care of all beasts, fowl, &c.; a groom of the stable; a messenger; four old women to tend the chambers.

4. "That the salaries of the above professors and officers amount to £3285 per annum, leaving £715 for keeping up the College and grounds. That the College be built to consist of three fair quadrangular courts, and three large grounds with gardens, just after the manner of the *Chartreux* beyond sea."

The author then enters into details respecting the uses of the various apartments, and recommends that out of the twenty professors, sixteen shall be always resident, and four travelling in the four quarters of the world, in order that they may "give a constant account of all things that belong to the learning, and especially Natural Experimental Philosophy, of those parts." A section is devoted to the duties of all the Officers, and another to the school, which was to contain two hundred boys, who were to be educated "by a method for the infusing knowledge and language at the same time into them." In his conclusion he says, "If I be not much abused by a natural fondness to my own conceptions, there was never any project thought upon which deserves to meet with so few adversaries as this; for the Colledge will weigh, examine, and prove all things of nature, delivered to us by former ages, and detect, explode, and strike a censure through all false moneys, with which the world has been paid and cheated so long, and (as I may say) set the mark of the Colledge upon all true coins, that they may pass hereafter without any farther tryal Secondly, it will recover the lost inventions, and, as it were, drown'd lands of the ancients. Thirdly, it will improve all arts which we now have, and lastly, discover others which we yet have not[14]."

Sprat says that Cowley's proposition, though not carried out, yet tended to accelerate the foundation of the Royal Society. It certainly drew the attention of the learned to the necessity of having some institution, in which science might be investigated in an unprejudiced manner, and where the absurd and fatal errors concerning philosophy might be permanently eradicated.

Amongst the schemes to establish a scientific academy or college, that of the ingenious Sir William Petty must not be overlooked. It is entitled, *Advice to Mr. Samuel Hartlib, for the advancement of some*

---

[14] For the whole of Cowley's plan, see his Works, fol. London, 1668. Cowley subsequently wrote an Ode in praise of the Royal Society, quoted in another chapter.

*particular parts of learning*, and was published in 1648. The leading features are: the establishment, in the first place, of a Gymnasium Mechanicum, or College of Tradesmen; where able mechanics, being elected Fellows, might reside, rent free. The labours and experiments of these mechanics, Sir W. Petty conceived, would be of great value "to active and philosophical heads, out of which to extract that interpretation of nature whereof there is so little, and that so bad, yet extant in the world." Within the Gymnasium he proposed to build a *Noscomium Academicum*, a *Theatrum Botanicum*, an Observatory, Ménagerie, &c.; in short, that an Institution or Academy should be founded, whose members "would be as careful to advance arts, as the Jesuits are to propagate their religion." He further recommended that a work should be compiled, to be entitled *Vellus Aureum, sive Facultatum luciferarum descriptio magna*, in which "all practised ways of subsistence, and whereby men raise their fortunes, may be at large declared. There would not then be," he adds, "so many unworthy *fustian* preachers in divinity; in the law so many *pettyfoggers;* in physic so many *quacksalvers*, and in country schools so many *grammaticasters*." It is worthy of remark, that in this scheme Sir W. Petty recommends writings to be multiplied by means of an instrument which he invented, and for which Parliament granted him a patent for seventeen years. He called it his art of double writing, and described the instrument as being of "small bulk and price, easily made, and very durable." This is the prototype of the "manifold letter-writer" of modern times, which has merely accomplished what Sir William Petty effected in 1648[15].

Concurrently with these contemplated plans for building philosophical institutions, another scheme was entertained to establish an institution at Vauxhall, for the advancement of science.

In a curious letter from Hartlib to Boyle, dated Amsterdam, May 18, 1649, and preserved in the archives of the Society, is the following Memorandum: "Fauxhall is to be sett apart for publick uses, by which is meant making it a place of resort for artists, mechanicks, &c., and a dépôt for models and philosophicall apparatus." It is further proposed, that "experiments and trials of profitable inventions should be carried on," which, says the writer, "will be of great use to the Commonwealth."

Hartlib adds, that the late King (Charles I.) "designed Fauxhall for such an use."

In another letter to Boyle, dated May 1654, Hartlib[16] says, "The Earl of Worcester is buying Fauxhall from Mr. Trenchard, to bestow the use of that house upon Gaspar Calehof and his son, as long as they shall live, for he intends to make it a College of Artisans. Yesterday," he adds, "I was invited by the famous Thomas Bushel to Lambeth Marsh, to see part of that foundation."

The attention of Parliament was called to the state of learning at this period, as appears from the

---

[15] For a full account of this invention, see Ward's *Lives of the Gresham Professors*, p. 218.

[16] Evelyn states that "Hartlib was a public-spirited and ingenious person, honest and learned, and has propagated many useful things and arts." Milton's *Tractate of Education* is addressed to him.

Journals of the House of Commons, which record that on the 20th July 1653, a committee was appointed "for the advancement of learning[17]," which consisted of eighteen members. They met in the Duchy Chamber, but did not present any Report[18].

The unsettled state of public affairs presented, as before observed, an insurmountable obstacle to the establishment of any permanent institution for philosophical purposes. "The progress of all the sciences," writes Dr. Whewell, alluding to this period, "became languid for a while; and one reason of this interruption was, the wars and troubles which prevailed over almost the whole of Europe. The baser spirits were brutalized; the better were occupied by high practical aims and struggles of their moral nature. Amid such storms the intellectual powers of man could not work with their due calmness, nor his intellectual objects shine with their proper lustre[19]."

---

[17] Vol. VII. p. 288.

[18] The establishment of a large public Library in St. James's Park was also thought of.

[19] *Hist. Ind. Sc.*, Vol. III. p. 327.

# CHAPTER III.

The Restoration favourable to the establishment of a Philosophical Society—Burnet's Account of the Founders of the Royal Society—Bacon's Philosophy—His *Instauration of the Sciences*—*New Atalantis*—High opinion entertained of him by the eminent early Fellows of the Society—First Official Record of Royal Society—Rules and Regulations—Original Members—Design of the Society approved by Charles II.—Experiments proposed—Reporters of Experiments—Manner of conducting Elections—Officers and Servants of the Society—Meetings contemplated at the College of Physicians.

---

1655—60.

AS the last and darkest thunder-cloud is often succeeded by calm and sunshine, so was the "fatal year 1659" followed by the "wonderful pacifick year 1660"—a year standing prominently forth in the page of English history, as that of the Restoration of the house of Stuart after a series of civil wars which extended over a period of twenty years. "Then," says Dr. Sprat, "did these gentlemen (alluding to the philosophers who had been in the habit of meeting in Gresham College), finding the hearts of their countrymen inlarg'd by their joys, and fitted for any noble proposition; and meeting with the concurrence of many worthy men, who, to their immortal honour, had follow'd the king in his banishment, *Mr. Erskins*, *Sir Robert Moray*, *Sir Gilbert Talbot*, &c., began now to imagine some greater thing; and to bring out experimental knowledge from the *retreats* in which it had long hid itself, to take its part in the triumphs of that universal jubilee. And, indeed, philosophy did very well deserve that reward; having always been loyal in the worst of times; for though the king's enemies had gain'd all other advantages, though they had all the garrisons, and fleets, and ammunitions, and treasures, and armies, on their side, yet they could never, by all their victories, bring over the reason of men to their party[1]."

"The men that formed the Royal Society," says Bishop Burnet, "were Sir Robert Moray, Lord Brouncker, a profound mathematician, and Dr. Ward. Ward was a man of great search, went deep in mathematical studies, and was a very dexterous man, if not too dexterous; for his sincerity was much questioned. Many physicians and other ingenious men went into the Society for natural philosophy. But he who laboured most, at the greatest charge, and with the most success at experiments, was the Hon. Robert Boyle. He was a very devout Christian, humble, and modest almost to a fault, of a most spotless and exemplary life in all respects. The Society for philosophy grew so considerably, that they thought fit to take out a patent, which constituted them a body, by the name of the Royal Society[2]."

The year of the Restoration was peculiarly favourable to the establishment of a scientific society, and the study and investigation of science. During a long period, the country had been torn by political revolutions, which, after the death of Cromwell, threatened to end in complete anarchy, when the Restoration, though far from realising all that was expected, relieved

[1] Dr. Johnson observes: "It has been suggested that the Royal Society was instituted soon after the Restoration, to direct the attention of the people from public discontent." Works, Vol. x. p. 86.

[2] *Hist. Own Times*, Vol. I. p. 192.

men's minds from the pressure of political matters, and left them more at liberty for other pursuits[3].

"There arose at this time," says Dr. Whewell, alluding to the period antecedent to the epoch of Newton, "a group of philosophers, who began to knock at the door where truth was to be found, although it was left for Newton to force it open. These were the founders of the Royal Society[4]." We can readily suppose these men adopting the language in which Cicero addresses philosophy: *ad te confugimus; a te opem petimus; tibi nos, ut antea magnâ ex parte, sic nunc penitus totosque tradimus*[5]." But it must not be forgotten how much is due to Lord Bacon, who died only thirty-six years before the incorporation of the Royal Society. With a comprehensive and commanding mind, patient in inquiry, subtile in discrimination, neither affecting novelty, nor idolizing antiquity, Bacon formed, and in a great measure executed, his great work, on the *Instauration of the Sciences*, which being clearly connected in its main features with the Royal Society, connects itself with our inquiry. The design was divided into six capital divisions. The first proposes a general survey of human knowledge, and is executed in the admirable treatise, *The Advancement of Learning*. In this Lord Bacon critically examines the state of learning in its various branches at that period, observes and points out defects and errors, and then suggests proper means for supplying omissions and rectifying mistakes.

The second, and the most considerable part, is the *Novum Organum*, in which the author, rejecting syllogism as a mere instrument of disputation, and

---

[3] Sprat's *Hist.*, p. 58. [4] *Hist. Ind. Sci.* Vol. II. p. 145. [5] *Tusc. Disp.*

putting no trust in the hypothetical systems of ancient philosophy, recommends the more slow but more satisfactory method of induction, which subjects natural objects to the test of observation and experience, and subdues nature by experiment and inquiry.

It will be seen how rigidly the early Fellows of the Royal Society followed Bacon's advice.

The third part of the work is the *Sylva Sylvarum*, or history of nature, which furnishes materials for a natural and experimental history, embracing all the phenomena of the universe.

The fourth part, or *Scala Intellectus*, sets forth the steps or gradations by which the understanding may regularly ascend in philosophical inquiries; and is evidently intended as a particular application and illustration of the author's method of philosophizing.

The fifth part, or *Anticipationes Philosophiæ Secundæ*, was designed to contain philosophical hints and suggestions; but nothing of this remains except the title and scheme.

The sixth portion was intended to exhibit the universal principles of natural knowledge deduced from experiments, in a regular and complete system; but this the author despaired of being himself able to accomplish. Having laid the foundation of a grand and noble edifice, he left the superstructure to be completed by the labours of future philosophers.

Indeed, he tells us, "I have done enough, if I have constructed the machine itself and the fabric, though I may not have employed, or moved it[6]." No writer seems ever to have felt more deeply than Bacon, that he properly belonged to a later and more enlightened

---

[6] *Interpretation of Nature.* Works, Vol. xv. p. 105.

age; a sentiment which he has touchingly expressed in that clause of his testament where he "bequeaths his name to posterity, after some time be past over."

It is, however, in his *New Atalantis* that we have the plan of such an institution as the Royal Society more distinctly set forth.

Describing this imaginary establishment, he says, "The end of our foundation is the knowledge of causes, and secret motions of things; and the enlarging of the bounds of human empire, to the effecting of all things possible. The preparations and instruments are—large and deep caves for coagulations, indurations, refrigerations, and conservation of bodies,—high towers for meteorological phenomena; great lakes, both salt and fresh, whereof we have use for the fish and fowl; violent streams and cataracts which serve us for many motions; artificial wells and fountains; great and spacious houses for experiments; certain chambers of health, where we qualify the air as we think good and proper, for the cure of divers diseases and preservation of health; large and various orchards and gardens; parks and inclosures of all sorts of beasts and birds; brewhouses, bakehouses, kitchens, dispensatories or shops of medicines, furnaces, perspective houses, sound houses, where we practise and demonstrate all sounds and their generation; perfume-houses, engine-houses, mathematical-houses, &c." These are called the riches of Solomon's house. The employments of the Fellows are then described: "We have twelve that sail into foreign countries, who bring in the books and patterns of experiments of all other parts. These we call merchants of light. We have three that collect the experiments which are in all books. These we call depredators. We have three that collect experiments of all mechanical arts, and also of liberal sciences; and also of practices which are not brought into arts. These we call mystery-men. We have three that try new experiments, such as themselves think good. These we call pioneers, or miners. We have three that draw the experiments of the former four into titles and tablets, to give the better light for the drawing of observations and axioms out of them. These we call compilers. We have three that bind themselves looking into the experiments of their fellows, and cast about how to draw out of them things of use and practice for man's life and knowledge, as well for works as for plain demonstration of causes, means, natural divinations, and the easy and clear discovery of the virtues and parts of bodies. These we call doing men, or benefactors. Then after divers meetings and consults of our whole number, to consider of the former labours and collections, we have three that take care, out of them, to direct new experiments, of a higher light, more penetrating into nature than the former. These we call lamps. We have three others that do execute the experiments so directed, and report them. These we call inoculators. Lastly, we have three that raise the former discoveries by experiments into greater observations, axioms, and aphorisms. These we call interpreters of nature[7]."

Rawley in his Preface to the *Atalantis* says, "This fable my lord devised, to the end that hee might exhibite therein a modell or description of a college, instituted for the interpreting of nature, and the producing of great and marvellous works for the benefit

---

[7] Works, Vol. II. pp. 364—377.

of men, under the name of *Solomon's House, or the College of the Six Dayes Works;* and even so far his lordship hath proceeded as to finish that part. Certainly the model is more vast and high than can possibly be imitated in all things, notwithstanding most things therein are within men's power to effect. His lordship thought also, in this present fable, to have composed a frame of laws; or of the best state or mould of a commonwealth; but foreseeing it would be a long work, his desire of collecting the Natural History diverted him, which he preferred many degrees before it[8]."

Tennison observes, speaking of the *New Atalantis*, 'Neither do we here unfitly place this fable, for it is the model of a college, to be instituted by some king, who philosophizeth for the interpreting of nature and the improving of arts."

Sprat was so fully impressed with the wisdom of Bacon's design that, in alluding to him, he writes, "If my desires could have prevailed with some excellent friends of mine, who engaged me to write this work, there should have been no other preface to my account of the Royal Society but some of his writings."

---

[8] The *New Atalantis* is well deserving of perusal. In 1660, an attempt was made to complete Bacon's sketch, in a work published under the title, "*New Atalantis*, begun by the Lord Verulam, Viscount St. Alban's, and continued by R. H., Esquire. Wherein is set forth a platform of monarchial government, with a pleasant intermixture of divers rare inventions and wholsom customs, fit to be introduced into all kingdoms, states, and commonwealths. London, printed for John Crooke, at the signe of the Ship, in St. Paul's Church-yard, 1660." Concerning this work, which is curious, Tennison says, "This supplement has been lately made by another hand, a great and hardy adventure, to finish a piece after Lord Verulam's pencil."

The testimony of Oldenburg may be adduced to show that Bacon was held to be one of the greatest, if not the chief, founder of the experimental school of philosophy. "The enrichment of the storehouse of natural philosophy was a work," he says, "begun by the single care and conduct of the excellent Lord Verulam, and is now prosecuted by the joint undertakings of the Royal Society[9];" and, at a subsequent period, he observes, "When our renowned Lord Bacon had demonstrated the methods for a perfect restoration of all parts of real knowledge, *the success became on a sudden stupendous*, and effective philosophy began to sparkle, and even to flow into beams of bright shining light all over the world[10]."

Boyle, in his voluminous works, which extend to five large folios, frequently commemorates and honours the name of Bacon. In his treatise on the *Mechanical Origin of Heat and Cold*, he tells us that "Bacon was the first among the moderns who handled the doctrine of heat like an experimental philosopher;" in his *Considerations touching Experimental Essays in General*, that "he had made considerable collections, with the view of following up Bacon's plan of a natural history;" in his *Experiments and Observations touching Cold*, he extols Bacon as "the great ornament and guide of the philosophical historians of nature;" in his *Excellency of Theology*, he says that Bacon was "the great restorer of physics, and had traced out the most useful way to make discoveries;" and he writes in his *Essay on the Usefulness of Experimental Philosophy*, "it was owing to the

---

[9] *Phil. Trans.*, No. xx. p. 391.

[10] Ibid., Preface to *Phil. Trans. for* 1672.

sagacity and freedom of Lord Bacon, that men were then pretty well enabled both to make discoveries, and to remove the impediments that had hitherto kept physics from being useful."

Various other writers of this period might be quoted, who pay grateful homage to Bacon, for the service he rendered to science, some calling him the "Patriarch of Experimental Philosophy[11]." "If," says Dr. Whewell, "we must select some one philosopher as the hero of the revolution in scientific method, beyond all doubt, Francis Bacon must occupy the place of honour[12]."

"It has been attempted by some," writes a distinguished philosopher of our day, "to lessen the merit of Bacon's great achievement by showing that the inductive method had been practised in many instances, both ancient and modern, by the mere instinct of mankind; but it is not the introduction of inductive reasoning as a new and hitherto untried process which characterises the Baconian philosophy, but his keen perception, and his broad and spirit-stirring, almost enthusiastic announcement of its importance as the alpha and omega of science, as the grand and only chain for the linking together of physical truths, and the eventual key to every discovery and every application. It is on this account that Bacon, though his actual contributions to the stock of physical truths were small, is justly entitled to be looked upon as the great reformer of science[13]."

---

[11] See particularly Glanvill's *Plus Ultra, or The Progress and Advancement of Knowledge since the days of Aristotle*, 8vo. London, 1668.

[12] *Phil. Ind. Sci.*, Vol. II. p. 230.

[13] Herschel, *Nat. Phil.*, p. 114.

It is not a little singular that whilst some English writers have ascribed the origin of the Royal Society to foreign influences, there are several continental philosophers, who trace the rise of their academies to effects produced by the writings of Bacon[14]. Sorbière, who acted for some time as secretary to one of those associations of French *savans* which existed before the Academy of Sciences was founded, says, speaking of Bacon:—*Ce grand homme est, sans doute, celuy qui a le plus puissamment solicité les interests de la physique, et excité le monde à faire des experiences*[15]."

The Abbé Gallois says, *On peut dire que ce grand Chancelier est un de ceux qui ont le plus contribué a l'avancement des sciences*[16]*;* and Puffendorf has recorded, that "it was Bacon who raised the standard, and urged on the march of discovery; so that if any improvements have been made in philosophy in this age, there has been not a little owing to that eminent philosopher."

Whilst the memory of this great man was cherished, and the spirit of his philosophy abroad, the establishment of the Royal Society was accomplished. A great number of eminent men existed at that period in England, nearly all of whom were warmly interested in the progress of science; and it thus only required the cessation of domestic troubles, to cause their attention to be turned to experimental philosophy.

In 1660, the meetings at Gresham College were revived, and were, according to the authority of Dr. Wallis, attended by an increased number of persons;

---

14 See Buchneri, *Acad. Nat. Curi. Hist.*

15 *Rélation d'un Voyage en Angleterre*, p. 678.

16 *Journal des Savans*, 1666.

and on the 28th November in that year, the first Journal-book of the Society, a plain unpretending volume, bound in basil, yet destined to receive great names, and to be the record of important scientific experiments, was opened, with the following 'Memorandum,' which is the first official record of the Royal Society.

"Memorandum that Novemb. 28. 1660, These persons following, according to the usuall custom of most of them, mett together at Gresham Colledge to heare Mr. Wren's lecture, viz. The Lord Brouncker, Mr. Boyle, Mr. Bruce, Sir Robert Moray, Sir Paul Neile, Dr. Wilkins, Dr. Goddard, Dr. Petty, Mr. Ball, Mr. Rooke, Mr. Wren, Mr. Hill. And after the lecture was ended, they did, according to the usual manner, withdrawe for mutuall converse. Where amongst other matters that were discoursed of, something was offered about a designe of founding a Colledge for the promoting of Physico-Mathematicall Experimentall Learning. And because they had these frequent occasions of meeting with one another, it was proposed that some course might be thought of, to improve this meeting to a more regular way of debating things, and according to the manner in other countryes, where there were voluntary associations of men in academies, for the advancement of various parts of learning, so they might doe something answerable here for the promoting of experimentall philosophy.

"In order to which, it was agreed that this Company would continue their weekly meeting on Wednesday, at 3 of the clock in the tearme time, at Mr. Rooke's chamber at Gresham Colledge; in the vacation, at Mr. Ball's chamber in the Temple. And towards the defraying of occasionall expenses, every one should, at his first admission, pay downe ten shillings, and besides engage to pay one shilling weekly, whether present or absent, whilest he shall please to keep his relation to this Company. At this Meeting Dr. Wilkins was appointed to the chaire, Mr. Ball to be Treasurer, and Mr. Croone, though absent, was named for Register.

"And to the end that they might the better be enabled to make a conjecture of how many the elected number of this Society should consist, therefore it was desired that a list might be taken of the names of such persons as were known to those present, whom they judged willing and fit to joyne with them in their designe, who, if they should desire it, might be admitted before any other[17]."

Upon which, this following Catalogue was offered:

| | |
|---|---|
| Lord Hatton. | Dr. Ward. |
| Mr. Robert Boyle. | Dr. Wallis. |
| Mr. Jones. | Dr. Glisson. |
| Mr. Coventry. | Dr. Bates. |
| Mr. Brereton. | Dr. Ent. |
| Sir Kenelme Digby. | Dr. Scarburgh. |
| Sir Ant. Morgan. | Dr. Phrasier. |
| Mr. John Vaughan. | Dr. Coxe. |
| Mr. Evelyn. | Dr. Merrett. |
| Mr. Rawlins. | Dr. Whistler. |
| Mr. Matthew Wren. | Dr. Clarke. |
| Mr. Slingsby. | Dr. Bathurst. |
| Mr. Henshaw. | Dr. Cowley[18]. |
| Mr. Denham. | Dr. Willis. |
| Mr. Povey. | Dr. Henshaw. |
| Mr. Wilde. | Dr. Ffinch. |

---

[17] Jour. Book, Vol. I. p. 1.

[18] He had been created M.D. at Oxford, Dec. 2, 1657. Wood, *Fas. Oxon.*

Dr. Baines.
Dr. Wren.
Mr. Smith.
Mr. Ashmole.
Mr. Newburg.
Mr. Austen.
Mr. Oldenburg.
Mr. Pett.
Mr. Croone[19].

On the following Wednesday, being the 5th December, a Meeting was held, when, it is recorded in the Journal-book, "Sir Robert Moray brought in word from the court, that the King had been acquainted with the designe of this Meeting. And he did well approve of it, and would be ready to give encouragement to it."

"It was ordered that Mr. Wren be desired to prepare against the next meeting for the Pendulum Experiment.

"That Mr. Croone be desired to looke out for some discreet person skilled in short-hand writing, to be an amanuensis.

"It was then agreed that the number be not increased, but by consent of the Society who have already subscribed their names: till such time as the orders for the constitution be settled.

"That any three or more of this company (whose occasions will permit them,) are desired to meete as a Committee, at 3 of the clock on Fryday, to consult about such orders in reference to the constitution, as they shall think fitt to offer to the whole company, and so to adjourne *de die in diem*." Under the above date of the 5th Dec. 1660, the first page of the Journal-book contains the following obligation:

[19] Locke's name will be missed here: he joined the Society a few years after its Incorporation, and contributed papers to the Transactions.

"Wee whose names are underwritten, doe consent and agree that wee will meet together weekely (if not hindered by necessary occasions), to consult and debate concerning the promoting of experimentall learning. And that each of us will allowe one shilling weekely, towards the defraying of occasionall charges. Provided that if any one or more of us shall thinke fitt at any time to withdrawe, he or they shall, after notice thereof given to the Company at a meeting, be freed from this obligation for the future."

To this are attached the signatures of all those persons comprised in the Catalogue of names prepared at the meeting on the 28th of November, as also, of seventy-three others, who were subsequently elected into the Society, as may be seen in the Journal-book.

On the 12th December another Meeting was held, when the following business was transacted:—

"It was referred to my Lord Brouncker, Sir Robert Moray, Sir Paul Neil, Mr. Matthew Wren, Dr. Goddard, and Mr. Christopher Wren, to consult about a convenient place for the weekly meeting of the Society.

"It was then voted that no person shall be admitted into the Society without scrutiny, excepting only such as are of the degree of Barons or above.

"Sir Kenelme Digby, Mr. Austen, and Dr. Bates, were then by vote chosen into the Society.

"That the stated number of this Society be five and fifty. That twenty-one of the stated number of this Society be the *quorum* for Elections.

"That any person of the degree of Baron or above may be admitted as supernumerarys, if they shall desire it, and will conforme themselves to such orders as are or shall be established.

"Whereas it was suggested at the Committee that

the Colledge of Physitians would afford convenient accommodation for the meeting of this Society; uppon supposition that it be graunted and accepted of, it was thought reasonable, that any of the Fellowes of the said Colledge, if they shall desire it, be likewise admitted as Supernumerarys, they submitting to the Lawes of the Society, both as to the pay at their admission, and the weekly allowance; as likewise the particular works or tasks that may be allotted to them.

"That the Publick Professors of Mathematicks, Physick, and Naturall Philosophy, of both Universitys, have the same priviledge with the Colledge of Physitians, they paying as others at their admission, and contributing their weekely allowance and assistance, when their occasions do permitt them to be in London.

"That the *quorum* of this Society be nine for all matters excepting the Businesse of Elections.

"Concerning the Manner of Elections.

"That no man shall be elected the same day he is proposed. That at the least twenty-one shall be present at each election.

"That the Amanuensis doe provide severall little scroles of paper of an equall length and breadth, in number double to the Society present. One halfe of them shall be marked with a crosse, and being roled up shall be lay'd in a heap on the table, the other halfe shall be marked with cyphers, and being roled up shall be lay'd in another heap. Every person coming in his order shall take from each heap a role, and throwe which he please privately into an urne, and the other into a boxe. Then the Director, and two others of the Society, openly numbering the crossed roles in the urn, shall accordingly pronounce the election.

"That if two-thirds of the present number do consent uppon any scrutiny, that election to be good, and not otherwise.

"CONCERNING THE OFFICERS AND SERVANTS OF THE SOCIETY.

"The standing Officers of this Society to be three, that is to say, a President or Director, a Treasurer, and a Register. The President to be chosen monthly.

"The Treasurer to continue one yeare, as also the Register.

"That there be likewise two servants belonging to this Society, an Amanuensis, and an Operator.

"That the Treasurer doe every quarter give in an account of the Stock in his hand, and all disbursements made to the President or Director, and any three others to be appointed by the Society: who are to report it to the Society.

"That any bill of charges brought in by the Amanuensis and Operator, and subscribed by the President and Register for any experiment made, and subscribed by the Curators of the experiment, or the major part of them, be a sufficient warrant to the Treasurer for the payment of that sum.

"That the Register provide three bookes, one for the statutes and names of the Society, another for experiments and the result of debates: and a third for occasionall orders.

"That the salary of the Amanuensis be 40*l.* per annum, and his pay for particular business at the ordinary rate, either by the sheet or otherwise, as the President and Register can best agree with him.

"That the salary of the Operator be foure pounds by the yeare, and for any other service, as the Curators who employ him shall judge reasonable.

"That at every meeting, three or more of the Society be desired that they would please to be reporters for that meeting, to sitt at table with the Register and take notes of all that shall be materially offered to the Society and debated in it, who together may form a report against the next meeting to be filed by the Register.

"When the admission-money comes to 20*l.*, then to stop."

It had been contemplated to find apartments for the Society in the College of Physicians, which was then situated in Knight Rider Street. This plan, however, was not carried into effect, though there is every reason to believe that the members of the College were very favourably disposed towards the infant Society of Philosophers.

---

## CHAPTER IV.

Large proportion of Physicians amongst the early Members of the Society—Profession of Medicine much cultivated at that period—Account of College of Physicians—Harvey's Discovery of the Circulation of the Blood supported by the Royal Society—Gresham College chosen as a place of meeting—Sir Thomas Gresham's Will—Gresham Professors—Description of College—Manner of holding the Meetings—Superstitions still believed in—Witchcraft—Touching for the Evil—Greatrix the Stroker—Believed in by Boyle—May-Dew—Virgula Divina—Happy effect exercised on these Superstitions by the labours of the Society.

---

### 1660—65.

AMONG the names of persons recorded as likely to promote the objects of the Society, a large proportion, as may have been observed, were attached to the profession of Medicine.

Biology, or the Science of Life, more particularly as applied to man, was cultivated with considerable diligence at the period of the foundation of the Royal Society; having received an extraordinary impetus by Harvey's immortal discovery of the circulation of the blood. This went far towards destroying those extraordinary hypotheses of Paracelsus and others, described in Sprengel's *History of Medicine*, where spirits, good and evil, are made to work within man[1].

The science of medicine was honoured by having had, long antecedently to this period, a College spe-

---

[1] Paracelsus affirmed that digestion was carried on by the Demon *Archæus*, who lived in the stomach. See Spr. III. 468. It is remarkable that this doctrine was subsequently received and expanded by Van Helmont.

cially devoted to its high purposes; and as a great number of the members of this institution assisted materially in founding, and promoting the objects of the Royal Society, it will be desirable to give a brief account in this place of the Institution. On the 23rd September, 1518, the College was incorporated by letters patent, granted to Thomas Linacre and others, who were constituted a perpetual "Commonalty or Fellowship of the Faculty of Physic." To Linacre is due the merit of establishing the College. He was born at Canterbury about 1460. He studied at Oxford, Bologna, and Florence, and is said to have been the first Englishman who read Aristotle and Galen in the originals. He studied natural philosophy and medicine at Rome, graduated in physic at Padua, and on his return home received the degree of M. D. at Oxford, where he gained great reputation by his medical lectures and classical knowledge. "He acquired," says Dr. Elliotson, "immense practice, and stood without a rival at the head of his profession; becoming physician to Henry VII., and VIII., and to Edward VI.; and not through interest, accident, caprice, or subserviency,—which have raised so many without the education of the scholar and man of science, or more than a scanty amount of professional knowledge and skill, to such posts,—but through the force of his attainments. To him it could not be said, as it was to Piso by Cicero, *Obrepsisti ad honores errore hominum.* He was perfectly straightforward, a faithful friend, the ready promoter of all the meritorious young, and kind to every one. To such a man the spectacle of brutally-ignorant pretenders treating the sick all over the kingdom without restraint, must have been distressing; and the duty of exerting his great influence with the government to reform the practice of his profession, must have been felt by him overwhelming[2]." It was when the sweating sickness, as it was called, raged with such fearful violence as not only to alarm the people generally, but even the carefully protected court, that Linacre brought his plan of a College of Physicians before Cardinal Wolsey, who, at the time, exercised almost unlimited power. He regarded the scheme favourably, and its establishment followed as a matter of course. The first meeting of the new Society took place at Linacre's house, No. 5, Knight Rider Street, a building known as the Storehouse, which he gave to the College, and which still remains in their possession. But the science of Medicine was not advanced by Linacre. "We are indebted to him," says Dr. Elliotson in his interesting Oration, "for no original observation, no improvement in practice." Caius, who flourished fifty years after Linacre, was a great benefactor to the College, increasing its reputation by his scientific attainments. He studied anatomy at Padua under Vesalius of Brussels, whose great work *De Humani Corporis Fabricâ*, is yet considered a splendid monument of art, as well as science[3]. Caius erected a statue to Linacre's memory in St. Paul's, and endowed Gonville College at Cambridge with estates for the maintenance of three fellows and twenty scholars; two of the former were required to be physicians, and three of the latter medical students[4]. The heal-

[2] *Harveian Oration for* 1846, p. 39.

[3] The figures in this work are stated to have been designed by Titian. See Cuvier's *Leçons sur l'Hist. des Sci. Nat.*

[4] Caius is the original of the ridiculous French doctor in the *Merry Wives of Windsor.*

ing art, however, advanced but slowly. The extreme clumsiness and cruelty with which operations were performed, even subsequently to the above period, would scarcely be credited, had we not authentic descriptions of them by the operators, still in our possession. Thus Fabricius of Acquapendente, the eminent professor at Padua, and preceptor to the immortal Harvey, describes what he considered an improved and easy operation in the following terms: "If it be a moveable tumour, I cut it away with a red-hot knife, that sears as it cuts; but if it be adhered to the chest, I cut it without bleeding or *pain*, (!) with a wooden or horn knife soaked in aqua-fortis, with which, having cut the skin, I dig out the rest with my fingers"!! When the surgeons of Edinburgh were incorporated, it was required as a pre-requisite that they should be able to read and write, "to know the anatomie, nature, and complexion of everie member of humanis body, and lykewayes to know all vaynes of the same, that he may make flewbothemie in dew time." These were all the professional qualifications considered necessary at that period, so that we must not feel any surprise at the low state of medicine and surgery. Sir William Petty informs us, that even in his time the proportion of deaths to cures in the Hospitals of St. Bartholomew and St. Thomas was 1 to 7; whilst we know by subsequent documents, that in the latter establishment, during 1741, the mortality had diminished to 1 in 10; during 1780 to 1 in 14; during 1813 to 1 in 16; and in 1827, out of 12,494 patients under treatment, only 259 died, or 1 in 48. His Royal Highness the Duke of Sussex justly said, in one of his addresses from the chair of the Royal Society, "Such is the advantage which has already been derived from the improvement of medical science, that comparing the value of life as it is now calculated, to what it was a hundred years ago, it has absolutely doubled." And Sir Astley Cooper asserted that the human frame was better understood in his time by students, than it had been previously by professors.

The great Harvey was born in 1578, at Folkestone in Kent. After studying at Cambridge, he went to Padua, where the fame of Fabricius attracted medical students from all parts of Europe. There, "excited by the discovery of the valves of the veins, which his master had recently made, and reflecting on the direction of the valves, which are at the entrance of the veins into the heart, and at the exit of the arteries from it, he conceived the idea of making experiments in order to determine what is the course of the blood in its vessels. He found that when he tied up veins in various animals, they swelled below the ligature, or in the part furthest from the heart; while arteries, with a like ligature, swelled on the side next the heart. Combining these facts with the direction of the valves, he came to the conclusion, that the blood is impelled by the left side of the heart in the arteries to the extremities, and thence returns by the veins into the right side of the heart. He showed, too, how this was confirmed by the phenomena of the pulse, and by the results of opening the vessels. He proved, also, that the circulation of the lungs is a continuation of the larger circulation; and thus the whole doctrine of the double circulation was established. Harvey made his experiments in 1616 and 1618. It is commonly said that he first promulgated his opinion in 1619, but the manuscript of his lecture, which he delivered before the College of Physicians, and which is preserved in the British Museum, refers them to April 1616[5]." It is proper to mention that Caldwall, a Fellow of Brazenose College, and President of the College of Physicians, had endowed, in conjunction with Lord Lumley, an anatomical and surgical lectureship; and it was through this channel that the announcement of the circulation of the blood was made[6].

It is not in accordance with the plan of this work, to enter into any details respecting the violent opposition that Harvey's announcement of the circulation of the blood met with. The fact is well known, but he had nevertheless the singular pleasure, often denied to discoverers, of seeing his discovery generally adopted during his lifetime, though we find a candidate for a medical degree taking for the title of his inaugural dissertation in 1642, under the auspices of the President Chasles, *Ergo motus sanguinis* NON *circularis*. And in 1672, thirty years later, another candidate, under the patronage of another president, selected for the title of his dissertation, *Ergo sanguinis motus circularis* IMPOSSIBILIS[7]. Aubrey, in his Manuscript of the Natural History of Wiltshire, preserved in the library of the Royal Society, writes, alluding to Harvey, "He told me himself that upon his publishing that booke, he fell in his practice extremely;" and Dr. Elliotson informs us, that "the medical profession stigmatized Harvey as a fool."

---

[5] Whewell's *Hist. Ind. Sci.*, Vol. III. p. 439.

[6] Elliotson, *Har. Ora.*, p. 43.

[7] Ibid., *loc. cit.*, where a most interesting account is given of the difficulties which Harvey experienced.

But, as has been stated, truth eventually prevailed; and at the period of the establishment of the Royal Society, the circulation of the blood was generally admitted. It is right to record, that Harvey was always supported by the Fellows of the Society. Medical science, and the College of Physicians, had acquired great reputation, and the Fellows of the College were regarded as valuable accessions to the young Society of Philosophers. To render this notice of the College of Physicians more complete, it may be added, that about the period of the accession of Charles I. the Fellows removed from Knight Rider Street to Amen Corner, where they purchased the lease of a house from the Dean and Chapter of St. Paul's. Here Harvey erected an elegantly furnished convocation-room and museum, and in 1656, the year before his decease, he added to those gifts the assignment of a farm of the then value of £56 per annum, to defray the expenses of a monthly collation, as also of an anniversary feast, and for the establishment of an annual Latin oration[8]. The great fire of London entirely destroyed the premises of the College, and the greater portion of the library. In 1669, a piece of ground was purchased in Warwick Lane, for the purpose of building a College, the design for which was furnished by Sir Christopher Wren. The new

[8] "The monthly collation is kept up in the form of coffee and cakes, and the annual Oration is continued in Latin, but the general feast has not been celebrated for five-and-twenty years, the money being applied, as there is authority for doing, to the solid purposes of the College, whose means have never been ample, and are all spent in the performance of what are its imperative duties." Elliotson, *Har. Ora.*

edifice was opened in 1674, under the presidency of Sir George Ent. Here the Fellows remained until 1825, when, as Dr. Macmichael observes in *The Gold-headed Cane*, "the change of fashion having overcome the *genius loci*," they removed to their present residence at the corner of Pall Mall East and Trafalgar Square, which was erected by Sir R. Smirke, and opened on the 25th July, 1825, with a Latin oration delivered by the President, Sir Henry Halford[9].

We must now return to the Royal Society:—At a Meeting held on the 12th December, it was "ordered that the next Meeting should be at Gresham Colledge, and so from weeke to weeke till further order." By this it is evident that the idea of meeting at the College of Physicians was abandoned; and as Gresham College may be regarded as the cradle of the Royal Society, where they assembled for many years, some account of that building may here be very properly introduced[10].

---

[9] The old College in Warwick Lane is now occupied by braziers and brass-founders.

[10] The young Society had not met many months at Gresham College before a poem was written, entitled, *In praise of the choice company of Philosophers and Witts, who meet on Wednesdays, weekly, at Gresham College.* It is signed W. G. (probably William Glanvill). The poem exists in the form of a MS. in the British Museum. There are twenty-eight sextains. The following quotation is a fair specimen:—

> "The Merchants on the Exchange doe plott
> To encrease the Kingdom's wealthy trade;
> At Gresham College a learned knott,
> Unparallel'd designs have lay'd,
> To make themselves a corporation,
> And know all things by demonstration.

In 1575, Sir Thomas Gresham made a will, in which he left one moiety of the building of the Royal Exchange, with all pawns, shops, cellars, vaults, messuages, tenements, &c. unto the Mayor, Commonalty, and Citizens of London, and to their successors, upon trust, to perform certain payments, and other intents hereafter limited—and willed, that the said Mayor, Corporation, and their successors, should every year give and distribute for the sustentation of four persons, to be chosen by the said Mayor and Commonalty, meet to read lectures of divinity, astronomy, music, and geometry, within his mansion situated in Bishopsgate Street, the sum of 200*l.* yearly, that is, 50*l.* per annum to each of the said readers; and out of the other moiety of the buildings of the Royal Exchanges and their appurtenances, which he bequeathes to the "Commonalty of the mystery of the Mercers of London," he willed that they, and their successors,

---

This noble learned corporation,
Not for themselves are thus combin'd,
But for the publick good o' th' nation,
And general benefit of mankind.
These are not men of common mould;
They covet fame, but condemn gold.

This College will the whole world measure,
Which most impossible conclude,
And navigation make a pleasure,
By finding out the longitude:
Every Tarpaulian shall then with ease
Saile any ship to the Antipodes.

The College Gresham shall hereafter
Be the whole world's University;
Oxford and Cambridge are our laughter;
Their learning is but pedantry;
These new Collegiates do assure us,
Aristotle's an ass to Epicurus."

should give and pay for the finding and sustentation of three persons, by them from time to time to be chosen, meet to read lectures on Law, Physick, and Rhetorick, within his said Mansion House in Bishopsgate Street, the sum of 150*l.* per annum, that is, 50*l.* yearly to each of the said readers."

This noble gift was subsequently confirmed by act of parliament; and, after the decease of Lady Anne Gresham, which occurred in 1596, lecturers were chosen, who delivered lectures, "to the great delight of many, both learned, and lovers of learning."—It is remarkable that, although Sir Thomas Gresham evidently intended that a lecture should be delivered daily throughout the year, by one of the seven lecturers, yet they were given only in term time, which created so much disappointment amongst the public, that a petition was preferred to the Trustees for managing the affairs of the College, praying that the Founder's Will, which required the lectures to be read daily, (as the petitioners understood) might be put in execution. This petition was taken into consideration by the Trustees, and the Professors heard in their defence, who cited authorities to prove that the word "daily," in the Founder's Will was an "academick word," and therefore to be understood as meaning the days in the Term-weeks only.

The result was, that, although the Trustees were divided in opinion, as to the intention of the founder, they acceded to the prayer of the petitioners so far only as to *enjoin* the Professors to read, not only in the broken weeks, but also a few days before the terms commenced. The Professors complied with the first condition of this injunction, but refused to lecture out of term-time, to the great discontent of the petitioners[11].

The Gresham Professors had apartments assigned to them in the college, which were most commodious and comfortable[12]. "Here," says Sprat, "the Royal Society has one publick room to meet in, another for a repository to keep their instruments, books, rarities, papers, and whatever else belongs to them: making use besides, by permission, of several of the other lodgings, as their occasions do require. And when I consider the place itself, methinks it bears some likeness to their design; it is now a Colledge, but was once the mansion-house of one of the greatest merchants that ever was in England[13]."

The following more precise description of Gresham College is taken from a curious pamphlet in the British Museum, entitled, *Account of the Proceedings in the Council of the Royal Society, in order to remove from Gresham College.* "The great hall, to which the ascent from the court is by but a few steps, is 37 foot long, near 20 foot broad, and 25, or 30 foot high. This spacious room is a noble entrance to the rest of the apartments of the Royal Society. The next room is about 35 foot long, near 20 foot broad, and 13 foot high; and in this the Society always met upon St.

---

[11] A pamphlet was published in 1707, entitled, *An Account of the rise, foundation, progress, and present state of Gresham College, with the Life of the Founder; as also of some late endeavours for obtaining the revival and restitution of the Lectures there; with some Remarks thereupon.* 4to. London, 1707. See also Tooke's tract entitled, *An exact copy of the last will and testament of Sir Thomas Gresham.*

[12] Ward's *Lives of the Gresham Professors.*

[13] *Hist. Royal Society*, p. 93.

GRESHAM COLLEGE.

1 Gate into Bishopsgate Street.
2 Physic Professor's Lodgings.
3 Reading Hall.
4 Music Professor's Lodgings.
5 Observatory.
6 Geometry Professor's Lodgings.
7 Divinity Professor's Lodgings.
8 Physic Professor's Elaboratory.
9 Rhetoric Professor's Lodgings.
10 Astronomy Professor's Lodgings.
11 South, or Long Gallery.
12 West, or White Gallery.
13 Almshouses.
14 Law Professor's Lodgings.
15 Stable-Yard.

Andrew's Day, for their anniversary elections. The inner room, for their ordinary weekly meetings, is about 22 foot long and 18 foot broad. These three rooms are all upon the same floor; from the last, two or three steps conveys you into the gallery, which is 140 foot long and 13½ broad. Beyond this is the repository of their curiosities, which, with the two rooms adjoining, is about 90 foot long and 12 or 12½ broad. Besides all these commodious rooms within, they have the use of a fair colonnade under the gallery, and of a spacious area about 140 foot long and 107 foot broad."

Some idea of Gresham College may be gathered from the annexed drawing, copied from an authentic engraving in Ward's *Lives of the Gresham Professors*[14]. The concluding portion of the history of Gresham College is a melancholy detail. As the value of land in the city increased, the two trustee corporations thought much less about keeping up the lectures than of realizing large sums of money, by letting the ground on building leases.

We shall presently see, that when the Royal Society ceased to meet in the College, this institution was an object of contempt to the citizens, as previously to that period (1710) the lectures had become a mere empty name. It is true, that in 1706 some high-minded and intellectual individuals made great but unsuccessful exertions to revive the lectures. Petitions were now sent in to parliament, for leave to destroy the building; but though the government, in the reigns

[14] The yearly value of this mansion-house, with the gardens appertaining to it, was, at the time of Lady Gresham's death, 66*l*. 13*s*. 4*d*.

of William III. and George I. evinced their respect for the will of Sir Thomas Gresham, by rejecting these petitions, the legislature of 1767 (when, be it remembered, the patron of science, George III. was on the throne), passed an act, authorizing the destruction of the building, upon whose site was reared the *Excise-office!* For the poor sum of 500*l.* per annum the trustees agreed to demolish the College, and to part with all the land (the property really of the public), for the very unphilosophical purposes of an Excise-office[15].

"But," says Professor Taylor, "this was not all; not only were the citizens of London thus deprived of their College, with the spacious lecture-hall in which they had been accustomed to assemble, but another part of the Act compelled the trustees and guardians of this property to pay 1800*l.* for, and towards the expense of pulling down the same. That is, they were constrained, by an especial law, to commit a gross and flagrant violation of their trust, and to employ those very funds, which Sir Thomas Gresham had vested in them for the support and maintenance of his College, in demolishing and destroying it." "Am I wrong," pursues the Professor, "in asserting that this transaction has no parallel in any civilized country?

"Thus was this venerable seat of learning and science, founded by the munificence of one of your most eminent citizens, and hallowed by a thousand interesting associations,—the mansion in which successive

---

[15] It is right to mention, that the act empowered the Trustees to provide sufficient and proper places in which the seven Professors might read their Lectures.

monarchs had been entertained, in which princes had lodged and banqueted,—which, when London lay in ashes, had afforded shelter and refuge to its citizens, a residence to its chief magistrate, an Exchange for its merchants, and a home to the houseless;—thus was the hall in which Barrow, Briggs, Bull, and Wren had lectured, and the rooms where Newton, Locke, Petty, Boyle, Hooke, and Evelyn associated for the advancement of science,—rased to the ground[16]."

Prophetic indeed was the intimation which Sir T. Gresham received from his parent university of Cambridge: "If you design your institution to last, you will place it here."

According to Professor Taylor, the public journals of the time contain no account of any effort on the part of the citizens of London to perpetuate the existence of their College. The only notice of its destruction, beyond the bare record of the event, is contained in a short poetical satire, entitled *Gresham's Ghost, or a Tap at the Excise-office*, with this motto; *Is this house, which I have called by my name, become a den of robbers? Behold, even I have seen it, saith the Lord.* (Jer. vii. 11).

It is very much to be regretted that Dr. Sprat has not given more particulars of the early proceedings of the Society. His history is, unfortunately, more devoted to a defence of philosophy and the Society, than to any account of the labours of the first members. We have in consequence very scanty information of the mode, or, as Dr. Sprat calls it, "Ceremonies of their meetings." The little he gives is, however, worth quoting:—

[16] *Inaugural Lect.*, p. 58.

"Their time of meeting is every Wednesday, after the lecture of the Astronomy Professor; perhaps in memory of the first occasion of their rendezvouses.

"Their elections performed by ballotting[17], every Member having a vote; the candidates being named at one meeting, and put to the scrutiny at another.

"Their chief officer is the President[17]; to whom it belongs to call and dissolve their meetings; to propose the *subject*, to regulate the proceedings; to change the inquiry from one thing to another; to admit the members who are elected.

"Besides him they had at first a *Register*, who was to take notes of all that pass'd; which were afterwards to be reduced into their journals and register-books. This task was first performed by Dr. Croone. But they since thought it necessary to have two Secretaries[18].

"This," adds Dr. Sprat, "is all that I have to say concerning their *ceremonial part;* for the work which the Society proposes to itself being not so fine and easie as that of teaching is, but rather a painful digging and toiling in nature, it would be a great incumbrance to them to be straitened to many strict punctilioes, as much as it would be to an *artificer* to be loaded with many clothes while he is labouring in his shop[19]."

---

[17] The custom of ballotting had been introduced into England a short time only before this period. Wood, in the *Athenæ Oxonienses*, says, that a political club met in 1659, at Miles' Coffee-house, Westminster. "The gang had a ballotting box, and ballotted how things should be carried; which being not used or known in England before, upon this account the room every evening was very full. Vol. II. p. 439.

[18] This was written after the Society were incorporated.

[19] P. 94.

The Society being now located, addressed themselves to the great objects of their institution, with all the diligence and ardour of zealous seekers after truth. The almost unexplored storehouse of Nature was before them; every step revealed new wonders and new facts, which had great effect in demolishing the errors and superstitions of preceding centuries; for much as there was to do, as much almost remained to be undone.

Though past the middle of the seventeenth century, men calling themselves philosophers, and so styled by the multitude, yet cherished a belief in witchcraft, which, supported by Royal authority (for James I. wrote on demonology[20]), and countenanced by Bacon[21], was almost universally adopted by the

---

[20] See James VI. *Dæmonologie*, in which the royal author says "The fearfull abounding at this time in this country of these detestable slaues of the Dieul, the witches or enchanters, hath moved me (beloued reader) to despatch in part this following treatise of mine, not in any wise, as I protest, to serve for any show of my learning and ingine, but only (moued of conscience) to proceed thereby, so far as I can, to resolue the doubting hearts of many, both that such assaults of Satan are most certainly practised, and that the instruments thereof merit most severely to be punished."

[21] Bacon, though so wonderfully in advance of the age in which he lived, and one of those men, born *rebus agendis*, full of outward movement, had still too much of the mortal about him to shake off all the superstitions of the times. There is little doubt of his having been a believer in sympathetic cures as well as witchcraft, as the following curious extract from his works testifies: "The taking away of warts by rubbing them with somewhat that afterwards is put to waste and consume, is a common experiment; and I do apprehend it the rather, because of mine own experience. I had from my childhood a wart upon one of my fingers; afterwards, when I was about sixteen years old, being then at Paris, there grew upon both my hands a number of warts (at least an hundred),

people. During the civil wars, upwards of eighty individuals were executed in Suffolk alone for supposed witchcraft, upon the accusation of Hopkins, the notorious witch-finder; and such was the feeling in France, that Bodin the renowned lawyer, who assisted at the trial and condemnation of several witches, taught that the trial of this atrocious offence must not be conducted like other crimes, and that "whoever adheres to the ordinary course of justice, perverts the spirit of the law, both human and divine[22]."

Such an awful and fatal error, worthy of the darkest ages of superstition, could only be dissipated by the light of knowledge; and though the statutes against witchcraft were not repealed until 1735, it is cheering to know that (according to Hutchinson, who wrote on witchcraft), "there were but two

---

hundred), in a month's space. The English Ambassador's lady, who was a woman far from superstition, told me one day she would help me away with my warts; whereupon she got a piece of lard with the skin on, and rubbed the warts all over with the fat side, and amongst the rest, that wart which I had had from my childhood; then she nailed the piece of lard, with the fat towards the sun, upon a part of the chamber-window which was to the south. The success was, that within five weeks' space all the warts went quite away; and that wart which I had so long endured for company. But, at the rest I did little marvel, because they came in a short time, and might go away in short time again; but the going away of that which had stayed so long, doth yet stick with me. They say the like is done by the rubbing of warts with a green elder stick, and then burying the stick to rot in muck." *Sylva Sylvarum*.

[22] Whitelock states, that "in 1649 fourteen men and women were burnt at a little village near Berwick, for witchcraft, the entire population of the village consisting only of fourteen families." *Mem.* p. 450.

witches executed in England after the Royal Society published their *Transactions*, and one of these was in the year after their first publication[23]." Another absurd, though happily not so fatal belief, was that of curing scrofula by the Royal touch. This was practised from a very early date down to the reign of Queen Anne. Collier, in his *Ecclesiastical History*, says, "King Edward the Confessor was the first that cured this distemper, and from him it has descended as an hereditary miracle upon all his successors. To dispute the matter of fact, is to go to the excess of scepticism, to deny our senses, and to be incredulous even to ridiculousness." Charles II. entered London on the 29th of May 1660, and on the 6th of July following he began to touch for the evil. Evelyn gives a graphic sketch of the ceremony. "His Majesty sitting under his state in y^e^ banquetting house, the chirurgeons cause the sick to be brought or led up to the throne, where they kneeling, y^e^ King strokes their faces or cheekes with both his hands at once, at which instant a chaplaine in his formalities says, 'He put his hands upon them, and he healed them.' This is sayd to every one in particular. When they have been all touch'd, they come up againe in the same order, and the other chaplaine kneeling, and having angel[24] gold strung on white ribbon on his arme, delivers them one by one to his Ma^tie^, who puts them about the necks of the touched as they passe, whilst the first chaplaine repeats, 'That is y^e^ true light who

---

[23] The last execution of a witch took place in Sutherlandshire, in 1722.

[24] Pieces of money so called, from having the figure of an angel on them.

came into y^e world.' Then follows an Epistle (as at first a Gospel) with the Liturgy, prayers for the sick, with some alteration; lastly, y^e blessing; then the Lo. Chamberlaine and Comptroller of the Household bring a basin, ewer, and towell for his Ma^tie to wash[25]."

Aubrey, who believed implicitly in this and many other superstitions, has an anecdote in his *Miscellanies* which, as characteristic of the times, is worth quoting:—

"*Arise Evans*," he says, "had a fungous nose, and said it was reveal'd to him, that the King's hand would cure him: and at the first coming of King Charles II. in St. James's Park, he kiss'd the King's hand, and rubb'd his nose with it, which disturb'd the King, but cured him[26]."

There is something excessively ludicrous in this story. And we might fancy that the merry Monarch would cease for ever exercising his power when it was thus abused. Well might his voluptuous Majesty be "disturb'd," and much shocked,—and the more so, as we do not hear of any Lord Chamberlain being at hand with "basin, ewer, and towell."

It is worthy of record, that William Beckett, a surgeon, and Fellow of the Royal Society, in his *Free and impartial Enquiry into the antiquity and efficacy of touching for the King's Evil,* published in 1722, endeavours by the most rational arguments to confute this belief. That it existed amongst men possessed of high abilities, is proved by the fact that the Hon. Robert Boyle believed in the efficacy of the touch of Valentine Greatrix, who went by the name

---

[25] Evelyn's *Diary,* Vol. I. p. 312.
[26] *Miscellanies,* 8vo. London, 1696, p. 101.

of Greatrix the Stroker[27]; and who was said to cure the evil, when the King even failed[28]. The supposed cosmetic virtues of May-dew, collected before sun-rise, was another popular superstition; which there is reason to apprehend had its origin from an allegory, by which some village Zadig attempted to induce the

---

[27] See Boyle's Works, and *Phil. Trans.*, No. 256, where his name is spelt as above.

[28] Flamsteed went to Ireland for the purpose of being touched by Greatrix. In his *Autobiography* he says, "I was stroked by him all over my body, but found as yet no amends in anything, but what I had before." p. 16.

There is a letter in the archives of the Society, from Greatrix to the Archbishop of Dublin, in which he thus describes the circumstances that led him to undertake curing by touch: "I was moved by an impulse, which, sleeping or waking, in public or private, always dictated: I have given thee the gift of curing the King's evil. At first I wondered within myself what the meaning thereof should be, and was silent: at length, I told my wife thereof, and that I had no quiet within myself for this impulse, and that I did verily believe that God had given me the power of curing the evil. She little regarded what I said, telling me only I had conceived a rich fancy. Soon after, such was the providence of God, one William Maher, of the parish of Lismore, brought his son that had the evil in several places very grievous, and desired to know if I would cure him. Whereupon I went to my wife and told her she should now see whether my belief were a fancy or no, whereupon I put my hands on young Maher, desiring the help of the Lord Jesus, for his mercies' sake, whereupon the evil, which was as hard as possible for flesh and blood to be, dissolved and rotted within forty-four hours, run and healed, and so through God's mercy continues to this day." At first Greatrix merely touched the parts affected; but afterwards he made "passes," or stroked the limbs of his patients, which led to his being called 'Greatrix the stroker.'

We may appropriately add here, that the last person of note operated on in this way was Dr. Johnson; he was *touched* by Queen Anne in 1712.

maidens to attend to the wholesome observances of early rising and exercise.

That some, at least, of the Fellows of the Royal Society believed in this, appears from the fact, as will be seen in the sequel, that they were in the habit of going out at early morn, for the purpose of collecting the precious dew.

The *Virgula divina*, or divining rod, had also its votaries; and many were the miles of ground traversed by credulous men in quest of that, which the science of geology has now enabled us to find with almost unerring certainty[29].

We might swell this list, but have said enough to show how much ignorance and credulity prevailed at

---

[29] This superstition had its votaries long after the establishment of the Royal Society.

In the *Gentleman's Magazine*, Vol. XXI. 1751, is a long and very curious paper "On the manner of using the Divining Rod," from which the following is an extract: "The most convenient and handy method of holding the rod, is with the palms of the hands turned upwards, and the two ends of the rod coming outwards; the palms should be held horizontally as nearly as possible; the part of the rod in the hand ought to be straight, and not bent backward or forward. The upper part of the arm should be kept pretty close to the sides, and the elbows resting on them; the lower part of the arm making nearly a right angle with the upper, though rather a little more acute. The rod ought to be so held that in its working the sides may move clear of the little finger. But after all the directions that can be given, the adroit use of it can only be attained by practice and attention. A person who can use the rod tolerably may soon give the greatest sceptics sufficient satisfaction, *except they are determined* not to *be convinced*."

See also a curious work on this subject entitled, *Mémoire Physique et Médicinale, montrant les rapports evidens entre les Phénomènes de la Baguette Divinatoire du Magnétisme et de l'Electricité*, by M. Thouvenel. Paris, 1784.

the period of the institution of the Royal Society, amongst men of superior intellect and education[30]. It was a labour well worthy the men who met avowedly for the investigation and developement of truth, to inquire into these superstitions, and patiently and dispassionately to prosecute such experiments as should tend to eradicate them. It would indeed be difficult to over-estimate the great benefit that accrued to society by their destruction, and a lasting debt of gratitude is due to the Royal Society, for having been so essential an instrument in dispelling such fatal errors[31].

Let not the reader, therefore, when he smiles, as he assuredly will, at many of the seemingly absurd and ridiculous experiments tried by the Society, which he will find in the following pages, criticise them as mere folly, or the performances of empirics:—they were necessary to the welfare of science,—as much so as it is important to clear away a rotten foundation, ere a solid superstructure can be reared; and it will be seen, how year after year errors were blotted out, and new facts and truths developed[32].

---

[30] These superstitions were by no means confined to Englishmen. The great Tycho Brahe, at the close of the sixteenth century, was superstitiously anxious about presages. His biographers relate that, if he met an old woman on first going out of doors, or a hare on the road during a journey, he immediately turned back, from the persuasion of having met with an ill omen.

[31] Sir Walter Scott, in his *Demonology*, remarks, that the establishment of the Royal Society tended greatly to destroy the belief in witchcraft and superstition generally.

[32] England in the seventeenth century was happily, however, more liberal and enlightened than the South American Republics in the nineteenth. Mr. Darwin states, that a German Natural History

History Collector, left some caterpillars in a house at S. Fernando, under the charge of a girl to feed, that they might turn into butterflies. The authorities conceived that the Professor dealt in heresy and witchcraft; and accordingly, when he returned, he was arrested. *Jour.* p. 326. Had the English government looked with similar suspicion and ignorance on the early experiments of the Fellows of the Royal Society, half of them would have been tenants of the Tower.

---

## CHAPTER V.

Early Labours of the Society—Committee appointed to receive Experiments—Wren's Pendulum—Boyle's Air-Pump—Register Book opened—Questions sent to Teneriffe—Charles II. sends Loadstones to the Society—Evelyn's Communications—Experiments made at the Tower—Glass Bubbles sent from the King—Sir Robert Moray elected President—Memoir of him—Communication respecting Barnacles—Time of Meeting determined—Pecuniary Difficulties—Letter from Duke Leopold—Wren requested to make a Globe of the Moon for Charles II.—Boyle and Evelyn appointed Curators—Order to present Books to Society—Duke of Buckingham orders his Chemist to send Charcoal to the Society—Sir G. Talbot's Sympathetic Powder, and Cures—Virgula Divina—Charles II. puts questions concerning the Sensitive Plant—Wren's paper on Saturn—The King sends papers—Hour of Meeting changed from two to three p.m.—Genoese Ambassador visits the Society—Graunt dedicates his book to the Society—Unicorn's Horn—Society meet in Temple Church to see an Engine—Petition to the King—Incorporation of Society—Deputation to wait upon the King—Preamble to Charter.

---

### 1660—65.

THE Society's Journal-books contain such curious evidence of their early labours, and the state of science at that period, as to render no apology necessary for devoting a few pages to them in this place.

Under the date of December 19th, it is recorded, that—:

"Dr Petty and Mr. Wren be desired to consider the philosophy of Shipping, and bring in their thoughts to the company about it.

"That every man of this company be desired to bring in such experiments to a committee to be appoynted to that end, as he shall think most fit for the advancement of the general design of the Company.

"That Dr. Wilkins, and as many of the Professors of Gresham Colledge as are of this Society, or any three of them, be a Committee for the receiving of all such experiments.

"Mr. Wren, to bring in his accompt of the Pendulum experiment, with his explanation upon it, to be registered[1].

"December 26. It was ordered that Dr. Goddard be added to Dr. Petty and Mr. Wren for the experiments about Shipping; and that Sir Kenelme Digby be desired to afford them his assistance.

"That the persons above mentioned be desired to bring some experiment against the next day.

"Mr. Boyle, Mr. Oldenburg, Mr. Denham, Mr. Rawlins, Mr. Ashmole, Mr. Evelyn[2], and Mr. Henshaw, were proposed as members of the Society.

"January 2, 1660—1. The Society again met, when Lord Brouncker was desired to prosecute the experiments of the Recoyling of Gunns, and to bring it in against the next meeting, and Mr. Boyle, his Cylinder[3].

---

[1] Dr. Wilkins, in his *Essay towards a Real Character and a Philosophical Language*, published 1668, says, that Sir C. Wren was the first to suggest the determination of a standard measure of length, by the vibration of pendulum. Sprat also gives Wren the credit of making this invention, for he says, "it was never before attempted."

[2] Evelyn thus records his election into the Society: "I was now chosen by suffrage of the rest of the Members a Fellow of the Philosophic Society, now meeting at Gresham College, where was an assembly of divers learned gentlemen." *Diary*, Vol. I. p. 316.

[3] This refers to his Air-pump, which, according to Professor Powell, "he reduced to nearly its present construction." The reader will be interested to know that the original Air-pump alluded

"That Dr. Merritt be desired to bring in his history of Refineing.

"That Mr. Boyle be desired to shew his experiments of the Air; Dr. Goddard, his experiments of Colours.

"That Dr. Petty be desired to bring in diagrams of what he discoursed to the company this day; and likewise the history of the Building of Ships."

On this day the first volume of the Register-book was opened, and the following questions proposed by Lord Brouncker and Mr. Boyle, for transmission to Teneriffe, were agreed to, and duly registered.

"*Questions propounded and agreed upon to be sent to Teneriffe by the Lord Brouncker and Mr. Boyle.*

"1. Try the quicksilver experiment at the top, and at severall other ascents of the mountain, and at the end of the experiment upon the top of the hill, lift up the tube from the restagnant quicksilver somewhat hastily, and observe if the remaining mercury be impelled with the usual force, or not. And take by instrument (with what exactness may be), the true altitude of every place where the experiment is made, and observe at the same time the temperature of the air, as to heat and cold, by a weather-glass; and as to moistures and dryness with a hydroscope, and note what sense the experimenters have of the air at those times respectively.

"2. Carry up bladders, some very little blown, some more, and others full blown, and observe how they alter upon the several ascents.

---

to above, and constructed by Boyle, was presented to the Society by him in 1662, and still remains in their possession. It consists of two barrels.

"3. Take up a statera, two balls of like substances, differing in weight or bignesse, and an open empty bottle, to the highest part of the hill, and there stop the bottle exactly well, and then weigh that, and the balls (each severally) with the statera there, and at the several ascents, and also below; and likewise the bottle again filled with the air below and stopped as before, noting the different weight of the stopper, if not exactly the same.

"4. Try by an hour-glass, whether a pendulum-clock goeth faster or slower on the top of the hill than below.

"5. Try the power of a stone bow, or other spring, both above and below, and note well the difference.

"6. Make the experiment of two flat polisht marbles upon one another with a weight hanging at the lower, and carefully note the greatest weight that may be applyed on the top of the hill, and also below.

"7. Try whether birds that fly heavily, or others clog'd with as much weight as they can well fly with below, can fly as well, better, or worse above.

"8. Observe what alterations are to be found in living creatures carried thither, both before and after feeding, and what the experimenters doe find in themselves as to difficulties of breathing, faintness of spirits, inclinations to vomit, giddinesse, &c.

"9. Try to light a candle with a match, and fire some spirits of wine; and observe if they burn upon the top of the hill, as well as below, and of what figures, colours, &c., the flames are.

"10. Fire powder in a fusee, or otherwise, observe the manner of firing, the force of the powder, the motion of the smoak and the duration of it: the like of other combustible things as to flame, smoak, &c.

"11. Carry up a viall of aqua-fortis, (or other smoaking liquor) and there open it, and observe whether the fumes ascend as much as they doe below. Quench lime at the top of the hill, and observe the degree of heat and duration of it, in respect to the like quenched below.

"12. Observe whether any vapours fasten in little drops to the outside of a vessel filled with snow and salt, and try the experiment of freezing with it.

"13. Carry to the top two or three bright pieces of iron or copper, and observe there, whether the air doth cause any beginning of rust in them.

"14. Take some of the snow that lyes the highest upon the mountain, up to the top (if it may be), and observe what alteration is made thereupon by the air.

"15. Try whether a filter or a siphon will bring over liquors as well on the top of the hill as below.

"16. Observe the difference of sounds made by a bell, watch, gun, &c., on the top of the hill in respect to the same below.

"17. Observe diligently by a quadrant or double horizontall, what variation the same needle hath both above and below.

"18. Look upon the starrs (or the letters of a book at some certain distance) with a perspective glasse, as well above, as below (the air being clear), and observe accurately the best distance of the glasses in each place.

"19. Try if any difference may be found above in things to be smelt and tasted, from what they had below.

"20. Make an exact narrative of everything observable upon it, and where it is earthy, sandy, gravelly, &c.; what caves, precipices, windings and turnings, &c.; what

H 2

living creatures, plants, &c.; and send over a little of every remarkable vegetable that may be found thereon.

"21. Report the experiments, if conveniently they may, at both the solstices and equinoctes.

"22. Observe accurately the time of the sun's rising on the top of the hill and below, and note the difference[4]."

The preceding "questions," as they are called, but which are rather a series of instructions, have been printed at length, not only on account of their intrinsic general interest, but also because they are the first measures taken by the Royal Society to procure authentic information of the natural history and physical condition of foreign countries, respecting which the greatest ignorance prevailed.

On the 9th January, Dr. Goddard was desired to bring in writing, his experiments of colours, and to produce what he had done with relation to the anatomy of trees; and Mr. Evelyn was ordered to shew his catalogue of trades. At the meeting on the 16th January, the King sent two loadstones by Sir Robert Moray, with a message, "that he expected an account from the Society of some of the most considerable experiments of that nature upon them."

"The trial of these experiments was referred to Mr. Ball."

Dr. Goddard presented the following Paper:

*A Brief Experimental Account of the Production of some Colours by mixture of some liquors, either having little or no Colour, or being of different Colours from those produced.* This Paper was ordered to be entered in the Register-book.

---

[4] Register-book, Vol. I. pp. 1—3.

Mr. Evelyn was desired to bring in an history of engraving and etching[5].

It was ordered that the members of the Society belonging to Gresham College, together with Sir Robert Moray, and as many others as thought proper, be a committee for magnetical inquiries.

At the next Meeting, on the 23rd January, Evelyn was desired to communicate his observations of the anatomy of trees[6]. Sir Kenelm Digby, to bring in writing his discourse made this day concerning the vegetation of plants. Dr. Petty, to deliver in his thoughts concerning the trade of clothing. Mr. Slingsby, to communicate his remarks upon the business of the Mint; and Mr. Wilde, to show the experiment of the stone kindled by wetting. Several gentlemen were proposed as candidates—upon which it was:—

"Resolved, that no more be proposed till it be known whether any of those who were first named, and not in the list, were desirous of being admitted into the Society or not; and that their particular acquaintance be desired, to learn their minds in that respect."

The following Wednesday being the day appointed for a fast and humiliation for the death of King Charles I., there was no meeting of the Society.

On the 6th February a committee was appointed, "to consider of proper questions to be inquired of in the remotest parts of the world."

---

[5] Evelyn states in his *Diary*, that he was recommended to publish what he had written of Chalcography.

[6] These were communicated on the 29th Jan. to the Society, in a letter to Dr. Wilkins, which is preserved in the first volume of the Society's Letter-book.

Dr. Goddard appears to have presented his Paper at this meeting, entitled, *Some Observations concerning the Texture and similar parts of the Body of a Tree, which may hold also in Shrubs and other Woody Plants.* It will be found in Evelyn's *Sylva.*

At the next Meeting the Danish ambassador visited the Society. He was introduced by Evelyn; and entertained by experiments made with Boyle's engine, &c. The same evening the Earl of Sandwich, one of His Majesty's privy council, was admitted a Fellow of the Society.

"Feb. 20. Mr. Evelyn was desired to prepare oyl of sulphur, for the tryal of the experiment of its weight and bulk; Dr. Merret to bring in an appendix to his Paper on the art of Refining, about Cementation, and the Antimony Horne[7].

"Feb. 23. Experiments were made at the Tower of London on the weight of bodies increased in the fire; an account of which being drawn up by Lord Brouncker, was registered, and afterwards printed in Sprat's *History of the Royal Society.*

"On the 25th Feb. it was resolved, in consequence of the increased number of experiments, that the amanuensis should attend every meeting-day, and that his salary be increased from 2*l.* to 4*l.* a-year. An essay-furnace was ordered to be built, and an accurate beam provided for the use of the Society."

At the Meeting of the 4th March, the King sent five little glass bubbles, by the hand of Sir Paul Neile, in order to have the judgment of the Society concerning them. And, at a subsequent Meeting we find

---

[7] This Paper appears in the *Phil. Trans.*, No. 142.

that "the amanuensis produced the bubbles he was ordered to prepare, and they succeeded as those which the King sent of this kinde. Some of ours were sent by Sir Paul Neile to His Majesty."

"A committee was appointed to goe to the glass-house at Woolwich, to enquire into the experiment of those solid bubbles the King sent,—viz. Sir Paul Neile, my Lord Brouncker, Mr. Slingsby, Mr. Bruce."

At another Meeting, Sir Robert Moray laid before the Society a poisoned dagger, sent by the King, who had received it from the East Indies. "The dagger was warmed, and with it blood was drawn from a kitten, to see whether it would be killed thereby. The kitten not dying whilst the Society were together, the operator was appointed to observe what should become of it." At the Meeting in the following week the wounded kitten was produced alive.

Sir Kenelm Digby[8] related that "Dr. Dee, by a diligent observation of the weather for seven years together, acquired such a prognosticating skill of weather, that he was on that account accounted a witch." The next entry records the following experiment by Sir William Persall:—

"Take a handfull of the powder of Roman vitriol, put it into a galley-pot in a pint of water, put in two or three small irons the length of a span, and three or four times a-day constantly stir the water and powder, and move not the irons at all, but let them stand constantly in night and day; and within the space of three weekes

[8] Digby delighted in the marvellous, and was, probably, the most superstitious of the early Fellows of the Society. It is related of him, that he fed his wife on capons fattened with the flesh of vipers, in order to preserve her beauty.

there will be crusted about the irons, as farre as they are in the water, a substance purer than copper, which you may take off, and will be malleable."

The Minutes of the ensuing Meeting, which was held on the 6th March, inform us, that Sir Robert Moray was chosen President;—and he was continued in this office by several subsequent elections, though their dates are not always mentioned. In a letter addressed to M. de Montmor, dated July 22, 1661, Sir Robert Moray styles himself *Societatis ad Tempus Præses*[9]. It appears, however, that he was the sole President of the Society until its incorporation; and his name is inserted as presiding at every meeting, with the exception of those when Dr. Wilkins occupied the chair; but there is no mention of his having been elected President.

Sir Robert Moray was therefore the first President of the Society, and, in accordance with the design of this work, a brief memoir of him is here introduced.

Sir Robert Moray was descended from an ancient and noble family in the highlands of Scotland. He was educated partly in the University of St. Andrew's, and partly in France, where, according to Burnet's *History of his Own Time*, he had afterwards a military employment in the service of Louis XIII., and gained a high degree of favour with Cardinal Richelieu. He attained the rank of colonel in the French army, and came over to England for recruits. At this period Charles I. was with the Scottish army at Newcastle, and Moray became, by a variety of circum-

---

[9] Letter-book, Vol. I. p. 1.

stances, so much attached to that unfortunate monarch, that in December, 1646, he contrived a plan for his escape[10]. Burnet, in his *Memoirs of the Dukes of Hamilton*, states that it was thus conceived: "Mr. William Moray, afterwards Earl of Dysert, had provided a vessel which was to lie off Tynemouth, and to which Sir Robert Moray was to have conducted the King in disguise. Matters proceeded so far that the monarch dressed himself in the clothes provided for the purpose, and went down the back stairs of the house where he was living with Sir Robert; but, apprehending that it would be impossible to pass all the guards without being discovered, and conceiving that it would be highly indecent to be taken in such a condition, he changed his resolution and went back."

Upon the Restoration, Charles II. made Sir R. Moray one of the privy council, and frequently consulted him on affairs of state. He was one of the first and most active members of the Royal Society, and, according to the historian before quoted, "the life and soul of the body." He had the great pleasure of being the bearer of the message from the King, to the effect that his Majesty approved the objects of the Society, and was willing to encourage it, and was generally the organ of communication between the King and the Society. He was as assiduous as his friend Sir C. Wren in promoting its valuable objects; and their names are to be met with in almost every page of the early volumes of the Journal-books. He was nominated one of the council in the

[10] Wood in the *Athen. Oxon.* says, that Sir R. Moray had been general of the ordnance in Scotland against Charles I., when the presbyterians of that kingdom first set up their covenant.

first and second charters of the Society. Authorities concur in assigning to him an extensive knowledge of natural philosophy and mathematics. Dr. Birch adds, that he was "remarkably skilled in natural history." This may be true, applying the observation to the period in which he lived; but his knowledge as a naturalist would certainly gain him little credit at present. The reader will have an opportunity of judging of it, especially in one branch of zoology, in subsequent reports of the meetings. He made a great number of communications to the Society, ten of which are printed in the *Transactions.* Wood declares in the *Athen. Oxon.*, that Sir Robert Moray was "an abhorrer of women." "This," says Dr. Birch, "is a gross mistake, for he married the sister of Lord Belcarres." All accounts agree in describing him as universally beloved and esteemed, and eminent for his piety; spending many hours a day in devotion, in the midst of armies and courts. Evelyn, who knew him well, calls him "an excellent man and admirable philosopher;" and Dr. Birch, in his sketch of him, says, "He had an equality of temper that nothing could alter, and was in practice a Stoic, with a tincture of one of the principles of that sect, the persuasion of absolute decrees. He had a most diffused love to mankind, and delighted in every occasion of doing good, which he managed with great zeal and discretion."

He died suddenly in his garden at Whitehall, on July 4, 1673; and was buried at the charge of Charles II., in Westminster Abbey, near the monument of Sir William Davenant.

At the same meeting that Sir Robert Moray was elected President, he sent in a Paper entitled, *A Re-*

*lation concerning Barnacles*[11]. In this he declares, that when he was in the western islands of Scotland he saw multitudes of little shells adhering to trees, having within them little birds, perfectly shaped. He opened several of their shells, and found, as he states, nothing wanting for "making up a perfect sea-fowle." He honestly adds, however, that he never saw any of the birds alive, nor met with any person who did. Here we have the absurd notion of the *Lepas Anatifera* breeding geese, brought before the Society by their President, in a Paper which was subsequently printed in the 137th No. of the *Transactions*[12].

On the 15th March Evelyn communicated to the Society a curious relation concerning Teneriffe:—and on the 20th March, it was "voted that the number of the Society be enlarged, and that the Gentlemen of the Colledge be overseers for the accommoding the roome for the Society's meeting."

"March 25. Dr. Henshaw was desired to enquire of his brother concerning the boat that will not sink.

"Mr. Boyle was desir'd to bring in the name of the place in Brasill where that wood is that attracts fishes;

---

[11] Printed in the *Phil. Trans.*, No. 142.

[12] Hector Boyce, the Scottish historian, who died about the year 1535, believed firmly in the story of the Barnacles. Scaliger, in one of his *Epistolæ* published in 1627, ridicules Boyce's belief that the birds grew on trees. *Nam de conchis anatiferis fabula prorsus est. Nullæ enim anates ex conchis producuntur, sed ex putredine vetustorum navigiorum, quibus conchæ adhærent, anates quasdam nasci certum est. Etiam arbores anatiferas esse in ultima Scotia, ubi nullæ prorsus arbores sunt, hactenus mentita est scriptorum vernilitas.* It is just as easy to believe that birds grow upon trees, as that they are produced from rotten wood; so that Scaliger's philosophy seems to have conducted him but a little way beyond the region of absolute credulity.

and also of the fish that turns to the wind when suspended by a thread.

"March 27. To enquire whether the flakes of snow are bigger, or less in Teneriffe than here.

"That adders be provided to try the experiment of the stone.

"April 3. Dr. Petty was intreated to inquire in Ireland for the petrifaction of wood, the barnacles, the variation of the compass, and the ebbing and flowing of a brook.

"Mr. Boyle presented his book concerning glass-tubes to the Society."

At the next Meeting, on April 10, Lord Brouncker, Sir R. Moray, Sir Paul Neile, Dr. Wallis, Dr. Goddard, and Mr Wren, were appointed "a committee to consider about all sort of tooles and instruments for glasses for perspectives for the Society on Fridays."

"Ordered, that the Society meet no more but once a-weeke, and that it be on Wednesdays."

It was "propounded that henceforward every one shall pay his contribution and arrears every Wednesday." This is the first intimation of pecuniary difficulties, which subsequently harassed the Society for a long time.—At the next meeting it is recorded, that the "Treasurer, or his deputy, be desired to call every day on the company for arrears."

On the 8th May it was proposed that the Society write to Mr. Wren, and charge him from the King, to make a globe of the moone[13].

---

[13] This globe represented not only the spots and various degrees of whiteness upon the surface of the moon, but also the hills, eminences, and cavities, moulded in solid work. The King received the globe with peculiar satisfaction, and ordered it to be placed

A letter was received from the Duke Leopold, and a committee appointed to correspond constantly with that Prince.

"Sir Robert Moray was desired to write to the Jesuits at Liege about the making of copperas there.

"Dr. Clarke was intreated to lay before the Society Mr. Pellin's relation of the production of young vipers from the powder of the liver and lungs of vipers.—Sir K. Digby promised such another under my Lord .........s hand.

"Dr. Clarke and Mr. Boyle were intreated to procure an history of vipers."

Here we have an excellent illustration of the ignorance of scientific men respecting the zoology of their own country, and at the same time evidence of an anxious desire to procure authentic information.

At this Meeting a motion was made for the erection of a library.

"Mr. Boyle and Mr. Evelyn were appointed curators for the observing of insects, and to meet at Mr. Boyle's lodgings.

"May 15. It was resolved from henceforth that every member, as soon as he shall be admitted, shall pay, besides twenty shillings advancement, his weekly contribution from the time of his admission.

"May 22. Col. Tuke was desired to bring, in writing, an account of what he related to the Society of the Academy at Paris[14].

---

among the curiosities of his cabinet. See Ward's *Lives of the Gresham Professors*, p. 100.

[14] This paper is preserved in the Register-book (Vol. I. p. 25), and is curious as giving one of the earliest accounts of the French Academy and the mode of conducting their meetings.

"Mr. Povey was intreated to send to Bantam for that poyson, related to be so quick as to turne a man's blood suddenly to gelly.

"My Lord Northampton was intreated to make inquiry for Mr. Marshall's book of insects.

"The Amanuensis was ordered to go to-morrow to Rosemary-lane, to bespeake two or three hundred more of the solid glasse balls.

"May 28. It was *resolved* that every member, who hath published, or shall publish any work, give the Society one copy.

"Dr. Clarke was intreated to bring in the experiment of injections into the veins.

"June 5. Col. Tuke related the manner of the rain like corn at Norwich, and Mr. Boyle and Mr. Evelyn were intreated to sow some of those rained seeds to try their product.

"Magnetical cures were then discoursed of. Sir Gilbert Talbot promised to bring in what he knew of sympatheticall cures. Those that had any powder of sympathy were desired to bring some of it at the next meeting.

"Mr. Boyle related of a gentleman who having made some experiments of the ayre, essayed the quicksilver experiment at the top and bottom of a hill, when there was found three inches difference.

"Dr. Charleton promised to bring in that white powder, which, put into water, heates it.

"The Duke of Buckingham promised to cause charcoal to be distill'd by his chymist.

"His Grace promised to bring into the Society a piece of a unicorne's horne.

"Sir Kenelme Digby related that the calcined powder of toades reverberated, applyed in bagges upon the stomach of a pestiferate body, cures it by severall applications.

"June 13. Col. Tuke brought in the history of the rained seeds which were reported to have fallen downe from heaven in Warwickshire and Shropshire, &c.[15]

"That the dyving engine be goeing forward with all speed, and the Treasurer to procure the lead and moneys.

"Ordered, that Fryday next the engine be tried at Deptford[16].

"June 26. Dr. Ent, Dr. Clarke, Dr. Goddard, and Dr. Whistler, were appointed curators of the proposition made by Sir G. Talbot, to torment a man presently with the sympatheticall powder.

"Sir G. Talbot brought in his experiments of sympathetick cures."

These are entered in the first volume of the Register-book. They are exceedingly curious as emblematic of the superstition of the times. The following extract from the Paper, hitherto unpublished, will not be uninteresting.

"An English mariner was wounded at Venice in four severall places soe mortally, that the murderer took sanctuary; the wounded bled three days without intermission; fell into frequent convulsions and swounings, the chirurgeons, despayring of his recovery, forsook

---

[15] The supposed grains of wheat turned out, after due examination, to be the seeds of ivy-berries deposited by starlings; thus one popular superstition was destroyed.

[16] Evelyn remarks in his Diary, under the date of July 19, "We tried our Diving Bell in y^e Water Dock at Deptford, in which our Curator continued half-an-hour: it was made of cast lead, let down by a strong cable."

him. His comrade came to me, and desired me to demand justice from the Duke upon the murderer (as supposing him already dead); I sent for his bloud and dress'd it, and bad his comrade haste back and swathe up his wounds with cleane linnen. He lay a mile distant from my house, yet before he could gett to him, all his wounds were closed, and he began visibly to be comforted. The second day the mariner came to me, and told me his friend was perfectly well, but his spirits soe exhausted, he durst not adventure soe long a walke. The third day the patient came himself to give me thanks, but appeared like a ghost; noe bloud left in his body[17]."

Incredible as it seems, yet this relation of Sir G. Talbot, who, be it remembered, held high offices under the crown, was believed; and, as some proof of this, it may be adduced, that a minute in the Journal-book, under the same date as the above, informs us;—that "Mr. Evelyn was intreated to bring in next day that powder of simpathie he has of Sir Gilbert Talbot's making."

But such superstition soon gave way before the test of experiments, and it speaks well for the little band of philosophers that they were not backward in putting deep-rooted beliefs to the severest trials.

Thus, under the date of July 10, it is recorded, that "The fresh hazell-sticks were produced, wherewith the divining experiment was tried, and found faulty."

July 17. The King having desired to know "why the humble and sensitive plant stirs, or draws back, at the touching of it," a committee was appointed to report upon the fact. At this Meeting the first

[17] P. 31.

volume of the Letter-book was opened, to register a copy of a letter addressed by the Society to the French Academy, requesting the interchange of scientific communications[18]. At the next Meeting:—

July 24. "A circle was made with powder of unicorne's horn, and a spider set in the middle of it, but it immediately ran out severall times repeated. The spider once made some stay upon the powder.

July 31. "Dr. Wilkins made his experiments of blown bladders, the account of which was registered[19].

"Mr. Croune produced a glass jar full of the powder of the bodies of vipers, and a gallipot full of the powder of only the hearts and livers of vipers.

"Dr. Wilkins was desired to procure pipes of several bores to be made, for the experiment of blowing up weights.

August 7. "Mr. Palmer brought in a powder reported to be that of a plant which was brought a hundred miles beyond Moscow, said to be used there for ulcers[20].

August 14. "Sir Robert Moray brought in glass-drops, an account of which was ordered to be registered[21].

August 21. "Dr. Clarke brought in and read his account of the humble and sensible plants, which was ordered to be registered[22]."

The Meetings from this time to the close of the

---

[18] The letter, and answer of the Academy, will be found in the Letter-book.

[19] This relates to the power of raising heavy weights with bladders inflated by the breath.

[20] Register-book, Vol. I. p. 56.

[21] These glass-drops are the well-known Prince Rupert's Drops. The first volume of the Register-book contains a long account of them and their manufacture.

[22] Register-book, Vol. I. p. 90.

year, were occupied by several experiments, now of minor interest. At that on the 9th of October, the Society received a present of a living chameleon; and Wren read his hypothesis of Saturn, contained in a letter to Lord Brouncker, which is preserved in the first volume of the Letter-book. After the reading of Wren's letter, one was communicated from Christian Huygens, dated at the Hague, 24th July, 1661, containing some observations on Saturn: this is also preserved in the Letter-book. It is worthy of remark that, in this communication, Huygens states that the members of the French Academy were excited by emulation of the Society of London, and purposed applying themselves to philosophical experiments; and he adds, this is "a good effect produced by your example."

At the Meeting on the 21st August:—

"Mr. Boyle presented the Society with his booke of Aire in Latine.

August 28. "Mr. Oldenburg read a part of a letter come from Mr. Borry, concerning the making of Incombustible Wood.

Sept. 4. "Sir K. Digby brought in a letter from a friend of his in Florence, written in 1656, which treats of a petrified city and inhabitants[23].

---

[23] Register-book, Vol. I. This city was supposed to be in Africa. See MS. letter in the Royal Society archives, from Hartlib to Boyle, in which the writer expresses his strong belief in this story. Captain Smyth's private Journal, printed in Beechey's *African Expedition*, pp. 504—512, contains an account of this tradition. Mr. Hartlib was one of the most credulous, in that credulous period; for in the same letter he gives an account of a child, whose mother drank petrifying waters, being petrified in the womb, and adds, that the stone which was taken out was a lasting monument of the fact.

"Some papers were delivered in by Sir P. Neile from the King, and endorsed by His Majesty's own hand: they were about Mr. Hobbes' mean proportionalls[24].

Sept. 18. "It was voted that the petition Sir R. Moray read to the company, be delivered to the King in the Society's name[25].

"Mr Boyle brought in his account, in writing, of the experiment hee made of the compression of aire with quicksilver in a crooked glasse tube, and ordered to be registered."

At the two subsequent Meetings, various experiments were tried and discussed; it was also debated, whether the hour of meeting should be altered from 3 in the afternoon to 9 in the morning;—but although no entry respecting the alteration occurs in the Journal-book, it appears by a MS. letter to Boyle, preserved in the archives of the Society (No. 18 Coll. of Letters), that no change took place. On the 16th October, Sir R. Moray acquainted the Society that;—

"He and Sir Paul Neile kiss'd the King's hand in the Company's name, and he was desired by them to return their most humble thanks to his Majesty for the reference he was pleased to grant of their Petition, and for the favour and honour done them of offering to be entered one of the Society."

---

[24] The problems, with Lord Brouncker's solution, are entered in the Register-book, Vol. I. p. 99.

[25] This Petition, Evelyn tells us, was praying the King for his Royal grant authorizing the Society to meet as a corporation, with several privileges. A diligent search in the State Paper Office for this document, has been, I regret to say, unsuccessful, and Mr. Lechmere informs me, that no Petitions exist in that establishment of an earlier date than Oct. 1662.

Oct. 23. "Dr. Goddard was desired to give an account of his dissection of the Chameleon[26].

"Mr. Croune read an account of Mercuriall Experiments Mr. Power made at the bottom and top of Halifax hill.

Nov. 13. "Dr. Goddard communicated an account of the experiments of the stone called *Oculus Mundi*, which being transparent, was heavier than when it was cloudy[27].

"The Doctor was intreated to try more experiments upon it.

Nov. 20. "Dr. Wilkins read his paper for a Natural Standard.

Nov. 27. "Sir Wm. Petty[28] brought in his History of Clothing."

This paper appears at length in the Register-book; it gives a detailed description of the manufacturing of various cloths, and is illustrated by several curious drawings, which are in the archives of the Society.

Dec. 4. "Sir Robert Moray was desired to bring in his Engine for hearing.

"Mr. Powle promised to bring in an account of Iron, from the ore to the bar.

"Mr. Ellis promised to inquire into the making of Lead.

"The Amanuensis produced serpents, which being fired and cast in the water, burnt there to the bounce."

On the 11th Dec. an entry was made in the Journal-book, thanking Evelyn for having "done honour

---

[26] See *Phil. Trans.*, Vol. XI. p. 930.

[27] Printed in Dr. Sprat's *History*.

[28] He had been knighted on the 11th April.

to the company in an excellent panagyrick to the King's Majesty[29], and since in an epistle dedicatorie, address'd to the Lord Chancellorum[30], in which the Company and its design are most affectionately recommended to the King and his Lordship."

The Society adjourned for a fortnight at Christmas, and resumed their Meetings in January.

Jan. 8, 1661—2. "Mr. Evelyn read an account of the making of Marbled Paper, which was ordered to be registered.

Jan. 29. "Sir W. Petty promised to bring in against next day his observations on the Pendulum's Vibration for measure.

"The Lord Embassadour of Genoa gave the Society a visite, and was entertain'd with the sight of Mr. Boyle's Engine for the Exsuction of Aire.

Feb. 5. "Dr. Whistler brought in a book of Observations on the Bills of Mortality, and read the epistle dedicatory thereof (to Sir R. Moray, President)—from Mr. Graunt; who sent fifty copies of the book to be distributed among the members of the Society; for which thanks were returned to him, and he was proposed as a candidate."

Dr. Sprat, in his *History*, alludes to Mr. Graunt as having been recommended to the Society by the King. "In whose election," says he, "it was so farr from being a prejudice that he was a shopkeeper of London, that his Majesty gave this particular charge

---

[29] Evelyn presented this Panegyric (which was a poem upon His Majesty's coronation) to the King on the 24th April.

[30] This alludes to his translation of Naudæus concerning libraries, which Evelyn says, was "miserably false printed."

to his Society, that if they found any more such tradesmen, they should be sure to admit them all, without any more ado." Graunt's dedication is very curious; he styles the Royal Society "the King's Privy Council for Philosophy, and his great Council for the three Estates of Mathematics, Mechanics, and Physics."

May 14. "Mr. Southwell produced a great horn, said to be a unicorn's, and also shewed a little one that grew on a cock's head, being the spur of the fowle, cutt close till it bled, and sett on the head immediately after the comb was taken off (and it being squeez'd on, and a few ashes strow d thereon to quence y^e blood), when the cock was fresh capon'd.

June 11. "Mr. Evelyn presented the Society with a book call'd the *History of Chalcography*[31].

July 2. "Ordered, that the committee to view Mr. Toogood's engine doe meet on Saturday next, at two of the clock, in the Temple Church[32]."

On the 9th July the first intimation of the incorporation of the Society by Charles II., was given by Lord Brouncker announcing to the meeting, that Sir Heneage Finch, his Majesty's Solicitor-General, had refused the fees to which he was entitled for signing the docket of the bill prepared by him. A committee was appointed to wait upon the Solicitor-General, and present him with the thanks of the Society. On the 13th Aug. the Journal-book records that "Let-

[31] Its title is *Sculptura, or the History of Chalcography*, 8vo. London, 1662.

[32] What would the Benchers of the present day think of their beautiful church being made a place of meeting for the trial of experiments? Yet this occurred not unfrequently at the above period.

ters Patent for the incorporation of the Society were read by Mr. Oldenburg; and it was voted that the President, attended by the Council, and as many of the Society as can be obtained, shall wait upon his Majesty, after his coming from Hampton Court to London, to give him humble thanks for his grace and favour, and, in the mean time, the President is to acquaint the King with their intention; and that afterwards my Lord Chancellor is to be thanked likewise, and Sir Robert Moray, for his concern and care in promoting the constitution of the Society into a corporation."

The King had evinced considerable interest in the Society from the time that he was first apprised of its formation. "When," says Dr. Sprat, "the Society addressed themselves to his Majestie, he was pleas'd to express much satisfaction that this enterprize was begun in his reign: he then represented to them the gravity and difficulty of their work, and assured them of all the kind influence of his power and prerogative. Since that, he has frequently committed many things to their search: he has referr'd many forein rarities to their inspection: he has recommended many domestick improvements to their care: he has demanded the result of their trials, in many appearances of nature: he has been present, and assisted with his own hands at the performing of many of their experiments in his gardens, his parks, and on the river. And besides, I will not conceal that he has sometimes reprov'd them for the *slowness* of their *proceedings:* at which reproofs they have not so much cause to be afflicted, that they are the reprehensions of a King, as to be comforted that they are the reprehensions of his love and affection to their progress. For a testimony of which royal benignity, and to free them from all hindrances, he has given them the establishment of his *Letters Patents*[33]."

It is stated that the draught of the preamble to the Charter was prepared by Wren, at the request of his brother philosophers. It runs thus:—

"Whereas amongst our Royal hereditary titles, to which, by Divine Providence and the loyalty of our good subjects, we are now happily restored, nothing appears to us more august or more suitable to our pious disposition, than that of father of our country, a name of indulgence as well as dominion, wherein we would imitate the benignity of heaven, which in the same shower yields thunder and violets, and no sooner shakes the cedars, but dissolving the clouds, drops fatness:—We therefore, out of paternal care of our people, resolve, together with those laws which tend to the well administration of government and the people's allegiance to us, inseparably to join the supreme law of *Salus Populi*, that obedience may be manifestly not only the public, but the private felicity of every subject, and the great concern of his satisfactions and enjoyments in this life. The way to so happy a government we are sensible is in no manner more facilitated, than by promoting of useful arts and sciences, which, upon mature inspection, are found to be the basis of civil communities and free governments, and which gather multitudes by an Orphean charm into cities, and connect them in companies; that so, by laying in a stock as it were of several arts and methods of industry, the whole body may be supplied by a mutual commerce of each other's peculiar faculties, and, consequently, that the various miseries and toils of

---

[33] *Hist. Royal Society*, p. 133.

this frail life, may be, by as many various expedients ready at hand, remedied, or alleviated, and wealth and plenty diffused in just proportion to every one's industry, that is, to every one's deserts.

"And, whereas we are well informed, that a competent number of persons, of eminent learning, ingenuity, and honour, concording in their inclinations and studies towards this employment, have for some time accustomed themselves to meet weekly, and orderly, to confer about the hidden causes of things, with a design to establish certain, and correct uncertain theories in philosophy, and by their labours in the disquisition of nature, to prove themselves real benefactors to mankind; and that they have already made a considerable progress by divers useful and remarkable discoveries, inventions, and experiments in the improvement of mathematics, mechanics, astronomy, navigation, physic, and chemistry, we have determined to grant our Royal favour, patronage, and all due encouragement to this illustrious assembly, and so beneficial and laudable an enterprize."

It is not difficult to conceive that the King must have smiled at some parts of this preamble, but its language is, however, quite in keeping with the addresses, petitions, &c. of the day.

The Charter of Incorporation passed the great seal on the 15th of July, 1662, and was read before the Society on the 13th of August following[34].

A copy of this important document will be found

[34] Evelyn records in his *Diary* under this date, "Our Charter being now passed under the broad seal, constituting us a corporation under the name of the Royal Society, for the improvement of natural knowledge, was this day read, and was all that was done this afternoon, being very large." Vol. I. p. 337.

in the Appendix[35]. Sprat says, alluding to the Charter: "This is the *Legal Ratification* which the Royal Society has receiv'd. And in this place I am to render their publick thanks to the Right Honourable the Earl of Clarendon, Lord Chancellor of England, to Sir Jeffry Palmer, Attorney-General, and to Sir Heneage Finch, Solicitor-General; who by their cheerful concurrence, and free promotion of this confirmation, have wip'd away the aspersion that has been scandalously cast on the profession of the law, that it is an enemy to learning and the civil arts. To shew the falsehood of this reproach, I might instance many judges and counsellors of all ages, who have been the ornaments of the sciences, as well as of the bar and courts of justice. But it is enough to declare that my Lord Bacon was a lawyer, and that these eminent officers of the law have compleated this foundation of the Royal Society, which was a work well becoming the largeness of his wit to devise, and the greatness of their prudence to establish[36]."

---

[35] It will be necessary for the reader unacquainted with the Charter, to peruse it before proceeding to the next Chapter, in order that he may form a correct idea of the progress of the Society.

[36] *Hist. Royal Society*, p. 144.

# CHAPTER VI.

Memoir of Lord Brouncker—The Society founded for Improvement of Natural Knowledge—Explanation of the term—The Society return thanks to the King—Their Address—Cowley's Ode to the Society—Incorporation of Society, a claim of respect to the memory of Charles II.—The King proposes to endow the Society with lands in Ireland—Duke of Ormonde, the manager of Irish Affairs—His political intrigues—Lord Brouncker addresses his Grace respecting the Society's claims—Lands intended for the Society granted to other parties—Sir William Petty's estimate of their value—First Statutes enacted—Experiments vigorously prosecuted—Hooke appointed Curator—His great zeal and energy—Second Charter.

---

1660—65.

IT will be observed, that the Charter of Incorporation appoints Lord Brouncker President of the Society.

In various sketches of the life of this Nobleman, it is stated that he was born *about* the year 1622; but through the kindness of Richard Brouncker, Esquire, of Boveridge, in Dorsetshire, who has supplied the pedigree of the Brounckers, of which family he is the living representative, I am enabled to show that Lord Brouncker was born in 1620. He was the son of Sir William Brouncker, Knight, gentleman of the Privy Chamber to Charles I., and Vice-chancellor to Charles II., when Prince of Wales, who was created Baron Brouncker of Newcastle, and Viscount Brouncker of Castle Lyons, Ireland, on the 12th of Sept. 1645. He had a grant for his services, of the monastery of Clonnis, in the county

Monaghan. His son, the second Viscount, and the subject of this memoir, after receiving the usual preparatory education, was sent at the age of sixteen to the university of Oxford, where he made himself a proficient in several languages, and acquired an excellent knowledge of mathematics, and of other important sciences. He then entered on the study of natural philosophy and medicine, and displayed so much ability in the latter, as to obtain the degree of Doctor of Physic in the university of Oxford, in 1646. But his tastes led him to follow mathematical studies in preference to any other, and to these he applied himself with great diligence[1]. In 1657, he was engaged in a correspondence with Dr. Wallis, which appeared in the *Commercium Epistolicum*, Oxford 1658. 4to. He made two remarkable mathematical discoveries, being the first to introduce continued fractions[2], and to give a series for the quadrature of a portion of the equilateral hyperbola[3]. He contributed several valuable papers to the *Philosophical Transactions;* and translated an essay by Descartes, entitled *Musicæ Compendium*, which was published without the translator's name, but as the work of "a person of honour[4]." With others of the nobility and gentry who adhered to Charles I., he signed the memorable declaration in 1660, in which General

[1] Huygens, in a letter to Oldenburg, preserved in the archives, congratulates the Society on having so eminent a mathematician for their President as Lord Brouncker.

[2] Wallis, *Works*, Vol. I. p. 469.

[3] *Phil. Trans.*, No. 34.

[4] Hawkins' *Hist. of Music.*

Monk was acknowledged the restorer of the laws and privileges of these nations[5].

After the Restoration, he was appointed Chancellor of the Queen Consort[6], and Keeper of her great seal; he was also one of the Commissioners for executing the office of Lord High Admiral, and Master of St. Catherine's Hospital, near the Tower of London. This he obtained in 1681, after a long equity suit with Sir Robert Atkins, one of the Justices of the Common Pleas.

He was created President of the Royal Society by their first Charter, dated July 15, 1662, and held the office until 1677[7].

During his Presidency he exhibited great zeal in the performance of all the duties attached to that distinguished office[8]. He was always prepared to devote his time to experiments, and ready to make such improvements upon them as were suggested by his penetration and skill. His devotion to the Royal Society entitles him to the greater praise, as his high

---

5 Kennet's *Reg.*, fol. p. 120.

6 The warrant for this appointment is in the State Paper Office. It bears date April 18, 1662, and gives Lord Brouncker a salary of 50*l.* a year, and 4*l.* per annum for his livery.

7 Sprat says, "This office was annually renewed to him by election, out of the true judgment which the Society made of his great abilities in all natural and especially mathematical knowledge."

8 Lord Brouncker was one of the most active promoters of the Royal Society. Evelyn states, "That Brouncker, Boyle, and Sir R. Moray, were above all others the persons to whom the world stands obliged for the promoting of that generous and real knowledge which gave the ferment which has ever since obtained, and surmounted all those many discouragements which it at first encountered." *Diary*, Vol. II. p. 304.

state offices required a large amount of his attention and time.

He died at his house in St. James's Street, on the 5th April, 1684, and was buried in a vault in the hospital of St. Catherine. In default of issue, the title passed to his brother Henry, formerly Cofferer to Charles I., who died in January 1687. A portrait of Lord Brouncker by Sir Peter Lely, presented to the Society by his Lordship, hangs in their meeting-room.

The Charter states, that the Royal Society was founded for the improvement of *Natural Knowledge.* Dr. Paris, it is believed, was the first person who drew attention to the sense (in this instance) of the word *natural.* In his *Life of Sir Humphry Davy,* he says, "This epithet *natural* was intended to imply a meaning, of which very few persons, I believe, are aware. At the period of the establishment of the Society, the arts of witchcraft and divinations were very extensively encouraged; and the word natural was therefore introduced in contradistinction to supernatural." Sprat, in his *History of the Society,* confirms this view by the passages in which he mentions "*Experiments of natural things* as not darkening our eyes, nor deceiving our minds, nor depraving our hearts;" and in another place he talks of the Society "following the great precept of the Apostle, of *trying all things,* in order to separate superstition from truth[9]."

The Incorporation of the Society by Royal Charter gave the Fellows great satisfaction, nor were they

[9] It is worthy of notice, that in all the original drafts of letters, &c. existing in the British Museum, the word natural has been carefully added, when forgotten in the first instance.

long in testifying their sense of the King's gracious favour; for on the 29th August, the President, Council, and Fellows, went to Whitehall, and returned their thanks to his Majesty for the Patent granted to them.

The Address was as follows:—

"May it please your MAJESTY,

WE, your Majesty's most loyal subjects, newly incorporated by your Majesty's Charter, and honoured with the name of the ROYAL SOCIETY, do, with all humility, present ourselves before your Majesty the Royal founder thereof, to offer you our most hearty thanks, as the only way we have at present to express our deep sense of your Majesty's grace and favour to us, and to assure your Majesty of our constant veneration for your sacred person, our devotion to your Majesty's service, and our firm resolution to pursue sincerely and unanimously the end for which your Majesty hath founded this Society, the advancement of the knowledge of natural things, and all useful arts by experiments. A design, Sir, that is deservedly counted great and glorious, and is universally reputed to be of that advantage to mankind, that your Majesty is highly admired and extolled for setting it on foot; and this Society is already taken notice of, and famous throughout all the learned parts of Europe, and doubtless in time will be much more by the continuance of your Majesty's gracious favour, and the happy success of their endeavours, to the great increase of the fame of your Majesty's prudence, which hath justly intitled you to the honour of laying the first foundation of the greatest improvement of learning and arts, that they are capable of, and which hath never heretofore been attempted by any; so that men cannot now complain that the favour and assistance of a potent Monarch is wanting to this long wished for enterprize. And, Sir, our assurance of this your Majesty's favour and assistance is that which gives vigour to our resolutions, and is the life of our hopes, that in due season we shall be able to make your Majesty an acceptable present of choice and useful experiments, and accomplish your great design, being thereto engaged by so many powerful motives.

"And in the mean time, we shall daily pray that God will be eminently gracious to your Majesty, and accumulate upon you all the blessings answerable to the largeness of your heart, the height of your condition, the weight of your charge, the multitude of your virtues, and the desires and wishes of all your faithful subjects."

Evelyn states in his *Diary*, that his Majesty gave a gracious reply to the above Address, and that the President and Fellows kissed the King's hand. On the following day, the same deputation waited upon the Lord Chancellor, who was addressed by the President, and thanked for the great interest he had evinced towards the Society.

Thus, was the Invisible College of Boyle, and the ideal Philosophical College of Evelyn, incorporated by Royal Charter; and from a few philosophers and lovers of science, meeting here and there as the times permitted, grew a Society, that soon acquired a stability which two centuries have not weakened.

This royal recognition of philosophy gave sincere satisfaction to all lovers of science, and called forth the muse of Cowley, who wrote an Ode to the Royal Society; a few extracts from which will not be out of place here. It opens by lamenting the length of time that Philosophy lay neglected, until—

"Bacon at last, a mighty man, arose,
(Whom a wise King, and nature chose
Lord Chancellour, of both their lawes,)
And boldly undertook the injur'd pupils' cause."

The poet then goes on to state that the great philosopher,

"——led us forth at last,
The barren wilderness he past;
Did on the very border stand
Of the blest promis'd land;
And, from the mountain's top of his exalted wit,
Saw it himself, and shew'd us it."

The Society is then apostrophized:

"From you, great champions, we expect to get
The spacious countries but discover'd yet;
Countries where yet instead of nature, we
Her images, and idols, worshipp'd see:
These large and wealthy regions to subdue,
Though learning has whole armies at command,
Quarter'd about in every land,
A better troop she ne'er together drew;
Methinks, like Gideon's little band,
God with design has pick'd out you
To do these noble wonders by a few.
* * * * * *
Mischief and true dishonour fall on those
Who would to laughter or to scorn expose
So virtuous and so noble a design,
So human for its use, for knowledge so divine."

It has been asserted that the Incorporation of the Royal Society was the only wise act of Charles II. It is not our province to enter on this question; but it may truly be said, that this act entitles his Majesty to be remembered with great respect and esteem, as by it he testified his love for science, and his desire for the advancement of truth and knowledge.

But his interest in the young Institution did not terminate here; for on the 17th September, 1662, it appears by the Register-book, that the King addressed a letter, with his own hand, to the Duke of Ormond, then Lord Lieutenant of Ireland, recommending the Royal Society "for a liberall contribution from the adventurers and officers of Ireland, for the better encouragement of them in their designes."

As this was the first instance of any pecuniary support being promised to the Society, it will not be uninteresting to give some account of the manner in which Charles the Second proposed carrying his intentions into execution; his own funds being wholly inadequate for the purpose. It will be remembered that the soldiers and adventurers, as they were called, who had served in Ireland under Cromwell, became landed proprietors; but, upon the Restoration, the King published his declaration, which constituted a basis for the new settlement of the landed property of the country. This document, after vesting all the confiscated property in the King, confirmed the adventurers and soldiers in the lands already granted to them. The estates of Protestants were to be restored to the owners, as likewise lands formerly belonging to "innocent Papists;" but the qualifications necessary for a Roman Catholic to claim the benefit of this clause were such, as to make it almost impossible for any person of this religious persuasion to establish his rights. All estates belonging to persons who took any part in the trial or execution of Charles I. were declared confiscated.

Thus the settlement of landed property became a work of great difficulty, and one offering many opportunities for the exercise of bribery and corruption. Lord Ormond was appointed the principal manager of Irish affairs, but the contentions and jealousies of innumerable parties ran so high, that the interference of Parliament became necessary. The Irish Parliament, however, could only frame the heads of a general Bill for the apportioning of the lands, which was liable to be altered and modified by the King and Council in England.

Thither, therefore, all parties interested, repaired or sent agents, to urge and defend their respective claims[10]. London became the scene of controversy, intrigue, cabal, and even violence. A Bill was eventually passed; and Ormond, now elevated to the rank of Duke, was sent back to Ireland, to settle in the most amicable possible manner the conflicting claims of all parties. Justice, however, had little to do in the matter; and this is evident from the fact, that upwards of three thousand ancient Irish families were stripped of their fortunes, without even having their claims inquired into. And we are informed that the Duke of Ormond's estates, which before the breaking out of the civil war only yielded £7000 per annum, after the adjudication of the lands brought him in a yearly income of £80,000.

This brief statement will suffice to show how poor a chance the Royal Society had of coming in for a portion of these "fractions," as they were called, when high families were cheated of their rights. On the 3rd Jan. 1662—3, the President (Lord Brouncker)

---

[10] The Journals of the House of Commons of this period contain a great number of petitions presented by individuals for grants of lands in Ireland; I cannot, however, find any such document from the Royal Society.

addressed the Duke of Ormond respecting the King's grant of lands in Ireland to the Society. The letter is of considerable interest:

*London, Jan.* 3, 1662—3.

"MY LORD,

"I AM desired by the Royal Society in their names to entreat your Grace's favour and countenance in the effectual settlement of the Fractions of Adventures, Arrears, Lands, &c., which, by the Act, for the better execution of his Majesty's gracious declaration, &c., were vested in his Majesty in trust for, and the better to enable his Majesty to grant the same to them: so as his Majesty being their founder, might also be their chief benefactor. In pursuance whereof his Majesty was pleased, by his private letters under his own hand in October last, to recommend the same unto you: and thereafter, some in behalf of the clergy of that kingdom made their application unto his Majesty for the same: but his Majesty, well remembering his promise to grant the same to the Society, was pleased to put a stop unto their addresses, and by his letters in the same month of October last under his royal signet, was further graciously pleased to direct your Grace to pass a grant of the said Fractions unto Mr. Robert Boyle, and Sir Robert Moray, for the use of the said Society. And understanding that afterwards Sir Allen Brodericke, Col. William Legge, and Mr. Henry Coventry, without informing his Majesty of the interest of the Society, have procured letters to your Grace, to pass grants of the said Fractions unto them, contrary unto the end for which the same were originally designed, and his Majesty's gracious intentions; the Society being much troubled thereat, and fearing that the same might have been absolutely passed the Seal there, before your Grace could have been informed of the truth of the case, nothing doubting, but that, as you have been an eminent favourer to the Society in their foundation, you will be so far from obstructing or diverting his Majesty's bounty or favour unto them, that you will contribute your interest to make the same effectual. Much more, where the interest of private persons, though very deserving, comes in competition with the publick concern of a Society, whose designs, if protected and assisted by authority, may so much conduce to the greatness and honour of their Prince, the real good of his dominions, and the universal benefit of mankind: and more especially where their right and pretentions unto the thing in question, are every way more just and considerable than theirs, who would endeavour to gain the same out of their hands.

"To prevent which, 'twas desired that Sir Henry Bennet would procure his Majesty's letter unto your Grace, to put a stop unto the said grants, untill his Majesty were fully informed of the case: wherein he was not so forward as was desired, and therefore 'twas thought fit to acquaint the Council of the said Society therewith, who desired me and others of the Society to sollicite his Majesty, not only to put a stop unto former grants, but also for an absolute grant of all the said Fractions to the use of the said Society, which was accordingly done in an open and publick way, and fully satisfied in the case that he would advise and consider of the pretentions of all parties before his Council.

"Whereupon his Majesty, after a week's consideration thereof, was graciously pleased to sign the grant of which I have herewith sent your Grace a copy, untill the original can be transmitted.

"And as the Society has so far tasted of his Majesty's

justice and favour, they are very confident that your Grace will not only in all things cause the same to be effectually executed to the advantage of the Society, but also in all other their concernments continue your favour unto them, so as they may encounter no other difficulties: whereby they may be the better enabled to effect those ends for which his Majesty was pleased to incorporate them.

"I am, &c.,

"BROUNCKER, P.R.S."

But neither the King's private letters, nor that of Lord Brouncker, had any effect on the Duke of Ormond, who was evidently too much occupied in considering his own interests, and those of his immediate relations and friends, to trouble himself about the Royal Society. Indeed, at the time when Lord Brouncker's letter was written, the land destined by the King for the Society had actually been granted by the Duke of Ormond to Sir Allen Roderick, Colonel William Legge, and Mr. Henry Coventry, although, as Lord Brouncker states, "the rights and pretentions of the Royal Society unto the thing in question, are every way more just and considerable than theirs."

Sir William Petty, who was at this period in Ireland, was requested to send over an estimate of the value of the lands granted by the King to the Society. Sir William made a rough calculation, but did not transmit it to the Society, in consequence of the lands having passed into other hands. Subsequently, however, at the request of several eminent Fellows, he communicated it to Sir Robert Southwell, whose letter to Oldenburg on the subject, dated Dublin, May 15, 1663, is preserved in the archives of the Society[11]. It is a curious document, shewing the inextricable complication of the whole affair. There is no attempt at any exact estimate of the value of the lands, for Sir William Petty observes that, "if the odd money and odd measure be understood in an unlimited way, then, I say, it will amount unto a great matter, but I know not what." In another part of the letter, he continues: "Those who told you that thirteen millions of acres were yet to be disposed of, did not calculate well, for I cannot imagine that there is one; and the better to confirm you, I am assured that all the profitable land in the whole kingdome of Ireland exceeds not nine millions; all the lands lett out to adventurers and soldiers, not much above two millions. Nor does all the forfeited lands, intended to be disposed of by act of settlement, extend to three millions, and much of what was intended will fall short and return to the Irish."

The non-fulfilment of the King's favourable intentions towards the Society, did not damp the philosophic ardour of the Fellows[12]. Before adverting to

---

[11] Collection of Letters, No. 20.

[12] It was about this period that Oldenburg, who had been acting as Secretary since the passing of the first Charter, wrote the following Memorandum, with the view, apparently, of procuring assistance. The original is preserved in the British Museum.

"The business of the Sec. of the R. S. He attends constantly the Meetings both of the Society and Councill, noteth the observables said and done there, digesteth them in private, takes care to have them entered in the Journal and Register-books, reads over and corrects all entrys, sollicites the performances of taskes, recommended, and undertaken, writes all letters abroad, and answers the returns made to them, entertaining a correspondence with at least fifty persons, employs a great deal of time, and takes much

their scientific labours, it is important to state that considerable time was devoted to the preparation of Statutes for the governance of the Fellows individually, and the Society at large, which were eventually, after receiving the King's approbation, passed into laws, in January, 1662—3. As they throw much light on the early history of the Society, a copy of them is inserted in the Appendix of this work[13].

The Journal-book contains the records of several experiments tried at the weekly meetings[14], and of papers read. Those relating to zoophytal or animal ingraftings, attracted considerable attention. Mr. Evelyn communicated a Paper from a friend of his on this subject, in which the author declares that "when his wife cuts the cocks for capons, by plucking the feathers, and applying them warm in an incision of the comb, and there holding them under her finger for some minutes till the gore-blood hath well cemented them, they grow without fail. Thus can she make any bright purple, or other beautiful feather, grow in the place of the crest. By the same address will the spur, taken fresh and warm from the heel of the cock, be made to grow in the place of the comb also."

---

much pains in satisfying forreign demands about philosophical matters, disperseth farr and neare, store of directions and enquiries for the Society's purpose, and sees them well recommended.

"Query: Whether such a person ought to be left vn-assisted?"

[13] They are contained in the Charter-book, with other Statutes subsequently enacted.

[14] These experiments were generally repetitions of experiments already made in private, and exhibited afterwards for the satisfaction and information of the Society. It is stated, that Boyle was so frequently engaged upon experiments that he used to write over his street-door when thus occupied, "Mr. Boyle cannot be spoken with to-day." See art. Boyle, *Biog. Britt.*

On the 15th Oct. 1662, the King gave further evidence of his goodwill towards the Society, by declaring his pleasure that "no patent should pass for any philosophical or mechanical invention, until examined by the Society." At the same meeting, Evelyn presented his Paper on Forest-trees, which was ordered to be printed[15].

Probably the most important scientific event in the history of the Society during this year, was the acquisition of the services of the celebrated Hooke, who on the 12th November was proposed by Sir Robert Moray, as willing to act in the capacity of Curator, and, in the words of the Journal-book, "to furnish

---

[15] It was published under the title of *Sylva, or a Discourse of Forest Trees, and the Propagation of Timber in His Majesty's Dominions.* Evelyn remarks in his *Diary*, that this was the first book printed by order of the Society, and by their printer, since it was a Corporation. In the Preface he says, "The reader is to know, if these dry sticks afford him any sap, it is one of the least and meanest of those pieces which are every day produced by that illustrious assembly, and which enrich their collections, as so many monuments of their accurate experiments and public endeavours in order to the production of real and useful theories, the propagation and improvement of Natural Science, and the honour of their Institution." In 1679 it reached a third edition, on which occasion Evelyn observes, "With no little conflict and force on my other business, I have yet at last, and as I was able, published a third edition of my *Sylva,* and with such additions as occurred, and this, in truth, only to pacifie the importunitie of very many (besides the printer), who quite tired me with calling on me for it; and above all, threatening to reprint it with all its former defects, if I did not speedily prevent it. I am only vexed that it proving so popular, as in so few yeares to passe so many impressions, and (as I heare), gratifie the avaricious printer with some hundreds of pounds, there had not been some course taken in it for the benefit of our Society. It is apparent that nere 500*l.* has been already gotten by it; but we are not yet œconomists." *Diary*, Vol. II. p. 106.

the Society every day they meete, with three or four considerable experiments, expecting no recompense till the Society gett a stock enabling them to give it." This proposition was received unanimously, and it was resolved that Mr. Hooke should at once be invited to take his seat with the Fellows of the Society; and that Mr. Boyle should have the thanks of the Society for dispensing with Mr. Hooke's services.

At this time Hooke was 27 years of age; he had been for many years in the habit of assisting Boyle in his experiments, and was acting as his assistant when that great philosopher generously relinquished his services in favour of the Royal Society. As Curator and Experimenter to the Society, Hooke laboured with extraordinary diligence. The Journal-books record the trial by him of several hundreds of experiments, mostly new, by which "facts multiplied, leading phenomena became prominent, laws began to emerge, and generalizations to commence[15]." Waller, in his *Life of Hooke*, states that it was "observed by several persons, that whatever apparatus he contrived for exhibiting any experiment before the Royal Society, it was performed with the least embarrassment, clearly, and evidently."

A short time after he entered into the service of the Society, he drew up a series of Papers entitled, *Proposals for the good of the Royal Society*, which are preserved in the archives. The leading features are:

"The Designe of the Royal Society being the promotion of naturall and usefull knowledge, is good: therefore all things tending to the advancement and per-

---

[15] Herschel, *Nat. Phil.*, p. 116.

petuating thereof ought to be promoted. To these ends tends;—such a constitution as will make it self-subsisting.

"All ages afford men enough inclin'd to the study of naturall knowledge. 'Twill be the interest of all such to endeavour to be members of this Society, provided the benefit received be greater than the expense and trouble will purchase elsewhere. Therefore, the benefit of every member thereof in this way, is the soul and life of the Society, and by all means to be advanced. Things tending hereunto are:

"1. That every member of the Society shall have equall freedom to be present at all meetings of it, and shall have free access to their Library, Repository, Instruments, &c., and if absent, shall receive an account of all experiments, observations, discourses, inventions, informations from foreign parts, or correspondencys here at home, querys or proposalls, &c., and whatever other benefit can be afforded them.

"2. That no other person whatsoever, upon any account, shall have any of the aforesaid benefits, before he be, by his earnest desire and the suffrage of the Society, made a Member thereof.

"3. That every Member of the Society shall be equally obliged to promote the ends thereof by paying 52*s.* yearly, and by doing some one duty that shall be charged on them by the Council once a year, or, if his occasions will not permit, to pay 52*s.* more per annum. The dutys may be various, as examining some subject by tryalls or experiments. Giving an account of authours;—giving a history of some trade, manufacture, country, operation, &c.; holding correspondence with some at home, or in foreign parts, about such matters as the Council shall desire, and taking care to provide some experiments for the Meetings.

"4. That these dutys may be more certainly performed, there should be two Secretarys and two Curators at least by office. That the Curators' salarys be but small, and that there be other encouragements given according to desert, upon each new invention, or discovery; either in money, plate, medals, or other gratuitys.

"5. That a certain number of the Society be appointed to manage the prosecution of any new invention, so as to bring it into use, and make it profitable for the Society, and the Inventor."

The remainder of the manuscript enters into details respecting the election of members, experiments, &c. and is not of sufficient importance to warrant insertion here. But it will be seen, that several of Hooke's propositions were ultimately carried into effect.

There was no change made in the Council at the Anniversary on St. Andrew's Day, and the Society continued their weekly meetings without any recess at Christmas.

In January 1662—3, a Committee, that had met several times at Arundel House, presented a Report, recommending the Society to "take measures to extend the growth of apple and pear-trees, for making cyder all over England;" and in March following another Committee strongly urged the Fellows of the Society who possessed land "to plant potatoes, and to persuade their friends to do the same, in order to alleviate the distress that would accompany a scarcity of food:" a recommendation, which, we are informed, was approved by the body generally.

A few months after the Charter of Incorporation had been granted, it was found, that, although thoroughly efficient and complete, as a Patent of Incorporation, it failed in giving the Society certain privileges essential to their welfare.

A summary of the required powers was given by Sir Robert Moray, to Sir Henry Bennett, then Secretary of State, through whose intercession a second Charter was granted to the Society, supplying the desired privileges, and retaining all the clauses of Incorporation contained in the first Charter.

At a meeting on the 25th March, 1663, Sir Robert Moray informed the Society, that the King had signed the second Charter, on which it was ordered that "the thanks of the Society should be given to Sir Henry Bennett for his favour and care." The Patent finally passed the Great Seal on the 22nd of April.

This document, which is of greater importance than the first Charter, is contained in the Appendix.

---

# CHAPTER VII.

Evelyn's Designs for the Society's Armorial Bearings—Grant of Arms by the King—Registered in the Herald's College—First Meeting of the Council—Obligation of Fellows—Business of the Society—Mace given by Charles II.—Described—Curious popular belief of its being the celebrated "Bauble"—Account of the Bauble-Mace—Letter from Mr. Swifte, Keeper of the Regalia—Warrant for making the Mace for the Royal Society.

1660—65.

IF the reader has referred to the second Charter, he will probably have noticed that a grant of Arms was made to the Society[1]. Smith, in his *Historical and Literary Curiosities*, gives a series of sketches copied from original drawings, of designs for the armorial ensigns and cyphers for the Royal Society, traced by Evelyn. The designs are headed, "Arms and Mottoes proposed for y^e Royal Society, 1660," and signed J. Evelyn. The first represents a vessel under sail, with the motto *Et Augebitur Scientia.* The second escutcheon is parted per fesse, Argent and Sable, a hand appears, issuing from clouds, holding a plumb-line, and underneath is the motto, *Omnia probate.* (1 Thess. v. 21.) In this sketch there appears to have been an intention of introducing the Royal Augmentation afterwards given to the Society, as there is an escutcheon in the dexter-chief. The third shield exhibits two telescopes extended in saltire, the object-glasses upwards; and on a chief argent, the earth and

[1] These are, a shield Argent, on a quarter Gules, and three lions of England in pale. A representation of them will be found elsewhere.

planets: the motto is *Quantum nescimus!* The fourth design is a shield, bearing the sun in his splendour, with the motto *Ad majorem Lumen;* on one side of the shield is written

Quis dicere Falsum
Audeat?——— GEOR. I.

The next design represents a shield bearing a canton only, with the motto, *Nullius in verba,* as at present used by the Society; it is probable, therefore, that this sketch was intended to show the disposition of the arms subsequently adopted. The last sketch represents a shield charged with a terrestrial globe, with a human eye in chief; above which are the words *Rerum cognoscere causas,* taken from the Second Book of the *Georgics* of Virgil. Besides these inscriptions appears the word *Experiendo,* and a repetition of the motto *Nullius in verba.*

With the exception of the last motto, none of these suggestions were adopted, as the King subsequently granted to the Society the very high and honourable Armorial Bearings described in the second Charter. In the official Volume, entitled *Royal Concessions in the College of Arms,* instead of the usual form in the instance of a grant of heraldic bearings, issuing from the principal and provincial kings of arms, the drawing is preceded by the following confirmation: "Whereas his Ma$^{tie}$, by his letters patent, under the Great Seal of England, bearing date 22nd day of April, in the 15th year of his reign, hath ordained and constituted a Society, consisting of a President, Council, and Fellows, called by the name of the President, Council, and Fellows of the Royal Society of London for the advancement of Natural Science, to whom, amongst other things, His said Sacred Ma$^{tie}$

hath therein granted a coat of arms, crest, and supporters: The said President, Council, and Fellows, being desirous to have the clause whereby the same are granted to them, together with a *trick* thereof, entered among the records of this office—It was this day, being the thirtieth of June, Anno Domini 1663, in full chapter, upon the motion of Elias Ashmole, Esq., Windsor Herald, and one of the Fellows of the said Society (by whom the said request was made, and the said Patent sent hither to be viewed) agreed and consented unto, and thereupon ordered to be entered as followeth." Then follows a repetition of that part of the charter, granting and describing the arms, which it is unnecessary to repeat.

On the 13th May, 1663, the Council of the Royal Society met for the first time, when the second charter was read, and after the new members of the Council were sworn before the President, the following important Resolutions were agreed to:—

"That the debate concerning those to be received and admitted into the Society, be kept under secrecy.

"That all persons that have been elected or admitted into the Royal Society, doe pay their whole arrears unto this day according to their subscription: and that the Treasurer, or Collector by him appointed, do repair to every such person, and demand the said arrears, showing unto him this order, together with the forme of the subscription hereunto annexed."

"Form of the Subscription.

"Wee whose names are underwritten do consent and agree that we will meet together weekly (if not hinder'd by necessary occasions), to consult and debate concerning the promoting of Experimental Learning, and that each of us will allow one shilling weekly, towards the defraying of occasional charges. Provided, that if any one or more of us shall think fit at any time to withdraw, he or they shall, after notice given thereof to the Company at a meeting, be freed from this obligation for the future."

A list was prepared of all the Fellows of the Society, who at this time amounted to 115[2].

On the 27th May, 1663, the Council resolved that the following obligation should be signed by every Fellow of the Society, and that any one refusing, should be ejected from the Society. This obligation is the same as that in force at the present time:—

"We who have hereunto subscribed, do hereby promise each for himself, that we will endeavour to promote the good of the Royal Society of London, for improving natural knowledge, and to pursue the ends

---

[2] The earliest Manuscript list of the Fellows of the Royal Society, is in the British Museum, MS. 4442. On the first fly-leaf is this epitaph:

"Underneath this stone is laid
Our neighbour Gaffer Thumb;
We trust, although full low his head,
He'll rise in the world to come.

This humble monument may shew,
Where rests an honest man,
Let kings, whose heads are laid as low,
Rise higher if they can."

On the 20th November, 1663, the Royal Society, according to another MS. also preserved in the Museum, consisted of 131 Fellows, of whom 18 were Noblemen, 22 Baronets and Knights, 47 Esquires, 32 Doctors, 2 Bachelors of Divinity, 2 Masters of Arts, and 8 Strangers, or Foreign Members.

for which the same was founded; that we will be present at the Meetings of the Society, as often as conveniently we can, especially at the anniversary elections, and upon extraordinary occasions; and that we will observe the Statutes and Orders of the said Society. Provided, that whensoever any of us shall signify to the President under his hand, that he desireth to withdraw from the Society, he shall be free from this Obligation for the future."

They also resolved that "the Ordinary Meetings of the Society should be held every Wednesday, at 3 o'clock, P. M., and continue until 6, unless the major part of the Fellows present shall resolve to rise sooner, or sit longer, and no Fellow shall depart without giving notice to the President;" and that "the President when in the chair is to be covered, notwithstanding the Fellows of the Society be uncovered."

On the 24th June, 1663, it was resolved, that Barons, and all persons of higher rank, and members of the King's privy council, coming forward as candidates, should be proposed and ballotted for on the same day.

In a manuscript volume of Hooke's *Papers*, in the British Museum, are the originals of the subjoined interesting documents, which were probably drawn up after the passing of the second Charter, as the date 1663 is appended to them.

"The business and design of the Royal Society is—

"To improve the knowledge of naturall things, and all useful Arts, Manufactures, Mechanick practises, Engynes and Inventions by Experiments—(not meddling with Divinity, Metaphysics, Moralls, Politicks, Grammar, Rhetorick, or Logick).

"To attempt the recovering of such allowable arts and inventions as are lost.

"To examine all systems, theories, principles, hypotheses, elements, histories, and experiments of things naturall, mathematicall, and mechanicall, invented, recorded, or practised, by any considerable author ancient or modern. In order to the compiling of a complete system of solid philosophy for explicating all phenomena produced by nature or art, and recording a rationall account of the causes of things.

"All to advance the glory of God, the honour of the King, the Royall founder of the Society, the benefit of his Kingdom, and the generall good of mankind.

"In the mean time this Society will not own any hypothesis, system, or doctrine of the principles of naturall philosophy, proposed or mentioned by any philosopher ancient or modern, nor the explication of any phenomena whose recourse must be had to originall causes (as not being explicable by heat, cold, weight, figure, and the like, as effects produced thereby); nor dogmatically define, nor fix axioms of scientificall things, but will question and canvass all opinions, adopting nor adhering to none, till by mature debate and clear arguments, chiefly such as are deduced from legitimate experiments, the truth of such experiments be demonstrated invincibly.

"And till there be a sufficient collection made of experiments, histories, and observations, there are no debates to be held at the weekly meetings of the Society, concerning any hypothesis or principal of philosophy, nor any discourses made for explicating any phenomena, except by speciall appointment of the Society or allowance of the President. But the time of the assembly is to be employed in proposing and making experiments, discoursing of the truth, manner, grounds and use thereof, reading and discoursing upon letters, reports and other papers concerning philosophicall and mechanicall matters, viewing and discoursing of curiosities of nature and art, and doing such other things as the Council or the President shall appoint[3]."

The other document runs thus:

"The designe of the Royal Society being the improvement of naturall knowledge, they pursue that designe by all means they conceive to conduce thereunto; and knowing that much of it lies dispersed here and there amongst learned and experienced men, when it is ofttimes little regarded because not enquired after, and too generally lost by the death or forgettfullness of the possessors, they conceive many usefull and excellent observations may be collected into a general repository, where inquisitive men may be sure to find them safely and carefully preserved, both for the honour of those that communicate them, and to the generall good of mankind; which is their principall and ultimate aim. And though a virtuous action be a sufficient reward to itself, and that it is ofttimes a greater pleasure to communicate than to concele an invention, yet they resolve to gratify all that communicate with suitable returns of such experiments, observations, and inventions of their own, or advertisements from others of their correspondents, as shall in some kind make them amends. And that you may understand what parts of naturall knowledge they are most inquisitive for at this present, they designe to print a Paper of advertisements once every week, or fortnight at furthest, wherein will be contained the heads or substance of the inquiries they are most

[3] Additional MSS. 4441.

solicitous about, together with the progress they have made and the information they have received from other hands, together with a short account of such other philosophicall matters as accidentally occur, and a brief discourse of what is new and considerable in their letters from all parts of the world, and what the learned and inquisitive are doing or have done in physick, mathematicks, mechanicks, opticks, astronomy, medicine, chymistry, anatomy, both abroad and at home;—First, it is earnestly desired that all observations that have been already made of the variation of the magneticall needle in any part of the world, might be communicated, together with all the circumstances remarkable in the making thereof; of the celestiall observations for knowing the true meridian, or by what other means it may be found, the time of making it, by whom and in what manner, with what kind of needle, whether a ship-board or upon land, &c. But from a considerable collection of such observations, Astronomy might be made of that admirable effect of the body of the earth upon a needle toacht by a loadstone, that if it will (as is probable it may) be usefull for the direction of seamen or others for finding the longitude of places, the observations collected together, with the theory thereof, may be published for the generall good of navigation, which they engage to doe soe soon as they have a sufficient number of such observations, wherein mention shall be made of every person soe making and communicating his observations[4]."

These comprehensive statements, although not strictly official, yet, being in the hand-writing of Hooke, who was so intimately connected with the

[4] Additional MSS. 4441.

Society, possess great value, as exhibiting the contemplated scope of operations.

In August, 1663, Charles II. presented the Society with the Mace at present in their possession[5]. His Majesty had probably resolved some time previously to honour the Society with this mark of his esteem, for, in the first and second Charters, permission is given to have two Sergeants-at-mace to attend upon the President (*duos servientes ad clavas, qui de tempore in tempus, super Præsidem attendant*). The Council-Book of the Society records, that "on the 3rd of August, 1663, the President (Lord Brouncker) informed the Society, that Sir Gilbert Talbot, Master of the Jewell House, had sent to him, without taking any fees, the Mace bestowed by his Majesty upon the Society; and that he, the said President, had, in the book of his Majesty's Jewell House, acknowledged the receipt thereof for the Society[6]."

This Mace, which fills so important an office in the Royal Society, as no Meeting can be legally held without it[7], is made of silver, richly gilt, and weighs 190

---

[5] The account of this Mace was read before the Royal Society, on the 30th April, 1846, and is printed in the Proceedings.

[6] Vol. I. p. 23.

[7] The same practice respecting this insignia exists at the Royal Society, as is observed in Parliament. In the House of Commons, when the Speaker is in the chair, and the Mace on the table, any member may rise to address the house. When the Speaker leaves the chair, the Mace is taken off the table; and when carried out of the building, the assembly is no longer a house. I may mention, however, that prior to the Presidency of Sir Hans Sloane, the Mace was only put on the table at the Society's Meetings when the *President* was in the chair. The first official act of Sir Hans Sloane, in his capacity of President, was to order the Mace to be used when a Vice-President occupied the chair, as well as when the President presided.

oz. avoirdupois. It consists of a stem, handsomely chased, with a running pattern of the thistle, terminated at the upper end by an urn-shaped head, surmounted by a crown, ball, and cross. On the head are embossed figures of a rose, harp, thistle, and fleur-de-lys, emblematic of England, Ireland, Scotland, and France, on each side of which are the letters C. R. Under the crown, and at the top of the head, the royal arms appear very richly chased; and at the other extremity of the stem are two shields, the one bearing the arms of the Society, the other the following inscription[8]:—

Ex Munificentiâ
Augustissimi Monarchæ
Caroli II.

---

[8] The Arms of the Society and the Inscription were engraved on the Mace by the Society's directions in the year 1663; it was cleaned and regilt in 1756 at the expense of Lord Macclesfield, who was at that period President of the Society, as appears by the following entry in the Council Minutes under the date of July 29, 1756:—

"The President having declared by letter to Mr. Watson, that he intended that the Mace shall be cleaned and repaired at his expense, it was

"Ordered, that Mr. Hawksbee do deliver the Mace to Messrs. Wyckes and Netherton, silversmiths in Panton Street, for that purpose."

The Mace accordingly was put into thorough repair, regilt, and registered in the Excise office as weighing 190 oz*. At a Meeting of the Society on the 25th of November, 1756, the thanks of the Society were unanimously voted to the President "for this obliging mark of his regard for them†."

In 1828, the Mace was again regilt and repaired, at an expense of 23*l*. 10*s*. It is now in excellent condition.

* *Council Minutes*, Vol. IV. pp. 177 and 178.
† *Journal-book*, Vol. XXIII. p. 418.

Dei Gra. Mag. Brit. Franc. et Hib.
Regis, &c.
Societatis Regalis ad Scientiam
Naturalem promouendā institutæ
Fundatoris et Patroni
An. Dni 1663.

To this Mace attaches a celebrity, which has long caused it to be regarded with extraordinary interest. It is almost superfluous to state, that this arises from the belief of its being the identical Mace turned out of the House of Commons by Oliver Cromwell when he dissolved the Long Parliament. So general has been this credence, that numberless visitors have come purposely to the apartments of the Society to see the Mace, having read, or been assured, that it is the famous "bauble;" and after minutely examining it, have departed, firmly persuaded that they have seen the Instrument so rudely dealt with by the Protector.

Nor has its fame been confined to oral tradition: Writers, professing to be authentic historians, have chronicled that the "bauble mace" is in the possession of the Royal Society; and it may be mentioned that the proprietors of the Abbotsford Edition of the *Waverley Novels* applied for permission to make a drawing of it to illustrate *Woodstock*, an engraving of which appears in the above work, accompanied by a statement, that it is a representation of the "bauble mace" formerly belonging to the Long Parliament, and now in the possession of the Royal Society.

It is difficult to conceive how this belief originated, and the more so, as there is not the slightest historical evidence in its favour; but, on the contrary, many facts which prove most indisputably, that the mace in question has no claim whatever to the designation of the "bauble" of the Long Parliament.

I confess that when the oft-repeated story, or legend as we may now call it, was imparted to me, I felt a strong desire to learn on what historical grounds it rested. As an officer of the Royal Society, I felt it to be almost my duty, when visitors came to see the "bauble," to be able to authenticate its history, though it may be observed, that I have never heard any doubts whatever cast upon its supposed authenticity: so true is it, that we willingly cling to whatever is interesting and marvellous.

It however frequently occurred to me, that the Society's Mace could not be that used in the House of Commons during the reign of Charles the First, and subsequently turned out by the Commonwealth Parliament; for when I thought of the democratic whirlwind that uprooted and swept away every vestige of royalty, it appeared that nothing short of a miracle could account for the preservation of so conspicuous and decisive an emblem of Sovereignty.

Researches for the purposes of this history led me, in the first place, to investigate the history of the famous "bauble;" and, secondly, that of the Mace of the Royal Society, in order to ascertain whether the latter and former are identical.

These researches were far more laborious than I anticipated; and although, unfortunately, they may have the effect of destroying a pleasing and long-cherished illusion, I feel confident nevertheless, that the Royal Society will not be displeased by having the real truth set before them.

On the 30th of January, 1649, Charles the First was beheaded; and on the 1st of February following, the Journal-books of the House of Commons inform us[9], that "a Committee appointed for securing the crown-jewells, and other things, late the King's, reported that they have disposed them in a room under several doors now locked up[10]." The royal Mace was doubtless among the articles of plate thus disposed of, as on the 17th of March, we find another entry that, "It be referred to the Committee for alteration of seals, to consider of a new form of Mace, and the special care thereof is committed to Mr. Love[11]."

On the 13th of April, 1649, "Mr. Love reported several forms of a new Mace," upon which it was "Resolved, that this shall be the form of the new Mace[12]."

* * * * *

Instead of a design appear a number of asterisks as above; but, fortunately, the *Parliamentary History* and Whitelock's *Memorials* enable us to fill the blank in the most satisfactory manner. On the 6th of June, we read, "It was ordered that the new Mace, made by Thomas Maundy of London, goldsmith, be delivered unto the charge of the sergeant-at-arms attending the Parliament; and that the said Mace be carried before the Speaker; and that all the Maces to be used in this Commonwealth be made according to the same form and pattern; and that the said Thomas Maundy have the making thereof, and none other[13]."

---

[9] Vol. vi. p. 164.

[10] Probably in the Tower; as Whitelock says, that he went at this period with others to see the Seals locked up in the Tower.

[11] Vol. vi. p. 166.

[12] Vol. vi. p. 184.

[13] Vol. vi. p. 226.

Now, according to the *Parliamentary History*, on the 6th of June (the same day, it will be observed, that the Journal-books of the House of Commons state "the new Mace was ordered to be delivered to the sergeant-at-arms,") a *new* Mace was brought into the House, ornamented with flowers instead of the cross and ball on the top, with the arms of England and Ireland, instead of the late King's[14]. Whitelock also records in his *Memorials*, that "on the 6th of June, 1649, a new Mace with the arms of England and Ireland, instead of the King's arms, was approv'd and delivered to Sergeant Birkhead, to be used for the House; and all other Maces for the Commonwealth to be of that form[15]." It is thus evident, that a new Mace was provided for the Commonwealth Parliament, differing essentially in form from that used in the time of Charles the First. The Journals of the Commons further inform us, that on the 11th of June, 1649, "the Committee of revenue was authorized and required to pay forthwith, unto Thomas Maundy of London, the sum of £137. 1*s.* 8*d.*, in discharge of his bill of charges for making the new Mace for the service of this House[16]." There appears to have been some error in this amount, as on the 7th of August 1649, it was "Ordered, that it be referred to the Committee of revenue to examine the particulars touching the charge for making the Mace for this House; and if they find the same was miscast, and that the sum of £9. 10*s.* remaineth yet due and unpaid for the same, that they forthwith make payment of the same unto Thomas Maundy[17]."

---

[14] Vol. III. p. 1314.

[15] Vol. VI. p. 228.

[16] P. 406.

[17] Vol. VI. p. 275.

Thus we have additional evidence, not only of the manufacture of a new Mace for the House, but even of its cost.

On the 9th of August, 1649, as appears from the Journals, it was "Ordered, that those gentlemen who were appointed by this House to have the custody of the Regalia, do deliver them over unto the trustees for sale of the goods of the late King, who are to cause the same to be totally broken; and that they melt down all the gold and silver, and sell the jewells to the best advantage of the Commonwealth, and to take the like care of them that are in the Tower[18]."

There is every reason to believe that this order was executed, and that not only the regalia, but all gold and silver articles (among which would be included the royal Mace) were melted down and sold. A curious MS., giving an account of the preparations for the coronation of Charles the Second, by Sir Edward Walker, Knt., Garter Principal King-at-arms, published in 1820, informs us, "that because through the rapine of the then late unhappy times, all the royal ornaments and regalia, theretofore preserved from age to age in the treasury of the Church of Westminster, had been taken away, sold, or destroyed, the Committee (appointed to order the ceremony) met divers times, not only to direct the re-making such royal ornaments and regalia, but even to settle the form and fashion of each particular, all which did then retain the old names and fashion, although they had been newly made and prepared, by orders given to the Earl of Sandwich, Master of the Great Ward-

---

[18] Vol. VI. p. 276.

robe, and Sir Gilbert Talbot, Knt., Master of the Jewell House."

The MS. then proceeds to enumerate the various articles ordered for the coronation of the King, rendering it still more evident that the former regalia had been destroyed[19].

The singular and fortunate discovery of the receipt of Sir Robert Vyner for 5500*l.*, being part payment of 31,978*l.* 9*s.* 11*d.*, the charge for making the Regalia and different gold and silver ornaments, destined as presents at the Coronation of Charles the Second,

---

[19] The following letter from Mr. Swifte, Keeper of Her Majesty's Jewel House, confirms the above statement:—

"*Her Majesty's Jewel House,*
*March* 15, 1846.

"DEAR SIR,

"YOU are but too right in your idea of the modern character of our Regalia. Whether as an Englishman, a Royalist, an Historian, or as a Gentleman, or in all these capacities, you must grieve over the wicked annihilation of its ancient memorials. The barbarous spirit which descended on the French revolutionists, when they destroyed even the tombs and the bones of their ancient monarchy, actuated our Puritans to break up and sell the old Crown Jewels of England.

"The two Jewel Houses (for then there were *two*, the upper and the lower) were betrayed by my predecessors, Sir Henry and Mr. Carew Mildmay, in 1649, and their precious contents transferred to the Usurper. The most shameful part of this afflicting transaction was the breaking up of King Alfred's wirework gold fillagree crown, and selling it for the weight of the metal, and what the stones would fetch.

"A new Regalia was ordered at the Restoration, to which additions or alterations have been made as requisite, constituting that which is now in my charge,

"Believe me, my dear Sir, very faithfully yours,

"EDMUND LENTHAL SWIFTE, K.C.J.

"*C. R. Weld, Esq.*"

lends additional and powerful weight to the presumption, that all the plate belonging to Charles the First was destroyed.

The receipt specifies the various articles made, among which are no less than "*eighteen maces*, and divers other parcells of guilt and white plate." It is worthy of mention, that this receipt was found by Mr. Robert Cole among the documents sold in 1838 by the then Lords of the Treasury as waste paper!! It forms the subject of a short communication made to the Society of Antiquaries in 1841, and is printed in the 29th volume of the *Archæologia*[19].

Between the period when the new Commonwealth Mace was first used, and the 23rd of April, 1653, the date of the celebrated dissolution, the Journals of the Commons frequently allude to the new Mace; and as there is no record whatever of any other mace having been ordered, we can only conclude that this was the celebrated Mace mentioned in all histories of that period as the "bauble," so called by Cromwell when he dissolved the Parliament. That the Mace was turned out of the House of Commons admits of no doubt, as all writers agree on the point, the only discrepancy being, that some say Cromwell, pointing to the Mace, ordered a musketeer to take away that "fool's bauble;" and others, that when all the Members had left the House, the doors were

---

[20] For another proof of the extraordinary want of judgment manifested by the Lords of the Treasury in selling several tons weight of national records, see a very curious pamphlet by Mr. Thomas Rodd, entitled, *Narrative of the Proceedings instituted in the Court of Common Pleas against Mr. T. Rodd, for the purpose of wresting from him a certain MS. roll, under the pretence of its being a document belonging to that Court.* 8vo. London, 1845.

locked, and the key, with the mace, carried away by Colonel Otley.

We may mention here, that West's famous picture of the Dissolution of the Long Parliament represents Cromwell in the act of pointing to the Mace as he uttered the words, "Take away that fool's bauble." The Mace, which occupies a most prominent position in the centre of the picture, agrees perfectly in its appearance with the description given of it in the *Parliamentary History* and Whitelock's *Memorials*, being nearly destitute of ornament, and without the crown and cross.

Had we no further evidence, this testimony from authentic documents would suffice to prove that the Mace turned out of the House of Commons by Cromwell was not that subsequently given to the Royal Society by Charles the Second, which differs totally in appearance from the Mace made for the Commonwealth Parliament, and, as we have seen, used by the House of Commons from 1649 to 1653. And when we reflect, that immediately after the King's execution, orders were issued to pull down, erase, and destroy every vestige of royalty throughout the length and breadth of the land[21], it is absurd to imagine that the individuals giving such orders, and exacting their most rigid execution, should, for a period of upwards of four years, have sat around a table on which lay a Mace, bearing not only the royal arms in the most conspicuous form, but also a crown and the letters

---

[21] The King's Arms over the Speaker's chair were taken down, and those of the Commonwealth substituted immediately after the execution of Charles the First. *Journals of the House of Commons.*

C. R. four times repeated: and this they must have done to make the story true, that the Mace given to the Royal Society by Charles the Second is the famous "bauble."

"The sacred Mace," as it has been called by some historians, though so unceremoniously expelled from the House of Commons, was, strange as it may seem, preserved and soon restored to its high office; for on the 7th of July 1653, only three days after the assembling of Cromwell's first Parliament, the Journals of the Commons inform us, that the Sergeant-at-arms was "Ordered to repair to Lieut.-Col. Worseley for *the* Mace, and to bring it to this House; and on the same day it was referred to a Committee to consider the use of the Mace, and with whom it shall remain, and report their opinion to the House[22]."

On the 12th of July the Committee reported, that "the Mace should be made use of as formerly;" upon which the House resolved, "That the Mace shall be used in the House as formerly; and it was ordered that the Mace be brought in, which was done accordingly[23]."

From this period to the Restoration, there is no record of a new Mace having been ordered; and by the Journals of the Commons it appears that the Mace was used on all occasions as heretofore, and sometimes even carried before the Speaker, when he went at the head of the House to attend service at St. Margaret's Church, on the days appointed for solemn fasts.

The Restoration, which put an end to every outward manifestation of republicanism, terminated the

---

[22] Vol. VII. p. 282. [23] Ibid., p. 284.

existence of the Commonwealth Mace; indeed, as much haste was shown to get rid of it, as was evinced—after the execution of the late King—in the ejection of the royal Mace from the House of Commons.

On the 27th of April 1660, the Journals further show, that E. Birkhead, Esq., late Sergeant-at-mace, was "Ordered forthwith to deliver the keys of the House, and the Mace belonging to the House, to Sergeant Northfolk;" and on the 21st of May it was resolved[24], "That two new Maces be forthwith provided, one for this House, and the other for the Council of State, with the cross and King's Majesty's arms, and such other ornaments as were formerly usual; and it was referred to the Council of State to take care that the same be provided accordingly[25]."

Here we have additional evidence that the Royal Society's Mace and the "bauble" are not identical, for we find the House of Commons ordering, a month before the Restoration, a new Mace, which is to be decorated "with the King's arms, cross, and other ornaments as were formerly usual."

Having thus clearly ascertained that the Mace presented to the Royal Society by Charles the Second is

[24] Vol. VIII. pp. 34 and 39.

[25] The Mace at present in the House of Commons corresponds in appearance to the above description, and is, I have every reason to believe, that made at the Restoration; it is very much like the mace in the possession of the Royal Society, with the exception that the chasing and ornaments are executed in a less elegant manner. It is 4 feet 8 inches long, and weighs 251 oz. 8 dwts. 2 grs. There is no inscription, date, or maker's name, but simply C. R. between the four shields, emblematic of England, Ireland, Scotland, and France; these letters are on all the maces made at the time of the Restoration.

not that expelled from the House of Commons by Cromwell, I turned my attention to discover, if possible, the history of the Mace belonging to the Royal Society.

It will be remembered that the archives of the Society throw no light whatever upon this point, nor is the Mace even described[26]. It is merely stated, that it was sent from the Jewel House in the year 1663. Under these circumstances, and by the advice of my friend Sir Henry Ellis, I addressed a letter to Edmund Lenthal Swifte, Esq., Keeper of Her Majesty's Jewel House, requesting permission to search the archives, which I presumed were kept in that office. In reply I received the following letter:—

"*Her Majesty's Jewel House,*
*March* 13, 1846.

"Dear Sir,

"It would have much gratified me to aid the wishes of any friend of Sir Henry Ellis.

"On your account too, your name and office would have been more than sufficient to claim attention. But I can only regret my inability in this matter. Since Edmund Burke's Bill, the Jewel House has undergone a radical change in its duties and functions. Previously, its Chief had the charge and presentation of the Royal gifts, whereof he had of course the accounts. Whatever entries there may be concerning the Mace, which was

[26] Evelyn says in his *Diary*, that "the King sent the Society a Mace of silver gilt of the same fashion and bigness as those carried before his Majestie, to be borne before the President on Meeting-daies," Vol. I. p. 338; and it is recorded in the Council-book of the Society, that Sir Richard Brown, through the medium of Evelyn, presented the Society with a velvet cushion, whereon the Mace was laid when placed before the President.

certainly given by Charles the Second to Lord Brouncker, as President of the Royal Society, in the old books of the Jewel House, they are most probably to be found in the Lord Chamberlain's Office, to whom the control of the Jewel House was transferred in (I believe) 1782. Not a single record is, or ever was, in my hands. Otherwise, to have accorded you fullest and freest access, would have been to me an especial pleasure.

"I am, my dear Sir,

"Very faithfully yours,

"EDMUND LENTHAL SWIFTE.

"*C. R. Weld, Esq.*"

The receipt of this letter caused me to write to the Lord Chamberlain for permission to examine the archives under his charge. This was immediately granted, and with the kind assistance of the chief clerks in Lord Delawarr's office, I fortunately, after a long search in a gloomy and damp apartment, which was formerly a stable, found the original warrant, ordering a Mace to be made for the Royal Society.

Subjoined is a copy of this important and valuable document. The book in which it exists is entitled, *The Book of Warrants of the Lord Chamberlain, Edward, Earl of Manchester, of His Majesty's Household, for the Years* 1663, 4, 5, 6, *&* 7, and the warrant is entered under the head of "Jewell House:"—

"A WARRANT TO PREPARE AND DELIVER TO THE RT. HON. WILLIAM LORD VISCOUNT BROUNCKER, PRESIDENT OF THE ROYALL SOCIETY OF LONDON, FOR THE IMPROVING OF NATURAL KNOWLEDGE BY EXPERIMENTS; ONE GUILT MACE, OF ONE HUNDRED AND

FIFTY OZ.[27], BEING A GUIFT FROM HIS MA$^{TIE}$ TO THE SAID SOCIETY."

This warrant is among those issued in 1663; and as several previous warrants exist, bearing the dates of January, February, March, and April; and others, entered subsequently, are dated May, June, and July, we may reasonably conclude, that the warrant for making the Society's Mace was issued early in 1663; and this is strengthened by the fact, that the Society received the Mace in the month of August in the same year[28].

This discovery not only destroys the long-entertained belief, that the Mace belonging to the Royal Society and the "bauble" are identical, but also affords conclusive evidence that the former was made expressly for the Royal Society.

On a minute examination of this Mace, in order to detect, if possible, the maker's name or a date, neither of which exist, I observed that the chasing on the stem consists entirely of thistle-leaves and flowers: at the time this fact passed unnoticed, but it is now evident that the thistle was employed as the principal ornament, on account of its being symbolical of St. Andrew, the patron saint of the Society,

---

27 Troy weight, which exceeds 150 oz. avoirdupois.

28 Since this account was read before the Society, Mr. Browell, the chief clerk in the Lord Chamberlain's Office, has been so obliging as to inform me, that the foregoing warrant is entered in another book of warrants, apparently a duplicate of that which I saw. The words of the warrant are similar to the above, but there is the important addition of the date, May 23, 1663; thus confirming my idea that the warrant was issued in the early part of the latter year.

in whose honour the Fellows of the Society were accustomed, at the early anniversary Meetings, to wear a St. Andrew's cross in their hats.

This use of the thistle is another proof that the Mace was made for the Society.

In conclusion, I cannot forbear observing, that although the Mace may not be as curious as before to the antiquary, divested as it now is, of its fictitious historical interest, yet it is much more to be respected; for surely a Mace designated a "bauble," and spurned from the House of Commons by a republican, would scarcely be an appropriate gift from a Sovereign to the Royal Society.

The Mace in its possession was expressly made for the Society, and given to it by its Royal Founder; and the associations appertaining to it, embracing the remembrance, that around it have been gathered men whose names not only shed imperishable lustre on the Royal Society, but on the civilized world, must hallow it to all lovers of science and truth.

---

# CHAPTER VIII.

Sorbière's Account of his visit to the Society—Sprat's Observations upon it.—Moncony's Description of the Society—Anniversary celebrated by Fellows dining together—Charles II. sends Venison—Exertions to increase the Income of the Society—Petition to the King for grant of Chelsea College—Society issue their Warrant for the bodies of executed Criminals—Notice of Dissection sent to the Fellows—Sir J. Cutler founds a Professorship of Mechanics—Hooke appointed Professor and Curator—Has apartments in Gresham College—His *Micrographia* printed by licence of Council—Dedicated to the Society—Appointment of Committees—Charter-book opened—Expected visit of Charles II.—Publication of Transactions by Oldenburg—Dedication of First Number—Contents—Their Sale—The Plague causes a suspension of the Meetings—Oldenburg remains in London—His alarm—Council-Meetings resumed—Purchase of Mr. Colwall's Collection of Curiosities—Formation of Museum—*Museum Tradescantium*—Transferred to Oxford—Coffee-House Museums—Oldenburg's extensive Correspondence—Presents of Rarities—Weekly Meetings resumed—Masters of Pest-house send their Observations on the Plague to the Society—Experiment of Transfusion of Blood—Great Fire of London interrupts the Meetings—The Society meet in Arundel House—Hooke's Model for rebuilding the City—The Duke of Norfolk presents his Library to the Society—Account of it—Duchess of Newcastle visits the Society—Arrest of Oldenburg—Warrant for his confinement in the Tower—His innocence and release—His Letter to Boyle—The Society obtain possession of Chelsea College—Its dilapidated state—Scheme of building a College—Contributions of the Fellows towards the building—Wren furnishes Design—Never carried out—Patent granting Chelsea College, and additional privileges.

---

## 1660—70.

IN the year 1663, M. Sorbière, historiographer to Louis XIV., visited England, and on his return to France, published an account of his travels, entitled,

*Rélation d'un Voyage en Angleterre, où sont touchées plusieurs choses qui regardent l'estat des Sciences, et de la Religion, et autres matières curieuses.* In this work the author gives the following interesting account of the Royal Society, to which he paid several visits:—

*L'Académie Royale des Physiciens de Londres est établie par des lettres du Roy, qui en est le fondateur, et qui luy a donné le College de Greshem, dans la ruë Biscop getstriidt*[1], *où elle s'assemble tous les Mercredis. Je ne sçay s'il n'y a point desia quelque reuenu affecté pour l'entretien des personnes qui gouernent les machines, et d'vn huissier, qui marche deuant le President auec vne grosse masse d'argent, laquelle il pose sur le Bureau de l'Assembleé, quand il y vient prendre place. Mais i'ay bien oüy dire, que l'on estoit apres à trouer vn fonds pour quatre mille liures de rente à deux personnes sçauantes, qui demeurerōt dans le College, et qui seront gagées pour rapporter à la compagnie ce dont elle voudra estre informée par la lecture des liures. Et à cet vsage il y a dé-ja vn commencement de Bibliotheque tout ioignant vne galerie, dans laquelle on passe au sortir de la sale de l'Assemblée: comme d'vn autre costé il y a au deuant de la mesme sale, vne anti-chambre fort honneste, et deux autres chambres en l'vne desquelles on tient le Conseil; sans conter le logement que l'on destine aux deux professeurs, qui recüeilleront des Autheurs les anciennes experiences Physiques et méchaniques, que l'on examinera pour s'en asseurer à l'auenir, tandis que l'on en fera aussi de nouuelles. La chambre de l'Academie est grande et lambrissée. Il y a vne longue table au deuant de la cheminée, sept au huict chaises à l'entour, couuertes de drap gris, et deux rangs de bancs de bois tout nud a dossier, le dernier estant plus esleué que l'autre en forme d'amphitheatre. Le President et les Conseillers sont electifs.*

[1] M. Sorbière thus curiously spells Bishopsgate Street.

*Ils ne gardent point de rang dans l'Assemblée; mais le President se met au milieu de la table dans vn fauteüil, tournant le dos a la cheminée. Le Secretaire est assis au bout à sa gauche, et ils ont chacun deuant eux du papier et une écritoire. Ie ne vis personne sur les chaises, et ie pense qu'elles sont reseruées pour les gens de haute qualité, ou pour ceux qui ont à s'approcher du President en certaines occasions. Tous les autres Académiciens prennent place indifferemment et sans cérémonie, et lorsque quelqu'vn suruiēt apres que l'Assemblée est formée, personne ne bransle, à peine est-il salué du President, et il prend place vistement là où il peut, afin de ne pas interrompre celuy qui parle. Le President tient vne petite masse de bois à la main, dont il frappe sur la table lors qu'il veut faire faire silence. On parle a luy découuert, iusques à ce qu'il fait signe que l'on se couure; et l'on rapporte en peu de mots ce que l'on trouue à propos de dire sur l'experience que le Secretaire a proposée. Personne ne se haste de parler, ny ne se picque de parler long-temps, et de dire tout ce qu'il sçait. On n'interrompt iamais celuy qui parle, et les dissentimens ne se poussent pas bien auant, ny d'vn ton qui puisse desobliger en aucune manière. Il ne se peut rien voir de plus ciuil, de plus respectueux, et de mieux conduit que cette Assemblée telle qu'elle me parut. S'il y a quelques entretiens particuliers qui se forment tandis que quelqu'vn parle, ils se passent a l'oreille, et l'on s'arreste tout court au moindre signal que le President fait; de sorte que l'on n'acheue pas mesme de dire sa pensée. Cette modestie me sembla fort remarquable en vn corps composé de tant de personnes, et de tant de sortes de nations*[2].

Sorbière laments that the natural reserve of the English, which extended to the Fellows of the Society; their repugnance to speak French, and their pronunciation of Latin, so foreign to his ears, gave him but

---

[2] pp. 86—91.

little opportunity of cultivating their acquaintance; but he frequently makes honourable mention of the Society, and their labours[3].

Monconys, in his *Journal des Voyages*, published in 1677, also gives an interesting account of the Royal Society, to which he was introduced by Sir R. Moray. He writes under the date of the 23rd May, 1663:—*Je fus à l'Academie de Gressin*[4], *où l'on s'assemble tous les Mercredis pour faire une infinité d'experiences sur lesquelles on ne raisonne point encore, mais on les rapporte à mesure que quelqu'vn sçait, et le Secretaire les escrit. Le President, qui est toujours une personne de condition, est assis contre*

---

[3] Dr. Sprat published in 1668, *Observations on M. Sorbière's Voyage into England, in the form of a Letter to Dr. Wren, Professor of Astronomy in Oxford.* I am led to mention this, as Dr. Sprat regarded M. Sorbière's book in the light of "an insolent libel on the English nation." With respect to his remarks upon the Royal Society, he says, "He has been utterly mistaken in the report of their main design. There are two things that they have most industriously avoided, which he attributes to them: the one is, a dividing into parties and sects; and the other, *a reliance on books for their intelligence of nature.* He first says, that they are not at all guided by the authority of Gassendus, or Des Cartes, but that the mathematicians are for Des Cartes, and the men of general learning for Gassendus. Whereas, neither of these two men bear any sway amongst them: they are never named there as dictators over men's reasons; nor is there any extraordinary reference to their judgments." He also asserts that the Royal Society, has appointed lodgings, and establish'd four thousand livres a year on two professors, who shall read to them out of authors, and that they have begun a Library for that purpose. Whereas they have as yet no Library, but only a Repository for their Instruments and Rarities, they never intend a Professorian Philosophy, but declare against it; with books they meddle not, farther than to see what experiments have been tried before: their revenue they designe for *operators*, and not for *lecturers*." pp. 206, 7.

[4] Gresham College.

*une grande table quarrée, et le Secretaire à un autre costé. Tous les academistes sont sur des bancs qu'il y a autour de la sale. Le President estoit le Milord Brunker, et le Secretaire M. Oldembourg. Le President a un petit maillet de bois à la main, dont il frappe sur la table, pour faire taire ceux qui veulent parler, lors qu'vn autre parle; ainsi il n'y a ny confusion ny crierie.*

Monconys paid several visits to the Society, and frequently alludes to the number of experiments tried at the meetings, adding: *le Secretaire escrit l'effet des experiences, soit qu'il ait reussi, ou qu'il ait manqué, afin qu'on puisse se détromper aussi bien des fausses propositions, que profiter des veritables.*

The evidence of Sorbière and Monconys is valuable and interesting, as the observations of foreigners of education written on the spot, and who were not likely to be prejudiced in favour of the Society.

On the 30th Nov. 1663, the Society, according to the Journal-book, "met in a solemn manner" to elect officers at 9 a.m., and celebrated their anniversary by dining together. Evelyn, in his *Diary*, says under the above date, "It being St. Andrew's day, who was our patron, each Fellow wore a St. Andrew's crosse of ribbon on the crown of his hat. After the election we dined together, his Majesty sending us venison." At the anniversary dinner in 1683, Evelyn records that the King sent the Society two does.

The accounts of the Treasurer, which were passed on this day, show that the amount received from the 28th Nov. 1660 to June 1663, was £527. 6*s.* 6*d.*, and the disbursements £479. 11*s.* 9*d.*, which included a sum of £11. 9*s.*, remitted by order of council to several members of the Society.

By a note it appears that the arrears due by Fellows amounted to no less than £158. 4*s.* 6*d.*

Great exertions were made to collect this sum, for the Society were in serious want of money to enable them to purchase instruments and apparatus for experiments. It is recorded that Mr. Colwall, who was elected one of the new council, presented the Society with £50, for which he was specially thanked, and Mr. Balle promised to make a donation to the Society of £100[5];—these sums, however, proved insufficient to meet the growing wants; and we find accordingly that at almost every Meeting of the Council various measures were brought forward for increasing the income of the Society.

Early in 1664, it was proposed "to solicit a grant from the King, of such lands as were left by the sea;" and it was moved "that the King might be spoken to, to confer such offices in the courts of Justice, or the Custom-house, as were in his Majesty's grant, upon some members of the Society for the use of the whole."

It was further resolved, "that every member of the Council should think on ways to raise a revenue for carrying on the design and work of the Society."

The result of these deliberations was, a petition to the King, praying his Majesty to grant Chelsea College, and the lands belonging to it, to the Royal Society[6].

---

[5] This he did subsequently, and he also gave an iron chest having three curious locks, which is still in the Society's possession.

[6] Chelsea College was founded by James I., and was built upon a piece of ground called "Thame Shot," containing about six acres, then belonging to the Earl of Nottingham, and leased by him to the College for the annual rent of 7*l.* 10*s.* The King laid the first

The petition was presented to the King in the month of June, and referred by his Majesty to Sir Henry Bennet, one of the principal Secretaries of State.

During the remainder of this year the Council used every exertion to procure a grant of the College, but there were too many difficulties in the way to hope for immediate success. Not the least of these was the circumstance that the College was claimed by two parties, who refused to give up their titles to the property without pecuniary satisfaction. Meanwhile, experiments were carried on with great vigour, and, in the beginning of 1664, the Society exercised their privilege of claiming the bodies of criminals executed at Tyburn, which were to be dissected in Gresham College.

The warrant demanding the bodies was as follows:

"These are to will and require you, that one Body, either Man or Woman, executed at Tyburne this present ——, being the —— day of ——, such as the bearer

---

first stone, and gave all the necessary timber from Windsor Forest, but the building was never completed according to the original design. The College was intended for "the defence of the true religion established within the realm, and for the refuting of errors and heresies repugnant to the same," and consisted of a Provost, and seventeen Fellows, all of whom were divines. In 1610, an Act of Parliament was passed empowering the Fellows to supply London with water, to be conveyed by pipes underground from the river Lea, the rents arising from which were to be devoted to the maintenance of the College. And in 1616, the King directed the Archbishop of Canterbury "to stir up all the Clergy for a liberal contribution in support of the College," but the sum thus collected was so small that the institution could not be maintained, and consequently fell to decay. See Fuller's *Church History,* Stow's *London*, and Collier's *Ecclesiastical History.*

hereof shall chuse, be delivered unto the said —— at the time and place of the said execution, for the use of the Royal Society; he paying the ordinary fees for the same.—Given under my hand, the day and year above written.

"Signed by the President.

"To all whom it may concern."

Dissecting instruments were provided at the expense of the Society, and when it was proposed to dissect a body, the following notice was sent to each Fellow[7].

"You are desired to take notice that there will be an anatomical administration at Gresham College to begin at —— day at ten of the clock precisely."

In the month of June 1664, Sir John Cutler founded a Professorship of Mechanics, and with the concurrence of the Council of the Royal Society, settled an annual stipend of £50 during life upon Hooke, empowering the President, Council, and Fellows of the Society to appoint the subjects and number of lectures. On the 23rd Nov. Hooke was proposed for the office of Curator to the Society, and on the 11th Jan. 1665, he was elected "for perpetuity, with a salary of £30 a year *pro tempore*." Apartments in Gresham College were also assigned as his residence.

In November 1664, the President, by order of the

---

[7] Dissections of the human body were at this period often witnessed by persons of rank. Pepys says in his *Diary*, that "Charles the Second saw Dr. Clark and Mr. Pierce dissect two bodies, a man and a woman, with which His Majesty was highly pleased." Vol. I. p. 217.

Council, signed a license for printing Hooke's *Micrographia*; which, whilst in manuscript, had frequently been brought before the Society. In a letter to Boyle, dated November 24, 1664, Hooke says: "My microscopical observations had been printed off above a month, but the stay that has retarded the publishing of them, has been the examination of them by several members of the Society."

The work was published in 1665, and is dedicated to the Society. Hooke was very desirous that his theories should not be supposed to represent the opinion of the Society. In the dedication he says: "The rules you have prescribed yourselves in your philosophical progress, do seem the best that have ever yet been practised; and particularly that of avoiding dogmatizing, and the espousal of any hypothesis not sufficiently grounded and confirmed by experiments. This way seems the most excellent, and may preserve both philosophy and natural history from its former corruptions. In saying which, I may seem to condemn my own course in this treatise, in which there may perhaps be some expressions which may seem more positive than your prescriptions will permit. And though I desire to have them understood only as conjectures and queries (which your method does not altogether disallow), yet if even in those I have exceeded, it is fit I should declare that it was not done by your directions."

Some idea may be formed of the activity of the Society at this period, by the following list of eight committees appointed on the 30th March, 1664. 1. Mechanical, consisting of 69 members. 2. Astronomical and Optical, consisting of 15 members. 3. Anatomical, consisting of Boyle, Hooke, Dr. Wilkins, and all the physicians of the Society. 4. Chymical, comprising all the physicians of the Society, and seven other Fellows. 5. Georgical, consisting of 32 members. 6. For histories of Trades, consisting of 35 members. 7. For collecting all the phænomena of Nature hitherto observed, and all experiments made and recorded, consisting of 21 members. 8. For Correspondence, consisting of 20 members[8].

It was expected during the early part of this year that the King would have honoured the Society with his presence, and consequently many experiments were prepared for the entertainment of his Majesty, who, however, does not appear to have paid the contemplated visit[9].

In this year the Charter-book was opened. This is a very handsome volume, bound in crimson velvet, with gold clasps and corners, having on one side a gold plate bearing the shield of the Society, and on the other side a corresponding plate showing the

---

[8] In a letter to Boyle, dated London, June 10, 1663, Oldenburg says, his Paris correspondents write to assure him that the English Philosophers do more for science than *toutes les autres peuples de l'Europe, nous ayans donné quantité de choses curieuses et particulières outre les grands ouvrages qu'ils ont donné au public.* Archives Royal Society, Letters. Oldenburg and an Italian named Galeazzo Victorio Villaro di Stato wrote several verses in praise of the Royal Society at this period.

[9] The reader will probably remember having read in more than one work that Charles II. visited the Society. It is evident that the Fellows expected this honour, and frequently made preparations to receive the King; but there is no Minute of his having attended any Meeting, and it is only reasonable to conclude that had he been present, the fact would have been duly recorded. Experiments were often made before the King at his palace and elsewhere, by Fellows of the Society, and this has probably led to the error of supposing that he visited Gresham College.

crest;—an eagle Or, holding a shield with the arms of England. The leaves of this book are of the finest vellum. The arms of England superbly emblazoned, adorn the first page, and those of the Society, equally well executed, appear on the next. A copy of the second Charter follows, occupying seventeen pages and a half. This is succeeded by sixteen pages, containing the third Charter; and this, again, by the Statutes of various dates, extending over sixty-six pages. Eleven blank leaves then intervene, after which the first page of the autograph portion of the volume exhibits, within an ornamented scroll-border headed by the Royal shield, the signatures CHARLES R., Founder[10],—JAMES, Fellow[11],—and GEORGE RUPERT, Fellow. All these autographs are in good preservation; that of Charles II. having been evidently written with a finely pointed pen, is not so distinct as the others, but is nevertheless quite legible. The next page is occupied with the autographs of various foreign ambassadors; and the third and succeeding pages contain the signatures of the Fellows beneath the obligation which heads each leaf.

It is impossible to regard the venerated pages of this most valuable Autograph-book without feelings of deep emotion. As we turn over the leaves, the eye is arrested by names glorious to the memory of Englishmen; Clarendon, Boyle, Wallis, Wren, Hooke,

---

[10] Written January 9, 1664—5.

[11] Pepys says in his *Diary*, under the date January 9, 1665, "I saw the Royal Society bring their new book, wherein is nobly writ their Charter and Laws, and comes to be signed by the Duke as a Fellow; and all the Fellows hands are to be entered there, and lie as a Monument; and the King hath put his with the word Founder." Vol. I. p. 324.

Evelyn, Pepys, Norfolk, Flamsteed, Newton[12], are here together with all their fellow-labourers in science.

In truth, it would be difficult to over-estimate the interest and value of this book, containing as it does the autographs of nearly all the illustrious patrons of philosophy, and scientific and literary individuals, from the time of the incorporation of the Royal Society, to the present period. Here are the autographs of the successive Kings and Queens of England, as well as many of the Sovereigns of foreign countries, who have visited England. Our gracious Queen has signed her name as Patron of the Society, and on the same page, which is richly illuminated, are the signatures of Prince Albert, and the Kings of Prussia and Saxony. Seventy-one pages are occupied by the autographs of the Fellows (including those on the Foreign list), and as these men represent the science of Europe, the volume, rich as it is at present, is annually becoming of greater value and interest.

On the 1st March, 1664—5, it was ordered at a Meeting of the Council, "That the *Philosophical Transactions*, to be composed by Mr. Oldenburg, be printed the first Monday of every month, if he have sufficient matter for it; and that the tract be licensed by the Council of the Society, being first reviewed by some of the members of the same; and that the

[12] It is worthy of remark, that the name immediately beneath that of Newton, though in characters four times the size of those of the illustrious Philosopher, is nearly obliterated, by the sad habit of touching. Individuals *will* persist in forgetting that the drop wears the stone away, and that each rub of the finger on the page (though the latter be vellum), will infallibly, at length, obliterate the autograph—and in time destroy the vellum itself.

President be now desired to license the first papers thereof, being written in four sheets in folio, to be printed by John Martyn and James Allestree[13]." In conformity with this order, the first number of the *Transactions* appeared on Monday, the 6th March. It consists of 16 quarto pages, at the end of which are the words, "Printed with license." The dedication is:—

"TO THE ROYAL SOCIETY.

"It will not become me to adde any attributes to a Title, which has a fulness of lustre from his Majestie's denomination.

"In these rude collections, which are only the gleanings of my private diversions in broken hours, it may appear, that many minds and hands are in many places industriously employed, under your countenance, and by your example, in the pursuit of those excellent ends, which belong to your heroical undertakings.

"Some of them are but the intimations of large compilements. And some eminent members of your Society, have obliged the learned world with incomparable Volumes, which are not herein mentioned, because they were finish't, and in great reputation abroad, before I entered upon this taske. And no small number are at present engaged for those weighty productions, which require both time and assistance for their due maturity. So that no man can, from these glimpses of light, take any just measure of your performances, or of your prosecutions; but every man may perhaps receive some benefit from these parcels, which I guessed to be somewhat conformable to your design.

---

[13] Printers to the Society.

"This is my solicitude; that, as I ought not to be unfaithful to those counsels you have committed to my trust, so also that I may not altogether waste any minutes of the leasure you afford me. And thus have I made the best use of some of them that I could devise; to spread abroad encouragements, inquiries, directions and patterns, that may animate and draw on universal assistances.

"The great God prosper you in the noble engagement of dispersing the true lustre of his glorious works, and the happy inventions of obliging men all over the world, to the general benefit of mankind! So wishes with real affections,

"Your humble and obedient Servant,

"HENRY OLDENBURG."

The contents of the first number are:

"An Introduction. An Accompt of the Improvement of Optick Glasses at Rome. Of the Observation made in England of a Spot in one of the Belts of the Planet Jupiter. Of the Motion of the late Comet predicted. The heads of many new Observations and Experiments, in order to an Experimental History of Cold, together with some thermometrical discourses and experiments. A relation of a very odd monstrous Calf. Of a peculiar Lead Ore in Germany, very useful for essays. Of an Hungarian Bolus, of the same effect with the Bolus Armenus. Of the new American Whale-fishing about the Bermudas. A Narrative concerning the success of the Pendulum-watches at sea for the Longitudes; and the grant of a Patent thereupon. A Catalogue of the Philosophical Books publisht by Monsieur de Fermat, Counsellour at Tholouse, lately dead."

The Introduction, written also by Oldenburg, comes next:—

"Whereas there is nothing more necessary for promoting the improvement of philosophical matters, than the communicating to such as apply their studies and endeavours that way, such things as are discovered and put in practice by others; it is therefore thought fit to employ the *press*, as the most proper way to gratifie those whose engagement in such studies, and delight in the advancement of learning and profitable discoveries, doth entitle them to the knowledge of what this kingdom, or other parts of the world, do from time to time, afford, as well as of the progress of the studies, labours, and attempts, of the curious and learned in things of this kind, as of their compleat discoveries and performances: to the end that such productions, being clearly and truly communicated, desires after solid and usefull knowledge may be further entertained, ingenious endeavours and undertakings cherished, and those addicted to or conversant in such matters may be invited and encouraged to search, try, and find out new things, impart their knowledge to one another, and contribute what they can to the grand design of improving natural knowledge, and perfecting all *Philosophical Arts and Sciences*. All for the glory of God, the honour and advantage of these kingdoms, and the universal good of mankind."

The foregoing extracts are not only interesting in themselves, but necessary to our purpose to show the manner in which the *Philosophical Transactions*, which have acquired a celebrity extending over the civilized world, were commenced[14]. Their publication,

---

[14] It is a curious and remarkable fact, that almost all the Philosophical Papers in the early numbers of the *Journal des Sçavans*, first published on the 5th January, 1665, are translations of the papers in the *Philosophical Transactions*.

as already explained, was undertaken by Oldenburg, who carried on the work until the period of his death in 1677. The Secretaries who succeeded Oldenburg continued, in like manner, to superintend the publication of the *Transactions* until the year 1750, when a Committee was appointed for this purpose. It is unnecessary to enter into any further details here, as a Table will be found in the Appendix, showing the Numbers, Volumes, and Editors of the *Philosophical Transactions* from their commencement.

Although the *Transactions* were published by the Secretaries, and a notice to this effect was inserted in the early numbers, yet it was generally believed that they emanated from the Royal Society. To correct this mistake, the following notification was inserted at the end of the 12th Number:

"Whereas 'tis taken notice of, that severall persons persuade themselves that these *Philosophical Transactions* are published by the Royal Society, notwithstanding many circumstances, to be met with in the already publish't ones, that import the contrary; the writer thereof hath thought fit expressly here to declare, that that perswasion, if there be any such indeed, is a meer mistake; and that he upon his PRIVATE account (as a well-wisher to the advancement of usefull knowledge, and a furtherer thereof by such communications, as he is capable to furnish by that philosophicall correspondency, which he entertains and hopes to enlarge,) hath begun and continues both the composure and publication thereof. Though he denies not, but that, having the honour and advantage of being a Fellow of the said Society, he inserts at times some of the particulars that are presented to them; to wit, such as he knows he may mention without offending them, or transgressing their orders; tending only to administer occasion to others also to consider and carry them further, or to observe and experiment the like, according as the nature of such things may require."

It may be remarked, however, that the intrinsic value of the *Transactions* depended very much on the various Presidents, and Councils, who had the power of licensing the publication of papers[15].

On the 28th June, 1665, the weekly Meetings of the Society were discontinued on account of the plague, which was then extending its fatal influence through-

---

[15] In a MS. letter of Oldenburg to Boyle, dated London, December 19, 1665, we have the following interesting account of the sale of the *Transactions:*

"Mr. Davis (the printer) wrote me the other day so heavy a letter, that it would very much slacken any man's pace in continuing such labor. For he tells me, that of the first *Transactions* he printed, he had not vended above 300, and that he fears there will hardly sell so many as to repay the charge of paper and printing, so that it seems my pains and trouble would be of no avayle to me. Yet he concludes that, notwithstanding these discouragements, if you and I doe think to have any more printed at Oxford, he will readily serve us in the managing thereof, and the present disbursing of the charge; intimating withall, that he undertook to print these papers, provided he might be secured from being a loser by it. What to say to this I know not; if Mr. Davis give over, it will look very ill, and if he continue, I must suffer very much. He thinks that London being like to be open now for commerce, if he do send to three or four active stationers in severall quarters of the town, and besides to Cambridge, Exeter, Brystoll, &c., item to Ireland and Scotland; a far greater number should then go off, but if he be not a man of a large and active correspondence, I had done much better never to have committed it to him. He should send some copies to a good bookseller about the Exchange (for there I find they are inquired after,) and to another about Dunstan's in Fleet Street, and to another about Westminster, and so dispose them to the chief parts of the city, especially now carriers begin to return hither." Archives Royal Society.

out London and Westminster. Most of the Fellows retired into the country, being previously "exhorted by the President to bear in mind the several tasks laid upon them, that they might give a good account of them at their return."

Oldenburg, however, remained at his house in Pall Mall during the entire period of the raging of the plague, and carried on a correspondence with Boyle and others, on scientific matters[16].

In a letter to Boyle, dated July 4, 1665, he says:

"If the plague should come into this row where I am, I think I should then change my thoughts and retire into the country, if I could find a sojourning corner. In the meantime, I am not a little perplexed concerning the books and papers belonging to the Society, that are in my custody; all I can think of to do in this case is, to make a list of them all, and to put them up by themselves in a box, and seal them together with a superscription, that so in case the Lord should visit me, as soon as I find myself not well, it may be sent away out of mine to a sound house, and *sic deinceps.*"

During the continuance of the plague, several Fellows of the Society remained at Oxford, where, according to the following interesting letter from Boyle to Oldenburg, they were in the habit of meeting in the lodgings of the former.

"*Oxford, Sept.* 30, 1665.

"To do any thing that is philosophical, I see I must withdraw from this place, where the making and receiving of visits takes up almost all my time. Yet I had the

---

[16] Several of Oldenburg's letters written during this period are in the archives of the Society.

honour of one that made me amends for all the rest, which you will easily believe when I tell you that it was made me by Sir Robert Moray, Sir Paul Neile, Sir William Petty, Dr. Wallis, Dr. Coxe, Cap. Graunt, Mr. Williamson, (and afterwards Mr. Secretary Morris, who yet knew nothing of the company he found). These gentlemen I had put in mind, that there being now at Oxford no inconsiderable number of the Royal Society, insomuch that the King seeing Sir Robert Moray and me with some others was pleased to take notice of it; I did not know why we might not, though not as a Society, yet as a company of virtuosi, renew our Meetings; and being put upon naming the day and place, I proposed Wednesday as an auspicious day, being, as you know, that of our former assemblies, and for the place till they could be better accommodated, I offered them my lodging, where over a dish of fruit we had a great deal of pleasing discourse and some experiments that I shew'd them, particularly one, which was thought odd enough, of turning a liquor like fair water in a moment into an inky substance, and presently changing that, first into a clear liquor, and then into a white one almost like milk. That Mr. Oldenburg was mentioned and drunk to by some of us, I have scarce time and paper to inform you, and that you were wish'd here, you will, I hope, easily believe, if you remember that there was in this company, besides Sir Robert Moray, Sir William Petty and Dr. Wallis,

"Your very affectionate friend,

"Robert Boyle."

During this melancholy period the seventh and eighth numbers of the *Philosophical Transactions* were printed at Oxford, in consequence of the impossibility of finding printers in London to execute the work. Hooke remained in London until the 15th July, when he accompanied Sir William Petty and Dr. Wilkins to Durdens, the seat of Lord Berkeley, near Epsom, where several experiments were made. Previous to his departure, he addressed a letter to Boyle, in which he communicated his ideas of the cause and nature of the plague. "I cannot," he says, "from any information I can learn of it, judge what its cause should be; but it seems to proceed only from infection or contagion, and that not catched but by some near approach to some infected person or stuff. Nor can I at all imagine it to be in the air, though yet there is one thing, which is very different to what is usual in other hot summers, and that is a very great scarcity of flies and insects. I know not whether it be universal, but it is here at London most manifest[17]."

It was not until February 1665—6, that there were a sufficient number of Fellows in London to enable the Meetings of the Council at Gresham College to be resumed. On the 3rd February, Hooke in a letter to Boyle says, "I hope we shall have again a Meeting, within this week or fortnight at farthest; and then I hope we shall prosecute experiments and observations much more vigorously, in order to which also I design, God willing, very speedily to make me an operatory, which I design to furnish with instruments and engines of all kinds, for making examination of the nature of bodies optical, chemical, mechanical," &c. He also observes in the same letter: "I am now making a collection of natural rarities, and hope within a short time to get as good as any that have been yet

---

[17] Boyle's *Works*, Vol. v. p. 543.

made in any part of the world, through the bounty of some of the noble-minded persons of the Royal Society[18]."

On the 21st Feb. 1665—6, the Council met again in Gresham College, Lord Brouncker the President being in the chair; at this Meeting it was resolved "that the donation of £100 presented by Mr. Colwall, should be expended in purchasing the collection of rarities formerly belonging to Mr. Hubbard;" and Oldenburg, writing to Boyle three days after the above Meeting, calls this "a very handsome collection of natural things;" and adds, "we are now undertaking several good things, as the collecting a repository[19], the setting up a chemical laboratory, a mechanical operatory, an astronomical observatory, and an optick chamber: but the paucity of the undertakers is such, that it must needs stick, unless more come in, and put their shoulders to the work. We know, Sir, you can and will do much to advance these attempts; and we hope the heavens are reconciled to us, to free us from the infection, and to return you to London."

The purchase of this collection was the first step of any magnitude towards the formation of a Museum, which eventually became the most extensive in London. The Fellows had already presented a few

---

[18] Boyle's *Works*, Vol. v. p. 545.

[19] Evelyn has in his *Diary*, Oct. 1666, "I made the Royal Society a present of the Table of Veins, Arteries, and Nerves, which great curiositie I had caused to be made in Italy, out of the natural human bodies, by a learned physitian, and the help of Vestlinguis (Professor at Padua), from which I brought them in 1646." Evelyn received the thanks of the Society for this present.

curiosities, but these were not sufficiently numerous to form a collection.

The only Museum, worthy the name at this time in London, was that formed by the Tradescants, a catalogue of which was published in 1656, entitled *Museum Tradescantium, or A Collection of Rarities preserved at South Lambeth, near London.* This museum contained not only stuffed animals and dried plants, but also minerals; implements of war and domestic use of various nations, also a collection of coins and medals. The Tradescants were very remarkable men[20]. John Tradescant, the elder, was a Dutchman; he travelled over Europe, Greece, Turkey, Egypt, and Barbary, and finally settled in England, as superintendent of the gardens of Charles I. During his travels, he procured specimens of whatever was rare and curious[21], amongst which we find enumerated in his Catalogue, "*Two feathers of the phœnix tayle.*" and "a *natural dragon.*" Tradescant's son, having imbibed his father's spirit, followed his example; and by their joint exertions a very large collection was brought together. Parkinson, in his *Paradisus terrestris*, mentions the father as "a painful industrious searcher and lover of all nature's varieties," and having "wonderfully laboured to obtain all the rarest fruits he can hear of in any place of Christendom, Turkey, yea, or the whole world."

In 1650, the younger Tradescant became acquainted

---

[20] Some exceedingly curious and interesting information respecting the Tradescants and Russia is contained in a work, by Dr. Hamel, entitled, *Tradescant der Aeltere in Russland.* 4to, St. Petersburg, 1847.

[21] The head of the dodo, (a bird now considered to be extinct,) in the Ashmolean Museum, was obtained by him.

with Elias Ashmole, who, with his wife, lived in the virtuoso's house during the summer of 1652. The result was so close a friendship, that Tradescant, at his death in 1662, left his museum to Ashmole, who, in 1682, bequeathed it to the University of Oxford, where it forms a portion of what is still called the Ashmolean Museum.

Dr. Hamel, in his *Tradescant in Russland*, states that Peter the Great visited the 'Tradescant Ashmolean Museum' in 1698[22].

The Tradescants possessed large gardens at South Lambeth, filled with various rare plants[23], the remains of which were still to be seen in 1749, when they were visited by Sir W. Watson, and described by him in the 46th Volume of the *Philosophical Transactions*.

Some coffee-houses in London possessed small museums, which served as attractions to the public. In 1664, an exhibition, opened in connexion with a house of entertainment, was thus advertised: "A Catalogue of Natural Rarities, collected with great industry, to be seen at the place called the Music House at the Mitre, near the west end of St. Paul's Church." Don Saltero's Coffee-house and Museum were celebrated things in their day. The Don had been a servant of Sir Hans Sloane, who gave him many curiosities for his Museum: this was situated in Cheyne Walk, Chelsea, where the site is still occupied by an Hotel. The catalogue ran:—"A Catalogue of Rarities. To be seen at Don Salter's Coffee-house in Chelsea;

[22] P. 175.

[23] The Tradescants introduced many new plants into Great Britain, some of which bear their name.

to which is added a complete list of the Donors thereof. Price Two Pence. O RARE!"

Immediately after the purchase of Mr. Hubbard's collection, the Society received so many accessions to their Museum, that application was made to the Trustees of Gresham College, praying "that they would be pleased to repair the floors and windows in the west gallery of the said College, where the Society's repository is to be."

Many presents were sent from abroad, at the request of Oldenburg, who was indefatigable in his correspondence with eminent foreigners. The first volume of the Letter-book contains a great number of letters written by him to persons on the Continent, who were in a situation to contribute to the Museum, or furnish useful information. The following extract from a letter to Mr. Henry Howard (afterwards Duke of Norfolk), then on the Continent, is a very fair specimen of the nature of these communications.

"It would be a great favour, if by your interest we might obtain some philosophical correspondents in the chief cities of Italy, and particularly at Florence, Pisa, Bologna, Milan, Venice, Naples and Rome. Riccio, Cassini, Fabri, Borelli and Campani, are no inconsiderable persons for such a commerce, and they may be assured that I shall make it a good part of my study to make some return.

"Methinks it were worth our knowledge whether there are not now some persons in Italy that know the old Roman way of plaistering, and the art of tempering tools to cut porphyry, the hardest of marbles.

"There is a curious artificial marble adorning the Elector of Bavaria's whole palace at Munchen, which we should be also glad to know the preparation of, and have a specimen.

"I am lately informed that there is a mineral salt plentifully to be found in the mines of Calabria, which has this particularity, that, being cast into the fire, cracks not, nor breaks in pieces. A specimen of that also would be acceptable."

The result of these letters was, that Oldenburg acquired a vast amount of information from foreign correspondents, which he duly imparted to the Society, and procured many valuable specimens for the Museum. Some of these are very curious. It is stated that "Sir R. Moray presented the stones taken out of Lord Belcarre's heart in a silver box;" and "a bottle full of stag's tears." Hooke gave "a petrified fish, the skin of an antelope which died in St. James's Park, a petrified fœtus," and other equally extraordinary things, which the Society not inappropriately called 'rarities.' The Trustees of Gresham College repaired the long gallery, and rendered it fit to contain the Museum, which increased rapidly.

On the 14th March, 1665—6, after an interruption of more than eight months, the weekly Meetings were resumed, and numerous investigations made respecting the late plague; the Masters of the Pest-house promising to send in their observations on it, at the request of Dr. Charlton[24]. The latter related at the

---

[24] Dr. Hodges, one of the city physicians during the plague, was of opinion, "that the true pestilential spots called the tokens, were gangrenated flesh, of a pyramidal figure, penetrating to the very bone, with its basis downwards, altogether mortified and insensible, though a pin or any other sharp body were thrust into it, and

Meeting on the 21st March, that the notion concerning "the vermination of the air as the cause of the plague, first started in England by Sir George Ent, and afterwards managed in Italy by Father Kircher, was so much farther advanced there, that by the relation of Dr. Bacon, who had long practised physic at Rome, it had been observed there that there was a kind of insect in the air, which being put upon a man's hand, would lay eggs hardly discernible without a microscope; which eggs being for an experiment given to be snuffed up by a dog, the dog fell into a distemper, accompanied with all the symptoms of the plague[25]."

The Annual Meeting of the Society was held on the 11th of April, the plague having prevented its being held on the preceding St. Andrew's day. The Treasurer's account showed that he had received £290. 7*s.* 4*d.*, and expended £256. 4*s.* 8*d.*, while the arrears of Fellows who had not paid their subscriptions amounted to £678. 5*s.*!

This awakens melancholy reflections, and is painful evidence of the crippled state of the Society. Royal in title, but almost paupers in fact, they had to struggle on, unassisted by the bounty of him who incorporated them. Deserving, assuredly, of great praise is that little band of truth-seekers, who despite all difficulties carried on the good work of promoting knowledge, and thereby increasing the happiness of mankind. It appears by the Minutes of Council, that

(what the Doctor thought especially remarkable), the next adjoining parts of the flesh though not discoloured, yet mortified as well as the discoloured ones." Register-book, Vol. II. p. 245.

[25] Register, Vol. II. p. 271.

great exertions were made at this period to recover the arrears, but with very indifferent success. Oldenburg writing to Boyle, says, "How to get them paid is the all-important question."

Meanwhile experiments were prosecuted, and several valuable Papers read. On the 20th June, the first notice of the remarkable operation of the transfusion of blood, is recorded in the Journal-book. Dr. Wallis related the success of the experiment made at Oxford by Dr. Lower, "of transfusing the bloud of one animal into the body of another, viz. that having opened the jugular artery of a mastiff, and injected by the means of quils the bloud thereof into the jugular vein of a greyhound, and opened also a vein in the same greyhound, to let out so much of his bloud as was requisite for the receiving that of the mastiff; the mastiff at last dyed, having lost almost all his bloud, and the greyhound having his vessels closed, survived, and ran away well[26]."

This account created considerable interest amongst the Fellows, and Boyle was requested to procure from Dr. Lower a detailed description of the manner in which the experiment was conducted, in order that it might be tried before the Society.

The desired account was subsequently laid before the Society, and on the 22nd August, Hooke was ordered to make the necessary preparations for the experiment.

The great Fire, which broke out on Sunday, September the 2nd, not only delayed the trial of this experiment, but interrupted the Meetings of the Society, as appears by the following entry in the Jour-

---

[26] Register, Vol. III. p. 4.

nal-book, under the date of Sept. 5: "The Society could not meet by reason of the late dreadfull Fire in London." And again, on the 12th Sept., "The Society being taken up with the considerations of the place for their future Meetings in this time of publick disorder and unsettlement, by reason of the late sad Fire, was thereby hindered from making experiments, and discoursing of philosophical matters, as they use to doe." The Fire did not extend to Gresham College, nor indeed beyond the end of Bishopsgate Street; but as the College was required for the use of the Lord Mayor and merchants, it became necessary to seek some other place of meeting[27]. In the meanwhile, it was resolved that the Society should assemble in Dr. Pope's lodgings in Gresham College. On the 19th September, the President reported that Mr. Howard had offered the Society rooms in Arundel House. The Council-minutes have no further notice of this report; than thanks to Mr. Howard for "his great respect and civility to the Society;" but we find from other sources, that his offer was subsequently accepted, although the Fellows did not meet in Arundel House until the following January. In November, "it was taken into consideration where the Society should meet for the future, Gresham College being too distant from the habitations of the Fellows generally, and the Earl of Northampton and the Bishop of Exeter were desired to speak to the Duke of Buckingham, that he would

[27] Oldenburg in a letter to Boyle, dated Sept. 18, 1666, writes, "all the hitherto printed *Transactions* which were carried into St. Faith's Church under Paul's were burnt." In a subsequent letter, dated Oct. 23, he complains of not being able to get any person to print the *Transactions*. No. 18 was printed for John Crook, in Duck Lane. Archives: Royal Society.

accommodate the Society with rooms in York House for their Meetings[28]."

It is worthy of remark, that on the 19th of September, Hooke exhibited to the Society, a model of his construction for rebuilding the city, which gave great satisfaction, not only to the Society, but also to the Lord Mayor and Aldermen, who, according to the minutes in the Journal-book, "expressed a desire that it might be shewn to his Majestie, they preferring it far before that which was drawn up by the surveyor of the city." Waller, in his *Life of Hooke*, remarks, that he had been given to understand, "that this model designed the chief streets, as from Leadenhall-corner to Newgate, and the like, to lie in a straight line, and all the other cross-streets turning out of them at right angles[29]."

At the Meeting on the 14th November the expe-

---

[28] Oldenburg says, writing to Boyle, "the Citty are striving hard to get the College totally into their hands for this time of distresse; which if they obtain, the Society are provided with another place to meet in, to wit, in Arundel House." *MS. Letters*, Archives: Royal Society.

[29] Wren had previously "drawn a modell for a new Citty, and presented it to the King, who produced it himselfe before his Councill, and manifested much approbation of it." Oldenburg was vexed that Wren had not submitted his model in the first instance to the Royal Society, who, on approving it, would have laid it before the King. Evelyn also prepared a plan, as he says in his *Diary*, under the date of Sept. 13. "I presented his Ma^ty^ with a survey of the ruines, and a plot for a new Citty, with a discourse on it." In a letter to Sir Samuel Tuke, dated Sept. 27, he writes, "I presented his Ma^ty^ with my own conceptions, which was the second within two days after the conflagration, but Dr. Wren got the start of me. We often coincided." *Diary*, Vol. I. p. 377. Part of Evelyn's plan consisted in lessening the declivities and filling up the shore of the river to low-water mark.

riment of transfusing the "bloud of one dog into another" was first made before the Society[30]. The operation was performed by Mr. King and Mr. Cox, upon a little mastiff and a spaniel, "with," says the Journal-book, "very good successe, the former bleeding to death, and the latter receiving the bloud of the other, and emitting so much of his own, as to make him capable of receiving that of the other." It was ordered at the next Meeting, that "the experiment of exchanging the bloud of animals be prosecuted and improved, by bleeding a sheep into a mastiff, and a young healthy dog into an old and sick one, and *vice versâ.*"

This experiment seems to have created great interest. Pepys writes in his *Diary*, under the date of Nov. 16, 1666, "This noon met Mr. Hooke, who tells me the dog which was filled with another dog's blood at the College the other day is very well, and like to be so ever, and doubts not its being found of great use to men, and so Dr. Whistler, who dined with us at the tavern."

The Journal-book contains the records of several experiments of the above nature made upon various animals, all of which appear to have been successful.

On the 2nd Jan. 1666-7, Mr. Henry Howard pre-

---

30 Wren had previously injected various liquors into living animals. In a letter to Sir W. Petty, (1656), he says, "The most considerable experiment I have made of late is this;—I injected wine and ale into the mass of blood in a living dog, by a vein, in good quantities, till he became extremely drunk, but soon after voided it by urine. It will be too long to tell you the effects of opium, scammony, and other things which I have tried in this way. I am in further pursuit of the experiment, which I take to be of great concernment, and what will give great light to the theory and practice of physic." *Parentalia*, p. 228.

sented the Society with "the Library of Arundel House, to dispose thereof as their property, desiring only that in case the Society should come to faile, it might return to Arundel House; and that this inscription, *Ex dono Henrici Howard Norfolciencis*, might be put upon every book given them. The Society," it is added, "received this noble donation with all thankfullnesse, and ordered that Mr. Howard should be registered as a benefactor[31]."

This gift may be regarded as the nucleus of the Society's valuable Library[32]. The history of the Arundel Library is interesting.—It formed originally a portion of the collection of Matthias Corvinus, King of Hungary, and after his death came into the possession of the celebrated Bilibaldus Pirckeimerus of Nuremberg, who died in 1530. It was purchased by Mr. Howard's grandfather, Thomas, Earl of Arundel, during

---

[31] Henry Howard was the second son of Henry Earl of Arundel, and became, on the death of his brother Thomas, sixth Duke of Norfolk. He was a great benefactor to the Royal Society, and presented the Arundel Marbles to the University of Oxford. He died in 1683—4.

[32] Evelyn says, it was at his instigation that Mr. Howard "granted the Society use of rooms in Arundel House, and also bestowed upon them that noble Library, which his grandfather especially, and his ancestors had collected." But this gentleman, he adds, "had so little inclination to bookes that this was the preservation of them from embezzlement." *Diary*, Vol. I. p. 380.

In another place he observes, "that many of the books had been presented by Popes, Cardinals, and great persons, including most of the Fathers printed at Basil, before the Jesuits abus'd them with their expurgatory Indexes. I should not," he concludes, "have persuaded the Duke to part with these, had I not seen how negligent he was of them, suffering the priests and everybody to carry away and dispose of what they pleas'd, so that abundance of rare things are irrecoverably gone." *Diary*, Vol. I. p. 471.

his embassy at Vienna; and it consisted of a great number of printed books and many rare and valuable manuscripts. Maitland, describing the Arundel portion of the Library of the Royal Society, informs us, that "this fine collection consists of 3287 printed books, in most languages and all faculties; and are chiefly the first editions of books, soon after the invention of printing. And the valuable and choice collection of Hebrew, Greek, Latin, Turkish, and other rare manuscripts, consists of 544 volumes[33]: which, together with the former, are thought to be of such value as cannot be paralleled for the smallness of their number."

In the *Collections of John Bagford concerning the History of Printing*, we have further evidence concerning this collection: we read, "Gresham College has a noble library, but it belongs not to the College, but to the Royal Society. These books, for the most part, were collected by the noble and learned antiquary, the Earl of Arundel, when he was ambassador at the court of Vienna, and some were presented to him by the Duke of Saxony." According to Pepys, the Society valued the books at 1000*l*.

At a Meeting of the Council, held on the 4th January, 1666—7, the printing of a circular announcing the change in the place of meeting was authorised: it runs thus:—

"These are to give notice, That the weekly Meetings of the Royal Society are appointed to be at Arundel House on Wednesday next, being the 9th

---

33 A portion of the Manuscripts in the Arundel Library was presented by the Duke to the College of Arms. A catalogue of these was made in 1829.

January, and thenceforward on the usual day and hour[34]."

Accordingly, on the 9th January, the Society met for the first time at Arundel House[35], when, to quote the Journal-book, "the President took notice of the great favour which Mr. Henry Howard had express'd to the Society, not only in accommodating them with convenient rooms for their Meetings, but also in presenting them with the Library of the said House[36]."

The Meetings were continued during this year at Arundel House. On the 30th May the Society were honoured by a visit from the Duchess of Newcastle, who had, according to the statement of Lord Berkeley, expressed a great desire to witness some experiments. The ceremonies, and subjects for her entertainment were referred to the Council, and arranged to be as follows: "Weighing the air; several experiments of mixing colours; dissolving flesh with a certain liquor; two cold liquors by mixture made hot; and making of a body swim in the middle of the water."

Pepys gives the following very interesting and amusing account of the visit of the Duchess.

"After dinner walked to Arundell House, the way

---

[34] Council Minutes, Vol. I. p. 120.

[35] Arundel House was situated to the West of Milford Lane, in the Strand: it was originally the Bishop of Bath's Palace. Its site is now marked by Arundel and Norfolk Street.

[36] Under the date of Jan. 9, 1667, Pepys writes; "To Arundell House, when first the Royal Society meet by the favour of Mr. Henry Howard, who was there. And here was a great Meeting of worthy noble persons: but my Lord Brouncker, who pretended to make a congratulatory speech upon their coming hither, and great thanks to Mr. Howard, did do it in the worst manner in the world." *Diary*, Vol. II. p. 4.

very dusty, where I find very much company in expectation of the Duchesse of Newcastle, who had desired to be invited to the Society; and was, after much debate *pro* and *con*, it seems many being against it; and we do believe the town will be full of ballads of it. Anon comes the Duchesse with her women attending her; among others the Ferabosco, of whom so much talk is that her lady would bid her show her face and kill the gallants. She is indeed black, and hath good black little eyes, but otherwise but a very ordinary woman, I do think, but they say sings well. The Duchesse hath been a good, comely woman, but her dress so antick, and her deportment so ordinary, that I do not like her at all, nor did I hear her say anything that was worth hearing, but that she was full of admiration, all admiration. Several fine experiments were shewn her of colours, loadstones, microscopes, and of liquors: among others of one that did, while she was there, turn a piece of roasted mutton into pure blood, which was very rare[37]. After they had shewn her many experiments, and she cried until she was full of admiration, she departed, being led out and in by several lords that were there[38]."

Evelyn also tells us, that he "went to London to wait on the Dutchess of Newcastle (who was a mighty pretender to learning, poetrie and philosophie, and had in both published divers books,) to the Royal Society, whither she came in great pomp, and being receiv'd by our Lord President at the doore of our Meeting-room, the mace, &c. carried before him, had several experiments shewed to her[39]."

---

[37] Query: Experiment, or Meat?

[38] *Diary*, Vol. II. p. 60.

[39] *Diary*, Vol. I. p. 383. This visit of the Duchess will call to remem-

It appears by the Journal-book, that the experiment of the transfusion of blood was tried frequently during this year upon various animals, as sheep, dogs, foxes, &c.[40] In the month of July, an account was received from Paris of two experiments made at the Academy of Sciences, on a youth and an adult, whose veins were opened and injected with the blood of lambs. The experiment, according to the account, succeeded so well, that the Society became anxious to perform it upon an individual, and Sir George Ent suggested that it would be most advisable "to try it upon some mad person in Bedlam." This proposal met with general approbation, and a Committee was forthwith appointed to communicate with Dr. Allen, the physician to the Hospital, and request him to furnish a lunatic for the experiment.

It will not excite surprise that Dr. Allen declined acceding to this request: the Committee reported, "That Dr. Allen scrupled to try the experiment of transfusion upon any of the mad people in Bedlam." They were then ordered "to consider together how this experiment might be most conveniently and most safely tryed."

In the interval Dr. King read a Paper to the Society, "On the Method of Transfusing Blood into Man," which was printed in the 28th Number of the *Transactions*.

A very remarkable event which occurred this year, seems to have had so much influence upon the Society

---

remembrance that of Queen Christina to the French Academy, mentioned by Pellisson, in his *History* of that learned body.

[40] See a curious Paper "On Injections into Veins, and the Transfusion of Bloud," by Dr. Clarke, in the 35th Number of the *Transactions*.

as to cause a suspension of the Meetings from the 30th May to the 3rd October. This was the imprisonment of Oldenburg in the Tower; a fact entirely unnoticed by Birch, or Thomson, and of which there is no record whatever in the Council or Journal-books of the Society.

He was arrested on suspicion of carrying on political correspondence with parties abroad, obnoxious to Charles II. and the Government.

By the kindness of Sir George Grey, Secretary of State, who permitted me to search the archives in the State Paper Office, I am enabled to produce a copy of the warrants for incarcerating the Society's Secretary.

"C. R.

"Warrant to seize the person of Henry Oldenburg for dangerous designes and practices, and to convey him to the Tower.

"*June* 20, 1667. "By order,

"ARLINGTON."

"Warrant to the Lieutenant of the Tower to take him into his custody and keep him close prisoner.

"*June* 20, 1667[41]. "By order,

"ARLINGTON."

It is probable that Oldenburg was arrested immediately on the issue of the foregoing warrants. He was certainly in the Tower seven days after, for Pepys writes, under the date of June 28, 1667, "I was told yesterday that Mr. Oldenburg, our Secretary at Gresham College, was put into the Tower, for writing news to a virtuoso in France, with whom

---

41 Vol. x. *Domes. Papers*, 1667.

he constantly corresponds on philosophical matters, which makes it very unsafe at this time to write, or almost do any thing[42]."

The suspicions of Government were doubtless excited by the voluminous correspondence carried on by Oldenburg with foreigners[43]; but it is not a little singular that the exercise of one of the privileges conceded in the Society's charter, viz. that of holding a correspondence on scientific subjects with all descriptions of foreigners (*cum omnibus, et omnimodis peregrinis et alienis*) should have led to the arrest and incarceration of the Secretary in the King's State Prison[44].

It is clear that the authorities were soon satisfied of Oldenburg's innocence, for, on the 26th August, 1667, a warrant was issued for his discharge. This document is also in the State Paper Office. It runs thus:

---

[42] *Diary*, Vol. III. p. 273.

[43] In a letter from Mr. Fairfax to Oldenburg, dated Sept. 28, 1667, he congratulates the latter upon his freedom and escape "from those mishaps arising from the frequency of exchanges by the pen." Archives: Royal Society.

[44] During the period that Oldenburg was engaged in the publication of the *Phil. Trans.*, he carried on a correspondence with above seventy individuals in various parts of the world. "I asked him," says Dr. Lister, in his *Journey to Paris*, "what method he used to answer so great a variety of subjects; and such a quantity of letters as he must receive weekly; for I knew he never failed, because I had the honour of his correspondence for ten or twelve years; he told me he made one letter answer another; and that to be always fresh, he never read a letter before he had pen, ink, and paper, ready to answer it forthwith; so that the multitude of his letters cloyed him not, or even lay upon his hands." p. 78.

"C. R.

"Discharge for ——— Oldenburg, prisoner in the Tower.

"*August* 26, 1667. "By order,

"ARLINGTON."

A few days after his discharge, Oldenburg went into the country, as appears by the following very interesting letter to his friend Boyle, which is preserved in the archives of the Society.

"*London, Sept.* 3, 1667.

"SIR,

"I was so stifled by the prison-air, that, as soon as I had my enlargement from the Tower, I widen'd it, and took it from London into the country, to fann myself for some days in the good air of Craford in Kent. Being now returned, and having recovered my stomack, which I had in a manner quite lost, I intend, if God will, to fall to my old trade, if I have any support to follow it. My late misfortune, I feare, will much prejudice me; many persons unacquainted with me, and hearing me to be a stranger, being apt to derive a suspicion upon me. Not a few came to the Tower, meerly to enquire after my crime and to see the warrant, in which when they found that it was for dangerous dessines and practices, they spread it over London, and made others have no good opinion of me. *In carcere audacter semper aliquid hæret.* Before I went into the country, I waited upon my Lord Arlington, kissing the rod. I hope I shall live fully to satisfy his Majesty and all honest Englishmen of my integrity, and of my reall zeal to spend the remainder of my life in doing faithfull service to the Nation to the very utmost of my abilities. I have learned during this comitment to know my reall friends.

God Almighty blesse them, and enable me to convince you all of my gratitude!

"Sir, I acknowledge and beg pardon for the importunities I gave you at the beginning, assuring you that you cannot lay any commands on me I shall not chearfully obey to the best of my power."

The rest of the letter relates to scientific subjects, a proof that, even during adversity, science had a share in the writer's thoughts.

Among those who went to see the Secretary in durance in the Tower was Evelyn, who records in his *Diary*, under the date of the 8th August, 1667, "Visited Mr. Oldenburg, now close prisoner in the Tower, being suspected of writing intelligence. I had an order from Lord Arlington, Secretary of State, which caused me to be admitted. The gentleman was Secretary to our Society, and I am confident will prove innocent."

Oldenburg's name is entered as having attended a Meeting of the Council on the 30th September, and from this period until his decease, he appears to have devoted all his time and abilities to the Society. His extraordinary labours will be better appreciated, when it is remembered that up to this date he only received a gratuity of 40*l*., which was voted to him "for the great pains he hath taken on behalf of the Society;" and it was not until June, 1669, that a salary of 40*l*. per annum was granted to him.

On the 27th September, 1667, the Society took possession of Chelsea College[45], which was in a most

---

[45] The College was delivered up to the Society by Evelyn, who writes in his *Diary*, Sept. 24, 1667, "I had orders to deliver the possession of Chelsea College (used as my prison during the

dilapidated condition; but, as the grant of the building and land had not passed the great seal, it was resolved, "that all repairs should be deferred until the Society obtained legal possession of the premises." This appears to have been regarded as extremely doubtful; and although Dr. Sprat speaks, in his history, of the College as intended to serve for the Society's meetings, laboratory, repository, and library, yet the Council resolved, that subscriptions should be set on foot for the purpose of building a College, "as the most probable way of the Society's establishment." With this view the subjoined form of subscription was drawn up:

"We whose names are underwritten being satisfied of the great usefullness of the institution of the Royal Society, and how requisite it is for attaining the ends designed thereby, to build a College for their Meetings, and to establish some revenue for discharging the expenses necessary for tryal of experiments, do heartily recommend it to the bounty of all generous and well-disposed persons for their assistance to a worke of such public usefullnesse. And we do each of us for ourselves, hereby promise to contribute to those good ends, the respective sums subscribed by each of us at four distinct quarterly payments, to be made to such persons as shall be authorized under the seal of the Royal Society for the receipt thereof; the first payment to begin at———."

A Committee was appointed, consisting of the President, the Bishop of Salisbury, Henry Howard, Mr. Boyle, Sir R. Moray, Sir John Lowther, Dr. Wilkins, Mr. Evelyn, Mr. Henshaw, Mr. Hoskyns, and Mr.

---

war with Holland, for such as were sent from the Fleete to London) to our Society, as a gift of His Majesty, our Founder."

Oldenburg; who were requested to solicit contributions for building the proposed College. The Committee, headed by the President, made great exertions to obtain donations; it is recorded even that they went expressly to Westminster Hall, for the purpose of asking those Members of Parliament who were Fellows of the Society, for their contributions.

Boyle was in Oxford at this period, where he principally resided until 1668, when he removed to London. All matters of importance relating to the Society were communicated to him by Oldenburg, who, in a letter written on this occasion, and printed in Boyle's *Works*, says,

"I cannot conclude this, without acquainting you of the design of the Society's Council, whereof you are a member, of building a college, as one of the most probable means of their establishment. They did on Saturday last consider of it in a full Council, when H. Howard of Norfolk was also present (who intends to continue his nobleness, either by contributing to this work a good sum of money, or by giving us ground about Arundel House to build upon), and proceeded so far as to view a draught and to select out of the Society's list such persons as they thought both willing and able for such contributions; and to carry it on the more effectually, they appointed a Committee of *Beggars* (as they were pleased to term it merrily) or solicitors, to speak to such as were called out for that purpose, and to recommend this work to their liberality. Our present Bishop of Salisbury, Mr. Howard, Dr. Wilkins and myself, are the principal beggars; we want Sir R. Moray exceedingly in this employment. I was particularly desired to acquaint you with this intention, and also Sir R. Moray, to engage him to beg in Scotland among the noblemen there, that are of the Society, on our behalf. I believe, Sir, you will not be displeased, if I am open and particular to you in this business, which I think myself obliged to be, as tending to your better direction and information. They guess there will be four classes of contributions; some of 100*l.*, some of 60*l.* or 50*l.*, some of 40*l.*, some of 20*l.* Our President hath already declared for 100*l.*, and I think the Bishop of Salisbury for the like sum. Dr. Wilkins for 50*l.*, Mr. Hayes for 40*l.*, Sir P. Carteret for 50*l.*, and, if there be occasion, for more.

"We begin with the Council, and proceed to the Society; that, when we go on to beg of others not of our body, they may not object, we would load others and draw our own necks out of the yoke. The Council, when they ran over the list of the Society, thought you, Sir, the fittest person to bespeak my Lord Clifford's generosity in this matter; thinking it superfluous to invite yourself otherwise than by a bare intimation. I am confident you take in good part both the Council's and my freedom in this discourse, which stops my pen, my saying more by way of excuse of it[46]."

Amongst the numerous manuscript letters of Oldenburg in the archives of the Royal Society, is one to Boyle, written at a later period, by which it appears that the latter was not very favourably disposed towards the proposed College, although he subsequently, as will be seen, contributed towards it. Oldenburg says,

"I find, by your last return, I am guilty of an unseasonable importunity expressed in my former, which yet, since I am not wont to be so in my own particular

[46] Vol. v. p. 381.

concerns, I am apt to believe I shall the more easily obtain your pardon for, the more it proceeded from a zeal to further a publick and useful work, as the building a College for the Royal Society is conceived to be by the Council, as being that which will in all likelihood establish our Institution and fix us (who are now looked upon but as wanderers and using precariously the lodgings of other men), in a certain place where we may meet, prepare and make our experiments and observations, lodge our curators and operators, have our laboratory, observatory, and operatory, alltogether. 'Tis a maxim I learned in my logick when a boy, *Qui amat finem, amat et media ad finem.* And I confess I am as averse from, and as much a bungler in, begging as any man; but I can deny myself, and go against the stream of my inclinations, when the prosecution of honour and publick usefulness is in the case, as here it is, and therefore am not ashamed to beg, when it is a good means to accommodate and promote the good ends of the Society's Institution. I think, since I have endeavoured to serve them to the utmost of my power these six years gratis, and am a beggar to boot as to my private fortune, I may extend my endeavours so much further as to go a begging for the Society's establishment, especially in so good company as our President, Mr. Henry Howard, Sir Robert Moray, and some others, have a mind to be."

Another letter from Oldenburg to Boyle, dated London, Jan. 18, 1667-8, gives us further information respecting the proposed College.

"I am almost persuaded that upon a mature consideration of the importance of building a College to establish the Society, you will not at all hesitate in giving your approbation and concurrence thereto. We had on Saturday last, at a Meeting of the Council, a draught of the structure before us, which was scanned sufficiently; Mr. Howard being there also, and giving no small proof of his understanding and skill in that art, as well as of his frankness and generosity in giving the foundation and forwarding the execution of the work. I think we shall at our next Council, which is summoned to meet on Thursday next in the evening, begin to make subscriptions in such a form as may be obligatory in law, and yet not severe in the expression. My Lord Anglesey, having the thing occasionally mentioned to him by me, shewed so much inclination to it, that he is like to be received into the Society, and to be both a contributor and a solicitor for this noblework. Sir, my opinion is, since you are pleased to demand it, that in case you think not yet fit to come to London, you may declare in a note to me the sum you intend to cast in, directing me to signify it to the Council, and giving order where the Society's treasurer shall receive it. Concerning the provision made by the Council to secure men's philosophical properties, they have ordered that, if any thing of that nature be brought in, and desired to be lodged with the Society—in case the author be not of their body—they should be obliged to shew it first to the President, (for fear of lodging unknownly ballads and buffooneries in these scoffing times), and that then it should be sealed up, both by the smaller seal of the Society, and the seal of the proposer: but if the authors were of the Society, that then they should not be obliged to shew it first to the President, but only to declare the general head of the matter to be laid up, and that then it should be sealed up as mentioned before."

At the request of the Council, Lord Brouncker

addressed the following letter to Sir Robert Moray, one of the Commissioners of His Majesty's Treasury in Scotland.

"The Council of the Royal Society having lately taken into serious consideration what might be the most probable means to establish the Society, and its design of improving useful knowledge to perpetuity, and having found, upon mature deliberation, that one of the ways most likely to effect the same may be the erecting of a College, fit to meet and to make their observations and experiments in; they have, accordingly, resolved to endeavour to engage as many of the members of the Society, and of others also, not of their body, as are able and willing to promote so noble and useful a work. In pursuance of which they have already begun to solicit divers of the Society, and found no ill success in this undertaking; in which they are more especially encouraged by the signal nobleness and bounty of the Hon. Henry Howard, of Norfolk, most generously bestowing on the Society a piece of ground in Arundel House, sufficient to build the College on; the raising of which they intend, God willing, to begin with this approaching Spring, and, if the design be seconded by cheerful contributions, to finish by Michaelmas next.

"And being persuaded, that those of the nobility in Scotland, whose names are here enrolled in the list of the Society, are, with many others, satisfied of the usefullness of this Institution, and of the necessity of making such an establishment as this; they thought fit to give you, of whose zeal for its prosperity they are well assured, notice of this their intention, that so you may be invited, as you have opportunity, to insinuate this undertaking to those of your noble countrymen as are of the Society, and to bespeak the concurrence of their generosity in contributing with all convenient speed what they may, to further so good an establishment: which being effectually done, as it cannot but redound to the immortal fame of the contributors, so it will certainly add to the reputation you have already so much gained and deserved of this Society."

There is a book in the archives of the Society entitled, *Contributions towards Building the College*, where the names of the Fellows of the Society are entered, with the various sums subscribed by each. These amount to 1075*l.*, contributed by 24 Fellows out of 207, a small proportion, showing that the proposition of the Council was not responded to very warmly. Indeed, we have evidence of this from Pepys, who says in his *Diary*, under the date of the 8th of April, 1668, "With Lord Brouncker to the Royall Society, when they had just done; but I was forced to subscribe to the building of a College, and did give 40*l.*; and several others did subscribe, some greater and some lesse sums; but several I saw hang off; and I doubt it will spoil the Society, for it breeds faction and ill-will, and becomes burdensome to some that cannot, or would not do it."

Lord Brouncker contributed 100*l.*, Boyle 50*l.*, and Evelyn, in addition to a similar sum, is recorded to have given 50,000 bricks. Mr. Henry Howard granted the Society a piece of ground in Arundel Gardens, 100 ft. long, by 40 ft. deep, for the proposed College, and he, as well as Hooke and Wren, drew designs for the building.

The latter communicated his plan in writing. The document is preserved in the archives of the Society. It is dated Oxford, June 7, 1668, and addressed to Oldenburg:—

"When I waited upon his Honour, Henry Howard of Norfolk, he took delight to shew me some designs he had thought of himself for your building, and commanded me to trace out to him what I had considered, the same in effect I shewed you at London. But this, at first appearance, seemed to him too chargeable a design, but afterwards he acquiesced in the reasons I gave him; and having taken the sketch with him, and delivered your letter with his own hand, he enjoyned me to give you an account of it.

"It contains in the foundations, first, a cellar and a fair laboratory; then a little shop or two, for forges and hammer-works, with a kitchen and little larder. In the first story it contains a vestibule, or passage-hall, leading through from both streets; a fair room for a library and repository, which may well be one room, placing the books after the modern way in glass presses; or, if you will divide the room with pillars, it will the better support the floor of the great room above it, and so place the presses for rarities in the other. Upon the same floor is a parlour for the housekeeper, and from the vestibule the great stairs lead you up to the ante-chamber of the great room, and not higher.

"The great room for the meeting is 40 feet long, and two stories high, divided from the ante-chamber by a skreen between columns, so that the whole length, in case of an entertainment, may be 55 feet. Upon the same floor is the Council-room, and a little closet for the Secretary.

"In the third story are two chambers with closets, for the Curators, and back stairs by them, which lead from the bottom to the top; one of the chambers being over the ante-room, looking down into the great room, very useful in case of solemnities.

"The fourth story is the timbers of the roof; which being 30 feet wide, and to be leaded, cannot be firm without bracing it by partitions to the floor below. These partitions are so ordered, as to leave you a little passage-gallery the whole length of the building, for trial of all glasses and other experiments that require length. On one side of the gallery are little shops all along for operators; on the other side are little chambers for operators and servants. The platform of lead is for traversing the tubes and instruments, and many experiments. In the middle rises a cupola for observations, and may be fitted, likewise, for an anatomy theatre; and the floors may be so ordered, that from the top into the cellar may be made all experiments for light.

"As for the charge of this fabric, I confess it is my opinion, that a fair building may easier be carried on by contribution, with time, than a sordid one. And, if I might advise, I could wish the foundations were laid of the whole, but then you need not build more than half at present; and this may be done for 2000*l.*, and will contain the necessary rooms, and so you will leave yourselves an opportunity of inlarging hereafter upon the same model. If you think to have a model made, I will willingly take care to have it done. I have so folded the papers, as to shew you what part I would have at present built; together with an extempore staircase of deal boards and laths. The cupola may be left till the finishing.

"Sir,<br>"I am your humble Servant,<br>"CHR. WREN."

Although it had been originally resolved that when the subscription amounted to 1000*l.*, the College should be commenced, yet some obstacles arose relating to the conveyance of the land, for on the 10th of August, 1668, it is recorded in the Minutes of Council, that "the building of the College should be deferred till Spring; and, in the mean time, good materials provided."

It may be as well to state here, that the College was never built, in consequence of legal difficulties, and want of funds. These might, however, have been surmounted; but the immediate necessity for so doing was removed by the grant of Chelsea College, which was conveyed to the Society by Royal Patent, dated April 8, 1669.

This document contains certain additional privileges, the most important of which is, that the President, Council, and Fellows, may hold their Assemblies anywhere within the kingdom of England.

A copy of the Patent will be found in the Appendix.

---

## CHAPTER IX.

Committee concerning Chelsea College—Proposition from Evelyn and a Nobleman respecting it—Resolution to let it—Cosmo III. Grand Duke of Tuscany visits the Society—Experiment of Transfusing the Blood of a Sheep into a Man—Popular Belief respecting Transfusion—Queen orders Thermometer to be made for her by the Society—Natural History Collections made by order of the Society—Letters of Recommendation given by the Council—Flamsteed's first Communication to the Society—Communications from Malpighi—Letter from Newton—Enemies of the Society—Glanvill's *Plus Ultra*—Poverty of the Society—Boyle lends Philosophical Apparatus—Wager of Charles II.—Newton proposed as Candidate by the Bishop of Salisbury—His Election—Sends his Reflecting Telescope to the Society—Supposed by some parties abroad to be a Maker of Telescopes—His Discoveries respecting Light—His gratitude to the Society—Controversies—Bishop Wilkins leaves a Legacy to the Society—The Society invited to return to Gresham College—Leuwenhoeck's Microscopical Communications—Presents his Microscopes to the Society—Society give his daughter a silver bowl—Pecuniary Difficulties—Means taken to collect Arrears—Obligation to furnish Scientific Communications and Experiments—Newton exempted from paying his Subscription—Erection of Greenwich Observatory—Flamsteed appointed Astronomer Royal—Society lend their Instruments to the Observatory—Peter the Great visits the Observatory—Astronomical Science neglected by Government—Valuable Communications from Travellers—Pains taken by the Society to procure information—Curious Account of Asbestos—Death of Oldenburg—Biographical Notice of him—Lord Brouncker resigns the Presidency.

---

1665—70.

THE grant of Chelsea College, and the lands appertaining to it, amounting to *about* thirty acres, was even in those days a munificent endowment; but, unfortunately for the Society, it presented a fairer appearance engrossed on a broad sheet of parchment, than warranted by the result; for various parties immediately claimed portions of the estate, which the Society, taking the Patent for their guide, conceived to be lawfully their property.

Within a month after the grant, a Committee consisting of Lord Brouncker, Mr. Charles Howard, Mr. Aerskine, Sir Robert Moray, Mr. Evelyn, Mr. Henshaw, and Mr. Hoskyns, was appointed "to consider of his Majesty's grant of Chelsea College, and *what may belong to it*[1]; and also to confer with Mr. Cheney about those acres which he held belonging to the College, and to commute parcels of land with the same, in case he should surrender his interest upon equitable terms to the Royal Society: and that the said Committee do meet at Lord Brereton's lodgings in Channel Row, beginning to do so on the Saturday following, at five in the evening; and that they make a report to the Council."

The Committee met frequently, but were unable for several months to come to any decision, in the absence of documents essential to the inquiry. Meanwhile, Evelyn proposed "that the College should be let as a prison-house during the war, he hoping to get 100*l.* per annum for it, besides some necessary repairs of the house."

The Council-minutes add, "It was ordered hereupon that the President, Treasurer, and Secretary that officiateth, should have power to agree, in the name of the Council, with Mr. Evelyn about the mat-

---

[1] Sir Joseph Sheldon assured Lord Brouncker that fifty acres of land adjoining the College belonged to the Society.—*Council Minutes.*

ter proposed, and conclude with him, if the abovementioned hundred pounds per annum, together with good repairs, could be obtained."

About the same time Sir Robert Moray communicated a proposition from a noble Member of the Society, who was willing to undertake the management of the College and lands, "and would plant the ground with all sorts of choice vegetables, exotick, and domestick, and in repairing the house, all upon his own charges, the Society always remaining proprietors and masters thereof, with a full power of ordering and directing what particulars they would have observed and done in the managing of this affair, the proposer only expecting to be perpetual steward of that place." This proposition, it is added, "was received with acclamations; only Sir Robert Moray was desired that he would employ his interest with the proposer, to have it put in writing for the prevention of mistakes."

Neither of these proposals was carried into effect; and the Committee were therefore earnestly requested to report on the best measures to be taken for rendering the College available to the Society. They eventually reported in favour of letting the building, and it was accordingly resolved, "that the influential Fellows of the Society be requested to use their interest to find a tenant." But the property was in so unsettled a state, that although great exertions were made to procure a tenant, and money expended in repairing the College, they met with no success, and the estate remained on the hands of the Society until finally disposed of to the Crown.

This transaction will be duly noticed in the proper place.

In 1669, Cosmo III., Grand Duke of Tuscany, visited England, accompanied by Count Lorenzo Magalatti, afterwards Secretary to the Academy del Cimento, and one of the most learned and eminent characters of the court of Ferdinand II. He wrote the Grand Duke's *Travels*, an English translation of which was published in 1821. The Duke visited the Royal Society on the 25th April. The following is the account of the visit, given by the Duke's chronicler:—

"He afterwards went, in his carriage, with his usual retinue, to Arundel House, where the Royal Society meets every Thursday, after dinner, to take cognizance of matters of natural philosophy, and for the study and examination of chemical, mechanical, and mathematical subjects . . . . . At their Meetings no precedence or distinction of place is observed, except by the President and Secretary; the first is in the middle of the table, and the latter at the head of it on his left hand, the other Academicians taking their seats indifferently on benches of wood with backs to them, arranged in two rows; and if any one enters unexpectedly after the Meeting has begun, every one remains seated, nor is his salutation returned, except by the President alone, who acknowledges it by an inclination of the head, that he may not interrupt the person who is speaking on the subject or experiment proposed by the Secretary. They observe the ceremony of speaking to the President uncovered, waiting from him for permission to be covered, and explaining their sentiments in few words, relative to the subject under discussion; and, to avoid confusion and disorder, one does not begin before the other has ended his speech; neither are opposite opinions maintained with obstinacy, but with temper; the language of civility and moderation being always adopted amongst them: which renders them so much the more praiseworthy, as they are a society composed of persons of different nations. It has for its coat of arms a field of silver, denoting a blank tablet, enlivened with the motto *Nullius in verba,* to shew that they do not suffer themselves to be induced by passion and prejudice to follow any particular opinions.

"The cabinet, which is under the care of Dr. Hooke, a man of genius, and of much esteem in experimental matters, was founded by Daniel Colwal, now Treasurer of the Academy, and is full of the greatest rarities, brought from the most distant parts; such as quadrupeds, birds, fishes, serpents, insects, shells, feathers, seeds, minerals, and many petrifactions, mummies, and gums; and every day, in order to enrich it still more, the Academicians contribute every thing of value which comes into their hands; so that in time it will be the most beautiful, the largest, and the most curious, in respect to natural productions, that is any where to be found. Amongst these curiosities, the most remarkable are: an ostrich, whose young were always born alive[2]; an herb which grew in the stomach of a thrush; and the skin of a moor, tanned, with the beard and hair white: but more worthy of observation than all the rest, is a clock, whose movements are derived from the vicinity of a loadstone, and it is so adjusted as to discover the distance of countries at sea by the longitude[3]."

---

[2] This truly was a *curiosity.* It is to be feared the Grand Duke's historian was not a profound zoologist.

[3] The analogy between the above clock and the electrical clock of the present day, is not a little remarkable. During 1669, the Journal-

The Grand Duke's physician, who was frequently present at the ordinary meetings, laid before the Society a number of philosophical and medical queries, which had been entrusted to him by some philosophers at Florence, for the purpose of obtaining the opinion and answers of the Society thereon.

Among the most remarkable experiments prosecuted at this period, was that of transfusing the blood of a sheep into a man, which was successfully performed for the first time in England, in the month of November, 1667. The subject of the experiment was a poor student, named Arthur Coga, who, hearing that the Society were very desirous to try the experiment of transfusion upon a man, and being in want of money, offered himself for a guinea, which was immediately accepted on the part of the Society[4]. The operation was performed by Drs. Lower and King at Arundel House, on the 23rd Nov., 1667, in the presence of several spectators, among whom were Mr. Henry Howard, the Bishop of Salisbury, and some members of Parliament. Oldenburg, in a letter to Boyle, giving an account of the experiment,

---

Journal-book contains many allusions to "Hooke's magnetic watch going slower or faster, according to the greater or less distance of the loadstone, and so moving regularly in any posture." On the occasion of the visit of illustrious strangers, this clock and Hooke's magnetic watches were always exhibited as great curiosities.

[4] Oldenburg, in a letter to Boyle, dated London, Nov. 25, 1667, observes, that Coga was looked upon as "a very freakish and extravagant man; that he had studied at Cambridge, and was said to be a bachelor of divinity, and an indigent person." And Dr. King in a letter to Boyle, dated the same day, remarks "that Coga was thirty-two years of age, that he spoke Latin well, but that his brain was *sometimes a little too warm.*"

observes, "Dr. King performed the chief part of it with great dexterity, and so much ease to the patient, that he made not the least complaint, nor so much as any grimace during the whole time of the operation; that he found himself very well upon it, his pulse and appetite being better than before, his sleep good, his body as soluble as usual, it being observed, that the same day he had three or four stools, as he used to have before."

Dr. King stated, that after the operation "the patient was well and merry, and drank a glass or two of canary, and took a pipe of tobacco, in the presence of forty or more persons; he then went home, and continued well all day, his pulse being stronger and fuller than before, and he very sober and quiet, more than before, as the people of the house said, who thought that he had only been let blood. In the night he slept well, but sweat two or three hours, and next day was very well, and so remained, and was very willing to have the experiment repeated, his arm being, he said, well. A person asking him why he had not the blood of some other creature instead of that of a sheep transfused into him, he answered, *Sanguis ovis symbolicam quandam facultatem habet cum sanguine Christi, quia Christus est Agnus Dei*[5]."

---

[5] Pepys says in his *Diary*, "Nov. 21, 1667. With Creed to a Tavern, where Dean Wilkins and others, and good discourse; among the rest of a man that is a little frantic (that hath been a kind of minister; Dr. Wilkins saying, that he hath read for him in his church), that is poor, and a debauched man, that the College have hired for 20*s*. to have some of the blood of a sheep let into his body, and it is to be done on Saturday next. They purpose to let in about 12 ounces." Pepys subsequently writes: "Nov. 30. I

The experiment was repeated on the 12th December following, when eight ounces of blood were taken from Coga, and about fourteen ounces of sheep's blood injected; the operation was performed at a public Meeting of the Society. An account of it is recorded in the third volume of the Register-book, by which it appears to have succeeded[6].

The most sanguine anticipations appear to have been indulged by uninquiring minds, and the new process was almost expected to realize the alchemical reveries of an elixir of life and immortality. Dr. Terne, physician to one of the London hospitals, expressed his willingness to try the experiment of transfusion upon morbid persons, but there is no record of any of his patients undergoing the operation. About this time, however, some papers were received from Dantzic, giving an account of trials made of injecting liquors into human veins; in two cases the individuals received great benefit, but a third person died[7]. In 1668, a lunatic in Paris, on

---

was pleased to see the person who had his blood taken out. He speaks well, and did this day give the Society a relation thereof in Latin, saying, that he finds himself much better since, and a new man. He is the first sound man that ever had it tried on him in England, and but one that we hear of in France." Vol. II. p. 160.

[6] Sir P. Skippon, in a letter to Ray, alluding to this experiment, says, "The effects of the transfusion are not seen, the coffee-houses having endeavoured to debauch the fellow, and so consequently discredit the Royal Society, and make the experiment ridiculous." *Phil. Let.*, p. 28.

[7] In a letter to Boyle, Oldenburg relates that a physician who was at the meeting when the above papers were read, "was so precipitate as to say, that he would engage that that one, viz. with the ill success, was the only true; but the other two both false

whom the experiment of transfusion had been tried without success, was again operated upon, but with fatal results, as explained in the annexed letter from M. Justel to Oldenburg.

"*Paris*, 3 *Févr.* 1668.

"*Il faut que je vous dise qu'on sçait que M. Denis et le Sieur Emerez voulant remedier à la folie du phrénétique, sur lequel ils avaient fait la transfusion, l'ont répétée, et lui ont ouvert la jugulaire et saigné au pied; mais il est mort entre leur bras. Sans le credit de M. de Montmor ils auroient été en peine, en ayant usé un peu hardiment*[8]. *Cette aventure decriera la transfusion, et on n'osera plus la faire sur les hommes*[9]."

These failures turned the current of public opinion, and led to the immediate abolition of the process, which was not practised again by the Society.

Scientific labours in the branches of physics and natural history, were prosecuted at this period with great diligence; the indefatigable Hooke is recorded in the Journal-books as having produced new experiments and inventions at almost every Meeting. The

---

"I could not," adds Oldenburg, "but take him afterwards aside, and represent to him how he would resent it, if he should communicate upon his own knowledge an unusual experiment to the curious at Dantzic, and they in public brand it with the mark of falsehood; that such expressions, in so public a place and in so mixed an assembly, would certainly prove very destructive to all philosophical commerce, if the curious abroad should be once informed how their symbolas were received at the Royal Society." Archives: Royal Society.

[8] They were tried for manslaughter, but acquitted.

[9] Two men died at Rome after undergoing the operation of transfusion. The Pope immediately issued an edict forbidding the practice. Merclin, *de Transfus. Sang*, p. 25.

mechanical science of the Society was even recognized at Court, for we find the Queen, Catharine of Braganza, requesting that a thermometer might be made for her, which was done by Hooke.

In Oct. 1669, Thomas Willisel, who had been engaged by the Society to collect zoological and botanical specimens in England and Scotland, returned to London with a large collection of "rare Scottish birds and fishes," and dried plants, which were laid before the Society, and preserved in their repository[10]. Evelyn having been at the Meeting of the Society when the specimens were shewn, says in his *Diary*: "Our English itinerant presented an account of his autumnal peregrination about England, for which we hired him, bringing dried fowls, fish, plants, animals, &c.[11]"

It was not an uncommon circumstance for the Council to grant to intelligent persons, whether Fellows of the Society or not, what are styled "Letters recommendatory." These were written in Latin, and bore the great Seal of the Society.

Their object was to request that all persons in authority abroad would kindly receive the bearer,

---

[10] Mr. Willisel's commission was as follows: "These are to certify all whom it may concerne, that the bearer hereof, Thomas Willisel, is employed by the President, Council, and Fellows of the Royal Society of London for improving natural knowledge, to go into several parts of His Majesty's dominions for purposes suitable to their Institution, according to authority unto them on this behalf given by his sacred Majesty that now is. And they earnestly recommend him to all generous and ingenuous spirits, desiring that as occasion shall require, they will assist him in promoting a work so generally beneficial to all mankind." To this the Seal of the Society was appended.

[11] Vol. I. p. 402.

who was desirous of cultivating science[12], and show him any attention in their power, particularly with reference to the nature of his scientific pursuits.

In November 1669, Flamsteed's name first appears in the Journal-books, as a contributor. He communicated a Paper *On Eclipses* in 1670; and inscribed the communication "To the Right Honourable William, Lord Brouncker, President of the illustrious Royal Society; also to the Right Worshipful, worthy, and truly ingenious Henry Oldenburg, Esq., Christopher Wren, M. D., and all other Astronomical Fellows of the said Society: J. F. humbly presents this epistle." At the close of it he writes: "Excuse, I pray you, this juvenile heat for the concerns of science, and want of better language from one who, from the sixteenth year of his age to this instant, hath only served one bare apprenticeship in these arts, under the discouragement of friends, the want of health, and all other instructors except his better genius. I crave the liberty to conceal my name, not to suppress it. I have composed the letters of it in Latin, in this sentence, *In Mathesi a sole fundes.* If I may understand that you accept of these, or think them worthy your notice, you shall certainly hear more from your's, J. F." Oldenburg's answer is too interesting to be omitted.

"*Jan$^y$*. 14, 1669—70.

Sir,

"Though you did what you could to hide your name from us, yet your ingenious and useful labors for the advancement of Astronomy, addressed to the noble President of the Royal Society, and some others

[12] These documents were often of service in adding to the Society's stores of information.

of that illustrious body, did soon discover you to us upon our solicitous enquiries after their worthy author. The said Society, having been made acquainted with your endeavours and performances too, and duly considered the importance and usefulness of these studies and astronomical predictions of yours, tending so much to state the motions of celestial bodies, especially that of the moon, have given me order to present you with their hearty thanks, both for your singular respect to them, and to congratulate you on the progress you have made in the excellent science of Astronomy; and withal to assure you, that you can do them no greater kindness than to continue this industry from year to year, and that, in compliance with your design, they will take what care they can to commit the province of observing those phenomena you have noted, to some of their most industrious and most skilful members. And, to the end that the better and ampler notice may be given of what you have so ingeniously and worthily begun to perform in this matter, it is intended that the most necessary part of your Papers shall be forthwith made public by the press; and that, perhaps, in the *Philosophical Transactions* for this month; reserving the rest, that cannot be conveniently concluded in the narrow bounds of those tracts, (which is to contain some variety of subjects), unto another opportunity. Which when done, I shall not fail (God willing) to see a copy of that book conveyed to you: whom I shall herewith desire to let me know the readiest and easiest way of sending things of this nature to you.

"What occasion you may have to employ my service in here, you need no more than signify it by a letter to me, sent by the ordinary post, addressing to me at my house in the middle of the Pal-mal, in St.

James' Fields, Westminster. Meantime, you must look upon me as the meanest of the Fellows of this Society, though I am, with all readiness and sincerity, Sir,

Your very affectionate friend, and real servant,

"H. OLDENBURG[13]."

Flamsteed writes, "in June, 1670, my father, taking notice of my correspondence with them whom I had never seen, would needs have me take a journey up to London[14], that I might become personally acquainted with the Fellows of the Society. I embraced the offer gladly, and there became first acquainted with Sir Jonas Moore, who presented me with Mr. Towneley's micrometer, and undertook to furnish me with telescope-glasses at moderate rates."

From this period, Flamsteed's name is of frequent occurrence in the Journal, and other books of the Society; and we shall often find him coming before connexion with its history.

It is deserving of record, that the celebrated Marcellus Malpighi, whilst holding the Professorship of Medicine at Messina[15], sent his work, *Dissertatio Epistolica de Bombyce*, to the Society, with a request that it might be published under their auspices. This was in 1669, and in the same year the Council-minutes record, "that the *History of the Silke Worme*, by Signor Malpighi, dedicated to the Royal Society, be printed forthwith by the printers of the same."

The work was published in 1669, with the following licence:

---

[13] Letter-book, Royal Society.

[14] He was living at Derby.

[15] He removed to Messina in 1666. See his *Opera Posthuma*, edited by Petrus Regus.

"*Tractatis cui titulus Marcello Malpighi Dissertatio Epistolica de Bombyce, Societatis Regiæ dicata, imprimatur à Johanne Martyn et Jacobo Allestry dictæ Societatis Typographis.*"

"BROUNCKER, P.R.S."

Malpighi addressed several of his best works to the Society. He had been elected a Member in 1668, and was afterwards a constant correspondent. There are several letters in the archives from learned foreigners, requesting the Society to allow their works to be published under their auspices, or dedicated to them[16]. In some instances the request was preferred through the medium of a Fellow, as exemplified by the subjoined letter, from Newton to Hooke, dated 'Trinity College, Cambridge, Dec. 3, 1670.'

"One Dominico Gasparini, an Italian Doctor of Physick, in the city of Lucca, has composed a Treatise of the method of administering the 'Cortex Peruviana,' in Fevers, in which he particularly discusses whether it may be administered in malignant fevers, and also whether in any fevers before the fourteenth day of the sicknesse.

"Upon the fame of the Royal Society, spread every where abroad, he is very ambitious to submit his discourse to so great and authentic a judgment as theirs is, and wishes therefore to dedicate his book to them."

Hooke replied, that "Dr. Gasparini needs no leave to dedicate his book to the Society, such things being very usually done without asking a consent; but doubtless they cannot but be very well pleased

[16] In 1671, Leibnitz dedicated his *Hypothesis Physica Nova*, to the Society.

with these testimonies of respect from learned and ingenious foreigners. And therefore, though they do not prompt any to such addresses, yet the author need not doubt finding such acceptance thereof by the Society, as may answer his expectations."

Circumstances like these attest how highly the Society was esteemed by foreign philosophers.

Yet it must not be supposed that they were without enemies and calumniators; among these was a Dr. Stubbe, a physician residing at Warwick, who made a fierce attack on the Society, in various works[17], written after the publication of Sprat's *History* and Joseph Glanvill's *Plus ultra*.

The language used by Dr. Stubbe is excessively scurrilous and violent, charging the Fellows not only with undermining the Universities, and destroying the established religion, but also of upsetting ancient and solid learning. Another detractor was the Rev. Robert Crosse, Vicar of Great Chew, in Somersetshire, who maintained that the Royal Society had done nothing to advance science, and that Aristotle had many more advantages for knowledge than the Royal Society, because he did *totam peragrare Asiam*. It was in consequence of Mr. Crosse's attacks, that Glanvill wrote the book entitled *Plus ultra, or the Progress and Advancement of Knowledge since the days of Aristotle, in an account of some of the most remarkable late improvements of practical useful learning, to encourage philosophical endeavours, occasioned*

---

[17] These are: *A Censure upon certain passages contained in the 'History of the Royal Society,' as being destructive to the Established Religion and Church of England;* and, *Legends no Histories, or a Specimen of some Animadversions upon the 'History of the Royal Society.'*

*by a conference with one of the notional way;* in which he affirms that "the impertinent taunts of those who accused the Society of doing nothing to advance knowledge, were no more to be regarded than the little chat of idiots and children." Crosse replied to Glanvill, but his book was refused a license at Oxford and Cambridge, on account of its scurrility[18].

Oldenburg alludes to these attacks in his Preface to the 5th Volume of the *Transactions*, "Let envy snarl," he says, "it cannot stop the wheels of active philosophy in no part of the known world. Not in France, either in Paris, or at Caen. Not in Italy, either in Rome, Naples, Milan, Florence, Venice, Bononia, or Padua. In none of the Universities, either on this or that side of the seas. Madrid and Lisbon, all the best spirits in Spain and Portugal, and the spacious and remote nations to them belonging; the Imperial Court, and the Princes of Germany; the northern Kings, and their best luminaries; and even the frozen Muscovite and Russian, have all taken the operative ferment, and it works high, and prevails every way to the encouragement of all sincere lovers of knowledge and virtue."

An enemy far more formidable to the Society was

[18] Evelyn in a letter to Glanvill, thanking him for a copy of his book *Plus ultra*, says, "I do not conceive why the Royall Society should any more concern themselves for the malicious and empty cavells of these delators, after what you haue said; but let the Moon-dogs bark on, till their throats are drie; the Society every day emerges, and her good genius will raise up one or other to judge and defend her; whilst there is nothing which does more confirme me in the noblenesse of the design, than this spirit of contradiction which the Devil (who hates all discoveries of those false and prœstigious ways that have hitherto obtain'd), does incite to stir up men against it." *Diary*, Vol. II. p. 234.

poverty, by which their resources were not only crippled, but their very existence rendered doubtful[19]. On the 30th November, 1671, it appears, by the Treasurer's accounts, which were audited on that day, that the arrears of subscriptions due to the Society from Fellows amounted to not less than 1696*l.*; while the receipts in quarterly payments during the year, were only 141*l.* 16*s.* This sum was wholly insufficient to purchase such philosophical apparatus as was required for the purpose of making experiments, and indeed scarcely sufficed to meet the ordinary annual expenses of the Society, who were frequently under the necessity of soliciting Boyle to lend them apparatus for experiments, which request, it is scarcely necessary to add, was never refused.

The name of the Royal Founder occurs so rarely at this period, as taking any interest in the Society, that I cannot forbear mentioning a curious wager, which Sir Robert Moray declares him to have made. The latter, at the request of the King, brought it forward at one of the Meetings in the early part of the year. It was to the effect, that his Majesty had wagered 50*l.* to 5*l.* "for the compression of air by water."

It was accordingly resolved, that "Mr. Hooke should prepare the necessary apparatus to try the experiment, which Sir Robert Moray said might be

---

[19] It is worthy of mention, that the Admiralty at this time requested the Society to raise some ships sunk off Woolwich. The Council replied, that their want of funds rendered it impossible for them to provide the necessary machinery, but that otherwise they would have great pleasure in complying with the request of the Admiralty.

done by a cane, contrived after such a manner, that it should take in more and more water, according as it should be sunk deeper and deeper into it."

The Minutes of a subsequent Meeting record the successful performance of the experiment, and it "was acknowledged that his Majesty had won the wager."

A curious pamphlet in the British Museum, entitled, *Propositions for the carrying on a Philosophical Correspondence already begun in the County of Sommerset, upon encouragement given from the Royal Society*, and published in 1670, affords interesting evidence of the gradual growth of philosophical inquiry throughout the country. A Society appears to have been organized, very similar in its constitution to the local Topographical and Archæological Associations of the present day, the head-quarters of which were at Bristol; and a remarkable instance of their desire to cultivate philosophy, is conveyed by the 12th section of their Rules, which orders, "That the Secretary set up a foot-post to go weekly from Bristol, the centre of intelligence, to various parts of the country, for the receiving and sending of letters."

It must have been highly gratifying to the Royal Society to see their endeavours to advance knowledge thus successful.

On the 21st December, 1671, the Journal-book records that "the Lord Bishop of Sarum (Seth Ward) proposed for candidate Mr. Isaac Newton, Professor of the Mathematicks at Cambridge." In a letter to Oldenburg, dated Jan. 6, 1671—2, Newton writes: "I am very sensible of the honour done me by the Bishop of Sarum, in proposing me candidate, and which I hope will be further conferred upon me by my election into the Society. And if so, I shall endeavour to testify my gratitude, by communicating what my poor and solitary endeavours can effect, towards the promoting their philosophical designs."

On the 11th Jan. 1671—2, Newton was elected a Fellow[20], and from this period until his decease, the name of this illustrious Philosopher is constantly found shedding lustre over the scientific history of the Society. Some time before his election, he transmitted his Reflecting Telescope to Mr. Oldenburg, with a description of the instrument. This telescope possesses so much interest, being the first perfect reflector invented, made by the hands of Newton, and, in his own words, "an epitome of what might be done," that a brief account of it will not be unacceptable.

In a communication to the Society in February, 1671—2, Newton tells us that when he "applyed himself to the grinding of optick-glasses of other figures than spherical, in the beginning of 1666," he discovered that "the perfection of telescopes was hitherto limited, not so much for want of glasses truly figured, according to the prescriptions of optick authors (which all men have hitherto imagined), as because that light itself is a heterogeneous mixture of differently refrangible rays. This made me take reflections into consideration, and finding them regular, so that the angle of reflection of all sorts of rays was equal to their angle of incidence: I understood that by their mediation optick instruments might be brought to any degree of perfection imaginable, provided a reflecting substance could be found which would polish as finely as glass, and reflect as much

[20] Newton was twenty-nine years old at the time of his election.

light as glass *transmits*, and the art of communicating to it a parabolick figure, be also attained. Amidst these thoughts, I was forced from Cambridge by the intervening plague, and it was more than two years before I proceeded further. But then having thought on a tender way of polishing, proper for metall, whereby, as I imagined, the figure also would be corrected to the last, I began to try what might be effected in this kind, and by degrees so far perfected an instrument (in the essential parts of it, like that I sent to London), by which I could discern Jupiter's four concomitants, and shewed them divers times to two others of my acquaintance. I could also discern the moon-like phase of Venus, but not very distinctly, nor without some niceness in disposing the instrument."

This extract shows that the year 1668 may be regarded as the date of the invention of Newton's reflecting telescope. I say Newton's reflecting telescope, because James Gregory, in his *Optica Promota*[21], published in 1663, describes the manner of constructing a reflecting telescope, with two concave

---

[21] It does not appear that Gregory ever constructed a Telescope. In 1664 or 1665, Sir D. Brewster states, "he attempted to make a Telescope, employing Messrs. Rives and Cox, who were celebrated glass-grinders of that time, to execute a concave speculum of six feet radius, and likewise a small one; but as they had failed in polishing the large one, and as Mr. Gregory was on the eve of going abroad, he troubled himself no farther about the experiment, and the tube of the Telescope was never made." *Life of Newton*, p. 28. Thus, Newton's Telescope was the first reflecting Telescope directed to the heavens. See a very interesting Paper "On the first invention of Telescopes," by Dr. Moll, in the first volume of the *Journal of the Royal Institution.*

specula; but Newton perceived the disadvantages to be so great, that, according to his statement, "he found it necessary, before attempting any thing in the practice, to alter the design, and place the eye-glass at the side of the tube, rather than at the middle."

On this improved principle he constructed the telescope now in the possession of the Royal Society, and of which the engraving at the end of this chapter is an accurate representation. The telescope was examined by Charles the Second, and the President and Fellows, who were so much pleased with it, that a letter describing it was drawn up by Oldenburg, which after being corrected by Newton, was sent to Paris, in order to secure the honour of the invention to its author[22].

The following letter from Newton to Oldenburg, describes the capabilities of this instrument.

"*Cambridge, March* 16, 1671.

"SIR,

"WITH the Telescope which I made, I have sometimes seen remote objects, and particularly the moon, very distinct, in those parts of it which were neare the sides of the visible angle. And at other times, when it hath been otherwise put together, it hath exhibited things not without some confusion; which difference I attributed chiefly to some imperfection that might possibly be either in the figures of the metalls

---

[22] Hooke informed the Society on the 18th January, 1671—2, that he possessed an infallible method of improving all sorts of optical instruments, so that "whatever almost hath been in notion and imagination, or desired in optics, may be performed with great facility and truth." Nevertheless, he did not communicate his method, but concealed it under the form of an anagram, which he never explained.

or eye-glasse; and once I found it caused by a little tarnishing of the metall in four or five days of moist weather.

"One of the ffellows of the College is making such another Telescope, with which last night I looked on Jupiter, and he seemed as distinct and sharply defined as I have seen him in other Telescopes. When he hath finished it, I will examine it more strictly, and send you an account of its performances, ffor it seems to be something better than that which I made.

"Your humble Servant,

"I. NEWTON[23]."

In the description of this telescope, entitled *An Accompt of a New Catadioptrical Telescope, invented by Mr. Newton*, published in the 81st Number of the *Transactions*, it is stated that "the objects are magnified about 38 times," whereas "an ordinary telescope of about two feet long only magnifies 13 or 14 times[24]."

Such is the instrument which, under the hands of Herschel and Rosse, has grown to proportions so gigantic, as to require the aid of machinery to elevate and depress the tube. Newton's first telescope is nine inches long, Lord Rosse's six-feet reflector is 60 feet in length[25]!!

---

[23] MS. Letters, Royal Society.

[24] It is not a little amusing to find that Newton's mechanical labours caused him to be regarded by some parties abroad as a maker of Telescopes. In a book of this period he is styled *Artifex quidam Anglus nomine Newton.*

[25] So exquisitely adjusted is the machinery connected with this gigantic instrument, that the tube is moved with all the ease and precision of that of a microscope. The delight which I experienced during a day spent in examining this wonderful piece of mechanism, under the guidance of Lord Rosse, will long be remembered by me.

On the 8th February, 1671—2, the Journal-book records a communication from Mr. Newton, "concerning his discovery about the nature of light, refractions, and colours, importing that light was not a similar, but a heterogeneous thing, consisting of difform rays, which had essentially different refractions, abstracted from bodies they pass through, and that colours are produced from such and such rays, whereof some in their own nature are disposed to produce red, others green, others blue, others purple, &c., and that whiteness is nothing but a mixture of all sorts of colours, or that 'tis produced by all sorts of colours blended together."

It was "Ordered that the Author be solemnly thanked, in the name of the Society, for this very ingenious discourse, and be made acquainted that the Society think very fit, if he consent, to have it forthwith published, as well for the greater conveniency of having it well considered by philosophers, as for securing the considerable notions thereof to the Author, against the arrogations of others."

These discoveries were the first of Newton's productions which saw the light. They are published in the 80th No. of the *Philosophical Transactions*. His experiments had been made in 1666, when he was only 23 years of age. Being employed in grinding glasses for telescopes, he had the curiosity to purchase a glass prism[26], in order to observe the celebrated phe-

---

[26] Although the following anecdote probably refers to a subsequent part of Newton's life, yet I insert it here as not entirely out of place. It forcibly exhibits the truthfulness and simplicity of the great philosopher's character, and at the same time the concentration of his thoughts upon philosophical subjects, to the

entire

nomena of the colours. This simple circumstance led to his important discoveries, which he designates in a letter to Oldenburg, as "the oddest, if not the most considerable detection, which had hitherto been made in the operations of nature." Newton accepted the proposed speedy and honourable method of publication, and in addressing his thanks to Oldenburg, says, "It was an esteem of the Royal Society for most candid and able judges in philosophical matters, that encouraged me to present them with that discourse of light and colours, which, since they have so favourably accepted of, I do earnestly desire you to return them my most cordial thanks. I before thought it a great favour to be made a member of that honourable body, but I am now more sensible of the advantage: for believe me, Sir, I not only esteem it a duty to concur with them in the promotion of real know-

---

entire forgetfulness of all beside. I found the anecdote in a very old number of the *Gentleman's Magazine*. It is written by an individual who states himself to have been personally acquainted with Newton.

"One of Sir I. Newton's philosophical friends abroad had sent him a curious prism, which was taken to the Custom-house, and was at that time a scarce commodity in this kingdom. Sir Isaac, laying claim to it, was asked by the officers what the value of the glass was, that they might accordingly regulate the duty. The great Newton, whose business was more with the universe than with duties and drawbacks, and who rated the prism according to his own idea of its use and excellence, answered, "That the value was so great, that he could not ascertain it." Being again pressed to set some fixed estimate upon it, he persisted in his reply, "That he could not say what it was worth, for that the value was inestimable." The honest Custom-house officers accordingly took him at his word, and made him pay a most exorbitant duty for the prism, which he might have taken away upon only paying a rate according to the weight of the glass!!"

ledge, but a great privilege, that, instead of exposing discourses to a prejudiced and censorious multitude (by which means many truths have been baffled and lost), I may with freedom apply myself to so judicious and impartial an assembly."

It was, however, the fate of Newton, in common with other great men, to have to submit to the law which ordains that merit, and more particularly success, shall give rise to envy. Biot has truly said, "By unveiling himself he obtained glory, but at the price of his repose."

No sooner were Newton's optical discoveries communicated to the world, than many individuals assailed, not only his conclusions, but the accuracy of the experiments from which they had been deduced. It is not within the province of this work to enter into any detailed account of the discussions which followed Newton's discoveries, but I may mention that Hooke, Pardies[27], Linus[28], and Huyghens, attacked him, impugning the accuracy of his theory of light. He replied in June 1672, in a manner which established his general doctrines on an impregnable basis.

Sir D. Brewster well says: "Harassing as such a controversy must have been to a philosopher like Newton, yet it did not touch those deep-seated feelings which characterize the noble and generous mind." The truth of this will appear by the following letter to Oldenburg, which is preserved in the archives.

---

[27] This was Father Pardies the Jesuit, Professor of Mathematics in the College of Clermont.

[28] He was a physician at Liege, and author of a Paper in the *Transactions*, entitled, *Optical Assertions concerning the Rainbow*. His objections to Newton's theory are contained in Letter-book, Vol. VII. p. 106.

"SIR, "*June* 11, 1672.

"I SEND you my answers to Mr. Hooke and P. Pardies, which I hope will bring with them that satisfaction which I promised. And as there is nothing in Mr. Hooke's considerations with which I am not well contented, so I presume there is as little in mine which he can except against, since you will easily see that I have industriously avoyded the intermixing of oblique and glancing expressions in my discourse. So that I hope it will be needless to trouble the R. Society to adjust matters. However, if there should possibly be any thing esteemed of that kind, I desire it may be interpreted candidly, and with respect to the contents of Mr. Hooke's considerations; and I shall readily give way to the mitigation of whatsoever the heads of the R. Society shall esteem personall. And, concerning my former answer to P. Pardies, I resigne to you the same liberty which he hath done for his objections, of mollifying any expressions that may have a shew of harshnesse."

"Your Servant,

"I. NEWTON."

Although these controversies terminated in the total defeat of all his opponents, yet Newton seems to have felt them most keenly. In a letter to Oldenburg, containing his reply to Huyghens, he says: "I intend to be no farther solicitous about matters of philosophy; and therefore I hope you will not take it ill, if you find me never doing any thing more in that kind[29]."

[29] In the original of this letter (which is preserved in the Royal Society), the lines quoted above are erased with a pen, but not so effectually as to prevent their being deciphered. It is probable that this was done by Oldenburg.

Fortunate, indeed, was it for science that such a body as the Royal Society existed, to whom Newton could make his scientific communications; otherwise, it is very possible that the *Principia* would never have seen the light.

At the Anniversary in 1672, John Evelyn was elected Secretary in the place of Thomas Henshaw; but retained office only one year. During 1672, the Society lost three eminent members: Matthew Wren, Francis Willughby, and Dr. Wilkins, Bishop of Chester. They were original Fellows of the Society; and the latter held office as Secretary, according to the terms of the First Charter, until raised to the Bench. He left a legacy of 400*l.* to the Society.

On the 26th April, 1673, Ward, in the Preface to his *Lives of the Gresham Professors*, states that "four gentlemen of figure, members of the Gresham Committee, Sir John Lawrence, Alderman, and Sir Thomas Player, Chamberlain, on behalf of the city, with Sir Richard Ford, Alderman, and Samuel Moyer, Esq., for the Mercers, were desired to attend the Lord Brouncker, President of the Royal Society, and, in the name of the Committee, to invite the Society to return, and hold their sessions in Gresham College, as they had been accustomed to do before the Fire. And the Professors of the College also waited on his Lordship, with the like request[30]." On the 9th October the President communicated this invitation to the Council; whereupon "the Council thought good to have their hearty thanks returned to the Committee of Gresham College for their kindness and respect; yet without saying any thing to them of acceptance, or not

[30] P. 15.

acceptance; only, in case they should give occasion for saying more, that then it might be intimated that this business was under consideration." It appears that the Council thought favourably of the proposition, as, on the 6th November, 1673, at a Council held that day, "the Lord Marshall was made acquainted with their thoughts of removing their weekly assemblyes to Gresham College, and of beginning to meet there again upon the next Anniversary; the Council being moved thereto, by considering the conveniency of makeing their experiments in the place where their Curator dwells, and the apparatus is at hand; as also, by the solemn invitation of the city of London and the Professors of Gresham College, and likewise from the hopes (they find ground to entertain), of meeting with some considerable benefactors at that end of the citty. ......That though the Society should thus remove their Meetings, yet they were full of hopes that his Lordship would be so far from removing his favours and kindnesses from them, that he would preserve them in the same degree he had done all along, and especially, during the many years he hath entertained them under his roof. To all which the Council added this humble request: that my Lord Marshall would be pleased to give the Council leave still to meet upon occasion in his Lordship's house; there to enjoy the honor and advantage of his counsel and directions, which they had always found so affectionate and considerable to them."

"Whereupon my Lord Marshall very obligeingly and generously declared: that though he always had esteemed, and still did esteem, it a great honour to his house, that the Royal Society kept their assemblies there, yet, understanding that the Council apprehended it really to be for the service and good of the Society to return to Gresham College, he could not but give up his reason to the reason of the Council: adding, further, that he should continue the same respect and concern for the Society wherever they mett, and be glad to receive the Council in his house upon any occasion of their meeting." The Society passed a vote of thanks to the Earl Marshall on the 6th November; and on the 1st December, 1673, assembled again in Gresham College[31]. As, however, the south and west galleries, which had been occupied by the merchants, were not in a state to be used as a repository and library, the Society continued occasionally to meet at Arundel House, until "November 12, 1674, when they re-settled themselves wholly at Gresham College[32]."

Amongst the many interesting communications made to the Society during 1673, must not be omit-

---

[31] The minutes of a sub-committee of the Gresham College Committee, held on Dec. 1, 1673, record that "The Committee now met, and welcomed the Royal Society into the same accommodations they enjoyed in this house before the late general Fyre in 1666." And Evelyn says in his *Diary*, under the date of Dec. 1673, "To Gresham College, whither the citty had invited the Royal Society by many of their chiefe Aldermen and Magistrates, who gave us a collation to welcome us to our first place of assembly, from whence we had been driven to give place to the citty, on their making it their Exchange on the dreadful conflagration, till their new Exchange was finished, which it now was." Vol. I. p. 441.

[32] Ward's *Lives*, Preface, p. 17, and Council-book, Vol. I. p. 222. At this meeting of the Council there were present, the Bishop of Salisbury in the chair, the Earl Marshall, Earl of Dorset, Lord Stafford, Sir John Lowther, Mr. Colwall, Mr. Hill, and Mr. Oldenburg.

ted those of the celebrated Leuwenhoeck, under whose hands the microscope became an instrument of infinite utility to science. The first notice of the Father of microscopical discoveries, as he may be called, occurs in a letter from Dr. Graaf to Oldenburg, dated April 28, 1673, in which "one Mr. Leuwenhoeck," he writes, "hath lately contrived microscopes, excelling those that have been hitherto made;" adding, that "he hath given a specimen of their excellency by divers observations, and is ready to receive difficult tasks for more, if the curious in London shall please to send him such; which they are not like to be wanting in." A short communication by Leuwenhoeck accompanied the letter, in which he described the structure of a bee and louse[33]. From this period, until his decease in 1723, he was in the habit of constantly transmitting to the Society all his microscopical observations and discoveries[34]. Some idea may be formed of his industry, by the fact, that there are 375 papers and letters from him, preserved in the archives, extending over a period of fifty years. His gratitude to the Society, for receiving and publishing his communications, will be seen by the following interesting extract from one of his letters:

"*Delft*, 2 *Aug.*, 1701.

"HON<sup>BLE</sup> GENTLEMEN,

"MY last to your Honours was dated the 21st June, wherein I humbly offer'd you my observations about spiders, since when I have received the

---

[33] Printed in *Transactions*, Vol. VIII. p. 6037.

[34] These, amounting to 125 Papers, are inserted in the *Transactions*. They have also been printed in Dutch at Delft and Leyden.

book which treats of fishes, and the whole set of *Philosophicall Transactions* for the year 1700, for which noble presents I return you my most hearty thanks.

"I have a small black cabinet, lacker'd and gilded, which has five little drawers in it, wherein are contained thirteen long and square tin boxes, covered with black leather. In each of these boxes are two ground microscopes, in all six and twenty; which I did grind myself, and set in silver; and most of the silver was what I had extracted from minerals, and separated from the gold that was mixed with it; and an account of each glass goes along with them[35].

"This Cabinet, with the aforesaid Microscopes, (which I shall make use of as long as I live), I have directed my only daughter to send to your Honors, as soon as I am dead, as a mark of my gratitude, and acknowledgment of the great honor which I have received from the Royal Society[36]."

It may be stated here that, in 1724, the Council presented Leuwenhoeck's daughter with a handsome silver bowl, bearing the arms of the Society, in testimony of their esteem for her deceased parent, and as an acknowledgment for his valuable legacy.

In 1674 the unhappy state of the Society's finances seriously engaged the attention of the Council. The Treasurer's accounts showed that the arrears, in November 1673, had increased to 1957*l.*; and, on care-

---

[35] These Microscopes were exhibited to Peter the Great, when he was at Delft in 1698. The Czar requested Leuwenhoeck to pay him a visit, and to bring some of his admirable microscopes, adding, that he would have gone to visit him at his residence, had it not been for the wish he had to escape the notice of the multitude.

[36] Letter-book, Vol. XIII. p. 183.

fully revising the list of Fellows, it was found that out of 146, the number on the books, "only 53 paid well, 79 did not, and 14 were absent in the country[37]." After several discussions as to the most advisable manner of proceeding, in order to collect a portion of the arrears, and enforce the payment of subscriptions, it was "Ordered, that there should be prepared a forme of a legal subscription for paying fifty-two shillings a year. That as many of the Fellows as are willing to further the business of the Society, shall be desired to advance a year's weekly contribution, for carrying on the work thereof with more vigour than hitherto."

The Attorney-General, at the request of the Council, drew up the following obligation, which was approved, and ordered to be signed by every newly-elected Fellow.

"I, ———, do grant and agree with the President, Council, and Fellows of the Royal Society of London for improving natural knowledge, that, so long as I shall continue a Fellow of the said Society, I will pay to the Treasurer of the said Society for the time being, or

---

37 In a Volume of MSS. in the British Museum, relating to the Royal Society, is a sheet containing the names of Fellows who will probably pay and give yearly one entertainment to the Society, "Amongst these are, Lord Brouncker, Bishop of Salisbury, Boyle, Petty, Wren, Evelyn, Wallis, and Goddard." The latter says, Aubrey "was a zealous member of the Royal Society for the improvement of natural knowledge. They made him their drudge, for when any curious experiment was to be donne, they would lay the task on him. He intended to have left all his papers and books to the Royal Society, had he made his will, and not dyed so suddenly." Opposite the names of Dr. Grew, Hooke, and Newton, in the MS. referred to, are the words, "No Pay, but will contribute experiments."

to his Deputy, the sum of Fifty Two Shillings per annum, by four equal quarterly payments, at the four usuall dayes of payment; that is to say, the Feast of the Nativity of our Lord, the Annunciation of the Blessed Virgin Mary, the Feast of St. John Baptist, and the Feast of St. Michael the Archangel: the first payment to be made upon the ——— next ensuing the date of these Presents; and I will pay in proportion, viz. One Shilling per week for any lesser time, after any the said dayes of payment, that I shall continue Fellow of the said Society. For the true payment whereof I bind myself and my heirs in the penal sum of Twenty Pounds. In witnesse whereof, I have hereunto set my hand and seal, this ———."

A collector was appointed, and some influential Fellows were requested to call upon several Noblemen who were in arrear with their subscriptions.

With the view of increasing the interest of the weekly Meetings, the Council "Ordered, that such of the Fellows as regard the welfare of the Society, should be desired to oblige themselves to entertain the Society, either *per se*, or, *per alios*, once a year, at least, with a philosophical discourse, grounded upon experiments made, or to be made; and, in case of failure, to forfeit 5*l*." And they passed a resolution, at a subsequent Meeting, rendering it imperative on every member of the existing Council "to provide an experimentall discourse for the Society, to be made at some publique Meeting within the year, either by himself, or by some other Member of the Society, or to pay forty shillings[38]." Besides the Members of

[38] The Earl of Aylesbury was the first who forfeited and paid the

Council, other Fellows were selected as "able and likely" to furnish "discourses" for the Society; and it was resolved, that the following letter should be written by the President, to every person so selected:

"Sir,

"The Council of the Royal Society, considering within themselves the great importance of having the Public Meetings of the said Society constantly provided with entertainments, suitable to the design of their Institution, have thought fit to undertake to contribute each of them one, not doubting but that many of the Fellows of the Society will join with them in carrying on such an undertaking. And, well persuaded of the approbation of this their purpose, so much tending to the reputation and support of the Society, they desire that you would be pleased to undertake for one, and to name any Thursday, as shall be most convenient to you, when you will present the Society, at one of the said public Meetings, by yourself, or by some other of the Fellows for you, with such a discourse, grounded upon, or leading to, philosophical experiments, on a subject of your own choice. In doing of which, you will benefit the Society, and oblige, Sir,

"Your humble Servant,

"Brouncker, P. R. S.[39]"

Oldenburg, with his habitual activity, obtained many papers for the Society: writing to Ray, under the date of September 15, 1674, he says:

---

the above sum: he engaged to provide the Society with an entertainment, but being called out of town, was unable to do so.

39 It is stated in the Council-book, that the draft of the above letter was prepared by Oldenburg, "but viewed and altered by the President."

"I cannot conclude this, without giving you notice, that the Council of the Royal Society intends to engage those of the Fellows of that body, that are able and willing to give them, once a year, each of them, an experimental entertainment at their ordinary Meetings: that is, some good discourse, grounded on experiments made, or to be made; that so their weekly Meetings may be more considerable and inviting than hitherto they have been, and the work of the Society not lie altogether on the shoulders of three or four of the Fellows. And this being to reach the absent as well as the present (I mean of those that have opportunity and ability), I do herewith intimate to you, that you are looked upon as one of those which the said Council have in their eye for such an exercise, desiring you, that you would think upon such a subject as yourself shall judge proper, for one entertainment of that company, after our Anniversary Election-day the next year: and if your occasions should not permit you to step to London, to present your discourse yourself, they have found an expedient, viz. to desire you, and such others as shall be in that case, to send it up to London to any of your friends, that may present and read it for you. It is farther intended, that such discourses shall be made publick if the author so think fit, and not otherwise[40]."

Although the Council used such strong measures to compel the Fellows of the Society, who were in arrears, to pay their subscription regularly, yet, it appears by the Minutes, that under extraordinary circumstances some Members were exempted from their pecuniary obligations; and it is not a little remark-

---

40 Phil. Letters between Ray and his Correspondents, p. 126.

able, that one of the first to request exemption was Newton. Under the date of January 28, 1673—4, the Council Minutes record: "It was mentioned by the Secretary, that Mr. Newton had intimated his being now in such circumstances that he desired to be excused from the weekly payments; it was agreed unto by the Council, that he should be dispensed with, as several others were." Mr. Baily, in his *Account of Flamsteed*, conceives that Newton's biographers are in error in attributing his wish to be excused his weekly payments, to his slender means at that time[41]. I cannot help differing from Mr. Baily on this point: my impression is, that it was no other than a temporary want of money which induced Newton to address the Council in terms carrying with them the conviction of pecuniary inability[42]. He had mentioned the subject at a much earlier period to Oldenburg, in a letter, dated "Cambridge, June 23, 1673:" "For your proffer," he says, "about my quarterly payments I thank you; but I would not have you trouble your-

---

41 P. 90.

42 Those who have read Baily's *Account of the Rev. John Flamsteed*, will probably remember, that the great aim of the work is to elevate Flamsteed, too often at the expense of Newton's character. But it should also be borne in mind, that the chief portion of the work consists of Flamsteed's *private* Journal or Diary, assuredly never intended by its author to meet the public eye. It is written, as all private diaries must necessarily be, very *ex parte*, and self-praising, and cannot be received by impartial persons as sufficient testimony with reference to the unfortunate dispute between himself and Newton. It is curious that Mr. Baily should have come to the conclusion, that Newton was able to pay his subscription to the Society with ease, when he prints in the same volume this assertion of Flamsteed's: "Newton was obliged to read mathematics for a salary at Cambridge."

self to get them excused, if you have not done so already[43]." In this reply to Oldenburg's offer, acquainting the Council with the young Pnilosopher's circumstances, we have satisfactory proof of the inconvenience he experienced in paying his subscription.

The generosity of the Council was not without its reward, as "the poor Cambridge Student," grateful for the consideration shown him, was, probably, incited to labour more zealously for science and the Royal Society, to whom he communicated all his noble discoveries. The great Philosopher praying to be excused from the payment of one shilling per week, contrasts curiously with his subsequent wealth.

In 1675 the Observatory at Greenwich was built, of which Flamsteed, who became a Fellow of the Society in 1676, was appointed the Astronomical Observator (the title still retained in official documents) by warrant, under the Royal sign-manual, with a salary of 100*l.* per annum.

The establishment of this highly-useful national institution, with which the Royal Society has been intimately associated from its foundation to the present time, forms so important an epoch in this history, as to merit some notice here.

Flamsteed thus describes the circumstances:

"Betwixt my coming up to London and Easter (1675), an accident happened that hastened, if it did not occasion, the building of the Observatory. A Frenchman, that called himself Le Sieur de St. Pierre, having

---

[43] Archives: Royal Society. This portion of the letter has been crossed by the pen, with the view of preventing its being printed.

some small skill in Astronomy, and made an interest with a French lady[44], then in favour at Court, proposed no less than the discovery of the longitude; and had procured a kind of commission from the King, to the Lord Brouncker, Dr. Ward (Bishop of Salisbury), Sir Christopher Wren, Sir Charles Scarborough, Sir Jonas Moore, Col. Titus, Dr. Pell, Sir Robert Moray, Mr. Hooke, and some other ingenious gentlemen about the town and court, to receive his proposals; with power to elect, and to receive into their number any other skilful persons; and, having heard them, to give the King an account of them, with their opinion, whether or no they were practicable, and would shew what he pretended. Sir Jonas Moore carried me with him to one of their meetings, when I was chosen into their number; and, after the Frenchman's proposals were read: which were,

"1. To have the year and day of the observations;

"2. The height of two stars, and on which side of the meridian they appeared;

"3. The height of the moon's two limbs;

"4. The height of the pole: all to degrees and minutes;

it was easy to perceive from these demands that the Sieur understood not that the best lunar tables differed from the heavens; and that therefore his demands were not sufficient for determining the longitude of the place where such observations were, or should be made, from that to which the lunar tables were fitted; which I represented immediately to the company. But they, considering the interests of his patroness at Court, desired to have him furnished according to his demand.

---

[44] In another place Flamsteed informs us that this lady was the Duchess of Portsmouth.

I undertook it, and gave him observations such as he demanded. The half-skilled man did not think they could have been given him; but cunningly answered, they were feigned. I then wrote a letter, in English, to the Commissioners, and another, in Latin, to the Sieur, to assure him they were not feigned. I heard no more of the Frenchman after this; but was told that, my letters being shewn King Charles, he startled at the assertion of the fixed stars' places being false in the catalogue[45]; said with some vehemence, 'he must have them anew observed, examined, and corrected, for the use of his seamen;' and, further, (when it was urged to him how necessary it was to have a good stock of observations taken, for correcting the motion of the moon and planets), with the same earnestness, 'he must have it done.' And when he was asked, Who could, or who should do it? 'The person,' says he, 'who informs you of them.' Whereupon I was appointed to it[46]."

Flamsteed was mainly indebted for his appointment to Sir Jonas Moore, at whose house in the Tower he carried on astronomical observations. Sir Jonas contemplated establishing a private observatory at Chelsea College, which was abandoned when the King expressed his intention of founding one at Greenwich.

According to Flamsteed, the site of Greenwich for the Observatory was determined by Sir Christopher Wren. In his *Autobiography* he says, "The next thing to be thought of was a place to fix in. Several were proposed, as Hyde Park and Chelsea

---

[45] Flamsteed stated in his letters that the places of the fixed stars were not truly given.

[46] Flamsteed's *History of his own Life*, p. 37.

College;—I went to view the ruins of this latter, and judged it might serve the turn, and the better, because it was near the Court. Sir Jonas rather inclined to Hyde Park: but Sir Christopher Wren mentioning Greenwich Hill, it was resolved on. The King allowed £500. in money, with bricks from Tilbury Fort, where there was a spare stock, and some wood, iron, and lead, from a gatehouse demolished in the Tower; and encouraged us further with a promise of affording what more should be requisite. The foundation was laid August 10, 1675, and the work carried on so well, that the roof was laid, and the building covered by Christmas."

Mr. Baily states, "that this Observatory was formerly a tower built by Humphrey, Duke of Gloucester, and repaired or rebuilt by Henry VIII. in 1526. That it was sometimes the habitation of the younger branches of the royal family; sometimes the residence of a favourite mistress; sometimes a prison, and sometimes a place of defence. Mary of York, fifth daughter of Edward IV., died at the tower in Greenwich Park in 1482. Henry VIII. visited "a fayre lady," whom he loved, here. In Queen Elizabeth's time it was called *Mirefleur*. In 1642, being then called Greenwich Castle, it was thought of so much consequence, as a place of strength, that immediate steps were ordered to be taken for securing it. After the Restoration, Charles II., in 1675, pulled down the old tower, and founded on its site the present Royal Observatory[47]."

The connexion of the Royal Society with the Observatory commenced by the Society lending to the

---

[47] *Account of Flamsteed*, p. 39.

new establishment some astronomical instruments, which Flamsteed required. The Minutes record:—"it was ordered, that the astronomical instruments belonging to the Society be lent to the Observatory at Greenwich, and that Mr. Hooke's new quadrant be forthwith finished at the charges of the Society."

Bearing in mind the apathetic conduct of the King towards the Royal Society, it will not appear extraordinary that the Observatory, so hurriedly established, was left for a period of nearly fifteen years without a single instrument being furnished by Government[48]. Sir Jonas Moore provided Flamsteed with a sextant, two clocks[49], a telescope, and some books; all the other instruments, excepting the foregoing, and those

---

[48] It is curious, though not gratifying, to contrast the noble manner in which Louis XIV. patronized science, with that of Charles the Second. In 1706, a solar eclipse occurred, when, says Fontenelle in the *Histoire de l'Académie Royale des Sciences*, "*Le Roi voulut voir faire les observations par des Astronomes de l'Académie, et pour cela M. Cassini le fils, et M. de la Hire le fils, allèrent à Marly avec tous les instruments nécessaires. Toute la maison royale et toute la cour furent témoins des operations, et Monseigneur le Duc de Bourgogne, qui fait bien voir que les sciences peuvent trouver leur place parmi les occupations des plus grands princes, détermina lui même plusieurs phases.*" p. 114.

[49] Under the date of Dec. 9, 1736, the following entry occurs in the Journal-book:—"Mr. Hodgson made a present of the late Mr. Flamsteed's Clock, which, in his time, stood in the great room of the Royal Observatory at Greenwich." It goes thirteen months without winding up, and was the donation of Sir Jonas Moore, whose name is inscribed on the dial-plate with this motto, "Sir Jonas Moore caused this movement with great care to be thus made, Anno 1676. Thomas Tompion." On the 3rd Nov. 1737, Mr. Hodgson also presented to the Society the object-glass of a 90-feet tube, "being that which Mr. Flamsteed designed to have used in the well of the Observatory at Greenwich, but was prevented in it by the damp of the place."

lent by the Royal Society, were made at Flamsteed's expense[50]. "It is true," says Mr. Baily, "that they had given him a house to live in, and had appropriated a *precarious* salary of 100*l.* a year; but, at the same time, although his employments were sufficiently laborious, the King had ordered that he should instruct monthly two boys from Christ Church Hospital, which was a great annoyance to him, and interfered with his proper avocations[51]."

It was well for astronomical science and the nation, that Flamsteed was selected to fill the important office of Astronomer Royal; for it is only paying a just tribute to his extraordinary energy and indefatigability to say, that any other man would probably have succumbed under the amount of drudgery appertaining to the office[52]; if, indeed, in the absence of encouragement, he would have continued in it at all, and particularly when the pecuniary reward was so insignificant.

Delambre in his *Histoire de l'Astronomie au 18^eme^ Siècle*, has justly remarked, *Ce n'est pas le tout, que de fonder un Observatoire, et de doter l'astronome; il faudroit faire un fonds annuel pour l'impression, et imposer à l'astronome l'obligation de n'être jamais en retard d'une année.* It is a curious and interesting fact, that the Observatory, so sadly neglected by its Royal Founder, was honoured by two visits from Peter the Great, to whom Russia is indebted for her Academy of Sciences, and first national Observatory[53].

---

[50] Baily's *Account*, p. 45. [51] Ibid., p. 27.

[52] In a letter to Sir Jonas Moore, Flamsteed talks of earning his salary by labour "*harder than thrashing.*"

[53] These were founded in 1724.

It appears by Flamsteed's *Historia Cœlestis*, that the Czar paid his first visit on the 6th Feb. 1697—8, opposite to which date Flamsteed has written, *Serenissimo* PETRUS MOSCOVIÆ CZARUS, *Observatorium primum visum venit, lustratisque instrumentis habitu privato abiit. Aderant secum Bruceus, parentibus Scotis Moscoviæ natus, legatus militaris; J. Wolfias et Stileus, mercatores Angli.* His second visit took place on the 8th March following, on which occasion his Imperial Majesty made a complete observation of Venus; thus recorded, *Observante Serenissimo* PETRO MOSCOVIÆ CZARO; proving that he not only saw the planet through the telescope, but made the observation.

It would be ungracious to pass from this subject, without remarking that the Observatory of Greenwich, although left so long unassisted by Government, has, from 1676 (when Flamsteed began his official labours) to the present time, continued to give the astronomical world a series of observations, unequalled for their extent, and unsurpassed by any in accuracy. *L'Observatoire de Greenwich*, says M. Struve, *a doté la science de cette série non interrompue d'observations, qui embrassent actuellement* 167 *ans, et qui, par rapport aux mouvements du soleil, de la lune et des planètes, et aux positions des étoiles fixés, doivent etre regardées comme la base des nos connaissances astronomiques. Il y a, dans l'histoire de l'observatoire de Greenwich, un point très remarquable, savoir que les astronomes ont travaillé sur un même plan, depuis l'origine de l'établissement jusqu'à l'époque actuelle*[54]. This is high praise from a high

[54] *Déscription de l'Observatoire de Poulkova*, 4to. Pétérsbourg, 1845, p. 5.

authority, and is honourable to the various Astronomers-Royal, of whom a list is subjoined[55]. The Observatory will be frequently brought under our notice in the course of this history.

In 1676 the Society received several communications from travellers, giving accounts of "the notables" observed in the countries they had visited. These communications were the results of the queries issued by the Society, and formed valuable additions to the very limited geographical knowledge of that period.

The Society were not satisfied with this mode alone of obtaining information: when a traveller arrived in London from remote countries, some Fellow was deputed to call on him, and invite him to the Meetings. Thus, Evelyn tells us, that he was "desired by the Royal Society to call upon, and salute a Mons. Jardine, who had been thrice in the East Indies and Persia, in their name; and to invite him to honour them with his company[56]." Evelyn adds, that he was accompanied by Sir C. Wren and Sir J. Hoskyns, and that they found Mons. Jardine "a very handsome person, extremely affable, and not inclined to talke wonders."

In the Minutes of the Oxford Philosophical Society it is recorded that Dr. Plot greatly entertained the

---

[55] Flamsteed ......1676 to 1719
Halley ..........1719 „ 1742
Bradley .........1742 „ 1762
Bliss.............1762 „ 1765
Maskelyne......1765 „ 1811
Pond ............1811 „ 1835
Airy.............1835 still continues.

[56] *Diary*, Vol. I. p. 490.

company by giving an account of what he had seen at a Meeting of the Royal Society in 1676. He stated "that a merchant, lately come from China, exhibited to the Society a handkerchief made of Salamander's wool, or *Linum Asbesti*, which, to try whether it were genuine or no, was put into a strong charcoal fire, in which not being injur'd, it was taken out, oil'd, and put in again; the oil being burnt off the handkerchief, was taken out again, and was altered only in two respects, it lost 2dr. 5gr. of its weight, and was more brittle than ordinary; for which reason it was not handled until it was grown cold, by which means it had recovered its former tenacity, and in a great measure its weight. The merchant who obliged the Society with the sight of so great a rarity, acquainted them that he received it from a Tartar, who told him that the Tartars, among whom this sort of cloth is, sold it at 80*l.* sterling the China ell, which is less than our ell; and that they greatly use this cloth in burning the bodyes (to preserve the ashes) of great persons, and that in Tartary it is affirmed to be made of the root of a tree[57]."

The year 1677 was attended by a melancholy event for the Society, the loss of Oldenburg, who died suddenly in September, at Charlton in Kent. Oldenburg, who sometimes wrote himself Grubbendol, was born at Bremen[58], and for several years acted as agent in England with the Protectorate, for the republic of Lower Saxony. In 1656 he entered as a student in the University of Oxford, under the

[57] MS. Minutes, Vol. I. p. 72.

[58] In the letters to him from Milton, printed in the *Epist. Familiares* of the latter, he is styled *Orator Bremensium.*

name and title of *Henricus Oldenburg, Bremensis nobilis Saxo*[59]. On leaving Oxford, he went to France with Mr. Richard Jones, son of Lord Ranalagh[60], with whom he travelled until 1661, in which year they returned to England. From 1662, when he became Secretary to the Society, to the period of his decease, he was indefatigable in the performance of all his secretarial duties[61], which, as we have seen, were unrewarded until 1669, when a salary of 40*l.* a year was allowed him[62]. This act of justice was most necessary to Oldenburg, as the *Philosophical Transactions* never yielded him a greater profit than 40*l.* a year, and generally fell considerably below this sum. In 1666, Boyle manifested great zeal in his favour, and joined Lord Brouncker in applying for the appointment of Latin Secretary to the King, which, however, they were unsuccessful in obtaining for him. He married a daughter of Mr. John Drury, a divine of considerable celebrity, by whom he had two children. His son, named Rupert, from his god-father Prince Rupert, received a present

[59] Wood, *Fasti. Oxon.*, Vol. II. p. 114.

[60] Milton, *Epist. Famil.* 24—25.

[61] Dr. Derham, in his *Life of Ray*, says, that "Oldenburg was in the habit of corresponding every month with that naturalist. He was a very diligent Secretary, and laboured very heartily to keep up the Society's correspondence, and get all the information he could about curious matters, from all persons that he knew or heard were able to furnish him with any: and the better to accomplish his ends, he would send his ingenious correspondents an account of matters that came to his knowledge, as well as expect a plentiful return from them." p. 44.

[62] An Ode in praise of Oldenburg, a copy of which exists in the British Museum, styles him the illustrious, laborious, and precise Oldenburg.

in 1717 from the Council of the Society, in consideration of his father's eminent services. These, indeed, it would be almost impossible to over-rate, and it may be mentioned that when attacked by Hooke on account of the *Philosophical Transactions*, he was at once justified by a declaration of the Council.

At the anniversary in 1677, Hooke and Dr. Grew were elected Secretaries, and Lord Brouncker retired from the Presidency, which he had held with great advantage to the Society for fifteen years. He was succeeded by Sir Joseph Williamson, a Memoir of whom commences the next Chapter.

*Inscription on the Stand.*

THE FIRST REFLECTING TELLESCOPE INVENTED BI $\text{S}^{\text{R}}$ ISAAC NEWTON, AND MADE WITH HIS OWN HANDS IN THE YEAR 1671.

# CHAPTER X.

Memoir of Sir J. Williamson—Norfolk Library removed to Gresham College—Rules for its preservation—Halley elected—Observatory at Gresham College—Monument used for Astronomical purposes—*Philosophical Collections* published by Hooke—His Salary increased—Sir J. Williamson resigns—Boyle chosen President—His reasons for declining the Office—Wren elected—Memoir of him—Grew's Catalogue of the Society's Museum—Chelsea College sold by Wren—Admission of Fellows made more difficult—Wren retires from the Presidency—Evelyn solicited to become President—Declines—Sir John Hoskyns elected—Memoir of him—Publication of *Transactions* resumed—Experiments made—Present of Curiosities from China—Limited state of Knowledge of Foreign Countries—Sir John Hoskyns resigns—Sir Cyril Wyche elected President—Memoir of him—Papin elected a Curator—His Bone-digester—Curious Account of a Supper prepared by him for the Fellows—His Steam-Engine Inventions—Croonian Lecture—Lady Sadleir's Legacy—Lister's Geological Maps—Lord Clarendon's Present of Minerals—Resignation of Sir Cyril Wyche—Pepys elected President—Sir Thomas Molyneux's Account of the Society.

---

1675—85.

SIR JOSEPH WILLIAMSON was the son of the Rev. Joseph Williamson, rector of Bridekirk in Cumberland. While yet a boy he visited London, in the capacity of clerk to Mr. Richard Folson, member of parliament for Cockermouth, who sent him to Westminster School, then presided over by Dr. Busby. His assiduity and talent gained for him a recommendation from Dr. Busby to Dr. Langbaine, provost of Queen's College, Oxford, by whom he was admitted on the foundation. He took the degree of B. A. in 1653, and immediately after went to France as tutor to a nobleman. In 1657 he took the degree of M.A., and was elected a fellow of his college.

After the Restoration he was appointed Secretary, successively to Sir Edward Nicholas, and Lord Arlington, Secretaries of State, and was made Keeper of the State-Paper Office in Whitehall. In 1667 he was chosen a clerk of the Council in Ordinary, and received the honour of knighthood. He was one of the plenipotentiaries at the treaty of Cologne, acting with the Earl of Sunderland and Sir Leoline Jenkins. On the 27th June, 1674, he became Secretary of State, in the room of Lord Arlington, to whom, according to the custom of the time, he paid 6000*l.* for the office. It is a curious fact, that two of the Secretaries of State in the reign of Charles II., Sir Leoline Jenkins and the subject of this memoir, had both been tutors. The former was of humble origin. It is reported of him, that when he rose to high office, he hung in his chamber the old pair of leathern breeches in which he first rode into Oxford, a poor scholar, to remind him of his former low estate[1].

Sir Joseph Williamson had a considerable share in establishing the *London Gazette.* This was originally called the *Oxford Gazette,* the first number being published Nov. 7, 1665, when Charles II. and the court were at Oxford. "But when the said Court removed to London, they were intituled and called the *London Gazette,* the first of which that was published there, came forth on the 5th of February following, the King being then at Whitehall. Mr. Joseph Williamson procured the writing of them for himself; and thereupon employed Charles Perrot, M.A.

---

[1] *Diary of the Times of Charles II.*, Vol. I. p. 306.

and fellow of Oriel College in Oxon, who had a good command of his pen, to do that office under him, and so he did, though not constantly, to about 1671[2]."

Sir Joseph Williamson was one of the first victims of the fear and excitement caused by the celebrated Popish plot. He was committed to the Tower by the House of Commons on the 18th November, 1678, on a charge of granting commissions to Popish officers, but was released by the King on the same day. On the 9th February following he resigned the Secretaryship of State, and was succeeded by the Earl of Sunderland. In December, 1679, he married the Baroness Clifton, widow of Henry Lord O'Brien, sister and sole heiress to Charles Stuart, Duke of Richmond, by whom he acquired large property, and the hereditary office of High-steward of Greenwich. Upon this marriage Evelyn remarks, "'Twas thought they lived not kindly after marriage, as they did before. She was much censured for marrying so meanly, being herself allied to the royal family." But Evelyn did not entertain a favourable opinion of Sir Joseph Williamson, as he calls him "Lord Arlington's creature, and ungrateful enough."

It is certain that he must have possessed considerable talents for business and courtiership, to have risen from so humble a beginning to the important situation of Secretary of State. His influence, too, must have been great, otherwise he would hardly have been selected to fill the chair of the Royal Society. Although deeply occupied by public affairs, he presided at every Meeting of the Council, and generally at the ordinary Meetings, in the proceedings of which

---

[2] Wood, *Athen. Oxon.*, Vol. II. p. 469.

he took much interest. He presented several curiosities to the Museum, and a large screw-press for stamping diplomas. He died in 1701, and left 6000*l*., and a valuable collection of heraldic manuscripts and memoirs relating to his foreign negociations, to Queen's College, Oxford, and 5000*l*., for the purpose of founding a Mathematical School at Rochester, for which town he had frequently sat in Parliament.

In a dedicatory epistle from Oldenburg to Sir Joseph Williamson, at the commencement of the ninth volume of the *Philosophical Transactions*, Oldenburg says, "Your merits raised you to that eminent place you are now possess'd of; and you are full of steady inclinations to on all occasions advance the ingenuous arts."

Sir Joseph presented the Society with his portrait, painted by Sir G. Kneller. It is suspended in the Meeting-room.

In June, 1678, the Duke of Norfolk commenced pulling down Arundel House, which occasioned the Library presented by his Grace to the Society to be removed to Gresham College.

The following Rules for its preservation and management were drawn up by the Council.

"Orders concerning the Government of the Bibliotheca Norfolciana.

"1. That the long gallery in Gresham College be the place for the Library, if it may be procured.

"2. That an Inscription in letters of gold be set up in some convenient place, in honour of the benefactor.

"3. That there be an exact Catalogue of all the Books of the Bibliotheca Norfolciana made apart; and also of all other books which shall accrue.

"4. That, for securing the Books and to hinder their being embezzled, no Book shall be lent out of the Library to any person whatsoever.

"5. That such person or persons as shall desire to use any Book in the Library, shall return it into the hands of the Library-keeper, entire, and unhurt.

"6. That the Library shall be surveyed once in the year, by a Committee chosen by the Council, to the number of six, any three of which to be a quorum."

Dr. Gale was requested to prepare the form of inscription mentioned in the second clause of the preceding orders; and, at the ensuing Meeting of Council, he submitted one, which was approved, and ordered to be set up on a Tablet in the gallery, where the books were placed. It is remarkable that there is no record of the inscription in the archives of the Society. According to Maitland, it was as follows:—

"BIBLIOTHECA NORFOLCIANA.

"*Excellentissimus Princeps* Henricus Howard, *Dux* Norfolciæ, *Comes Marechallus* Angliæ, *Comes* Arundeliæ, Suriæ, Norfolciæ, et Norwici, &c. *Heros, propter Familiæ Antiquitatem, Animi Dotes, Corporis Dignitatem, pene incomparabilis, Bibliothecam hanc instructissimam (quæ hactenus* Arundeliana *appellabatur) Regiæ Societati Dono dedit, et perpetuo sacram esse voluit.*

*Huic*<br>
*Pro eximia erga se Liberalitate, Societas*<br>
*Regia Tabulam hanc devotæ*<br>
*Mentis testem fixit;*<br>
*Præside* Josepho Williamson,<br>
*Equite Aurato,*<br>
A.D. MDCLXXIX.[3]

On the 30th November, 1678, Edmund Halley was elected a Fellow of the Society. He had just

[3] Maitland's *London*, p. 656.

returned from his voyage to St. Helena, the principal object of which was to make astronomical observations. The great result of the expedition remains in his *Catalogus Stellarum Australium;* but it is important to remember that he had attentively studied the variation of the magnet, and made such observations as materially assisted him in drawing up the remarkable papers upon magnetism, printed in the *Transactions,* which preceded his magnetic chart. These communications will come under notice in a subsequent part of this work[4].

"To Halley," says Sir John Herschel, "we owe the first appreciation of the real complexity of the subject of magnetism. It is wonderful indeed, and a striking proof of the penetration and sagacity of this extraordinary man, that with his means of information he should have been able to draw such conclusions, and to take so large and comprehensive view of the subject, as he appears to have done[5]."

Halley always evinced the greatest interest in the Society, and endeavoured by every means in his power to advance the objects for which it was founded. When he left England for the Continent, in 1680, he was furnished with letters of introduction to the eminent philosophers in Paris, of whose labours he gives some account in the following letter to Hooke. It is dated Paris, January 15, 1680—1.

"I got hither the 24th of last month, after the most unpleasant journey that you can imagine, having

---

[4] It will be sufficient to state here, that Halley supposed the earth to be one large magnet, having four magnetical poles or points of attraction, two in each hemisphere. See his Papers in the 180th and 195th number of the *Transactions.*

[5] Art. Terr. Mag. *Quar. Rev.*, Vol. LXVI. p. 277.

been forty hours between Dover and Calais, with wind enough.

"The letter you were pleased to intrust me with, did me the kindness to introduce me into the acquaintance of MM. Justel and Toynard, with whom is the rendezvous of all curious and philosophical matters. The general talk of the virtuosi here is about the Comet, which now appears, but the cloudy weather has permitted to be but seldom observed: whatsoever shall be made publick about him here I shall take care to send you. Whilst I am here I shall be able to serve you in procuring you what books you shall desire that are to be purchased for money; but those that have been published by the Academy of Sciences, amongst which is the book of plants Sir John Hoskyns desires, will be much more difficult to come by. However, I have hopes to get them for the Society's Library, at least, to get a sight of them, so as to give you some account of what they contain. There is just now finished the book of Astronomical Voyages, but I have not gotten a sight thereof. But Mr. Cassini, who seems my friend, will, I hope, grant it me. If I can but get it in my own possession, I will make hard shift to copy the most material things[6]."

A turret for astronomical purposes had been erected by the Gresham Committee, over the apartments of the Professor of Geometry, and Hooke availed himself of the facilities afforded him for carrying on several astronomical and meteorological observations[7]. It was on this account that the instruments

---

[6] Archives: Royal Society.

[7] "The Monument on Fish Street Hill, built by Wren, and completed in 1677, was used by Hooke and other members of the

lent to the Greenwich Observatory were recalled; the removal took place in January, 1679, by order of the Council. Flamsteed was evidently very much annoyed at this circumstance, as he talks of "Hooke wresting the quadrant from him;" but it must be borne in mind, that the Society had received no account whatever of the progress of Flamsteed, at which the President expressed his surprise, and, moreover, the Society who had, as we have seen, but very little money to purchase philosophical apparatus, were really desirous to have the instruments restored to them for scientific purposes. At the same time, however, that this order was issued, the Council directed a magnetical needle to be made at the cost of the Society, and lent to Flamsteed to make observations at Greenwich on variation[8].

In 1679 Hooke commenced publishing the *Philosophical Collections*, which are generally regarded as constituting a portion of the *Transactions*. These had ceased to appear in 1678; and it was in consequence of the great desire expressed by philosophers for their revival, that Hooke undertook to publish the *Collections*. At a Meeting of Council, in February, 1679, it was;—

"Resolved that Mr. Hooke be desired to publish

---

Royal Society for Astronomical purposes, but abandoned on account of the vibrations being too great for the nicety required in their observations." Elmes' *Life of Wren*, p. 289. It is curious that this gave rise to the report that it was unsafe. Captain Smyth, in his *Cycle of Celestial Objects*, tells us, that when taking observations on the summit of Pompey's pillar, the mercury was sensibly affected by tremor, although the pillar is a solid.

[8] It may interest the reader to know that the variation at the above period (1679) was between 4 and $4\frac{1}{2}$ degrees west.

(as he declares he is now ready to do,) a sheet or two every fortnight, of such philosophical matters as he shall meet with from his correspondents; not making use of any thing contained in the Register-books, without the leave of the Council and author."

At another Meeting it was;—

"Resolved, that, in consideration of propositions made by Mr. Hooke for a more sedulous prosecution of the experiments for the service of the Society, and particularly the drawing up into treatises several excellent things which he had formerly promised the world; the Council, as an encouragement, according to the poor abilities of the Society, have agreed to adde forty pounds for this year, ending at Christmas, to Mr. Hooke's salary."

The *Philosophical Collections*, which extend to seven numbers, comprise Papers by various individuals, giving, in the words of the title-page, *Accounts of Physical, Anatomical, Chymical, Mechanical, Astronomical, Optical, and other Mathematical and Philosophical Experiments and Observations*.

The second and third numbers were printed in 1681, and the fourth, fifth, sixth, and seventh, in the early part of 1682.

At the anniversary of the Society in 1680, Sir Joseph Williamson resigned, and the Hon. Robert Boyle was chosen President, but declined accepting office. The subjoined letter addressed to Hooke, makes it appear that he was actuated by conscientious motives.

"SIR, "*Pall Mall, Dec.* 18, 1680.

"THOUGH, since I last saw you, I met with a lawyer who has been a member of several Parliaments, and found him of the same opinion with my Council in reference to the obligation to take the test and oaths you and I discoursed of; yet, not content with this, and hearing that an acquaintance of mine was come to town, whose eminent skill in the law had made him a judge, if he himself had not declined to be one, I desired his advice, (which, because he would not send me till he had perused the Society's Charter, I received not till late last night,) and by it I found, that he concurred in opinion with the two lawyers already mentioned, and would not have me venture upon the supposition of my being unconcerned in an act of parliament, to whose breach such heavy penalties are annexed. His reasons I have not now time to tell you, but they are of such weight with me, who have a great (and perhaps peculiar) tenderness in point of oaths, that I must humbly beg the Royal Society to proceed to a new election, and do so easy a thing as, among so many worthy persons that compose that illustrious company, to choose a President that may be better qualified than I for so weighty an employment. You will oblige me, also, to assure them, that, though I cannot now receive the great honour they were pleased to design me, yet I have as much sense of it as if I actually enjoyed all the advantages belonging to it. And, accordingly, though I must not serve them in the honorable capacity they were pleased to think of for me; yet I hope that, God assisting, I shall not be an useless member of that learned Body, but shall manifest, in that capacity, both my zeal for their work, and my sense of their favours. This you will oblige me to represent in such a way as may persuade the virtuosi that you will discourse with, how concerned I am to retain the favourable opinion of persons that have so great a share in his esteem, who shall reckon your good offices, on so important an occasion, among the welcomest favours you can ever do[9],

"Sir,

"Your most affectionate Friend and humble Servant,

"Ro. Boyle."

Superscribed:

"*These for my much respected friend, Mr. R. Hooke, Professor of Mathematics at Gresham College.*"

The effect of Boyle's declining to accept the office of President, was the election of Sir Christopher Wren, who was sworn in at the Council held on the 12th January, 1680—1.

The family of the Wrens, according to tradition of Danish origin, had been long and honourably distinguished in England before the birth of this great architect. Sir Christopher was born at East Knoyle, Wilts, on the 20th October, 1632. His father was chaplain in ordinary to Charles I., and dean of Windsor, and his uncle (Dr. Matthew Wren) was successively Bishop of Hereford, Norwich, and Ely. All his biographers describe him as a small and weakly child, whose rearing required much care. His precocity was surprising. In fact, it almost partakes of the marvellous, when we are told, that at the age of thirteen he invented an astronomical instrument (which he dedicated to his father in Latin rhyme), a pneumatic engine, and "a peculiar instrument," says the author of *Parentalia*[10], "of use in Gnomonics, which he

---

[9] Archives: Royal Society.

[10] The MS. of the *Parentalia* is in the Library of the Royal Society.

explained in a treatise entitled, *Scistericon Catholicum.*" His mind rose early into maturity and strength. He loved classic lore; but mathematics and astronomy were from the first his favourite pursuits.

It would be impossible, in a sketch like the present, to do more than glance briefly over the principal events in the life of Sir Christopher Wren, and happily too many biographies of him exist, to render more than a rapid sketch necessary.

In his fourteenth year he was admitted as a gentleman-commoner at Wadham College, Oxford, where, by his acquirements and inventions, he gained the friendship of Dr. Wilkins, Seth Ward (Bishop of Salisbury), Hooke (whom he assisted in his *Micrographia*), and other scientific men, whose meetings laid the foundation of the Royal Society.

Thus Wren may be said to have been one of the earliest members of the club of Philosophers. In 1654 Evelyn visited Oxford, and went to All Souls', when he says, "I saw that miracle of a youth, Christopher Wren."

It deserves to be mentioned, that while Wren was at Oxford he ranked high in his knowledge of anatomical science. His abilities as a demonstrator, and his attainments in anatomy generally, are acknowledged with praise by Dr. Willis in his *Treatise on the Brain,* for which Wren made all the drawings; and he is allowed to have been the originator of the physiological experiment of injecting various liquors into the veins of living animals, which Sprat calls "a noble experiment," exhibited at the Meetings at Oxford[11].

---

[11] Dr. Clarke, in a Paper "On Anatomical Inventions and Observations,"

In 1653 he was elected a fellow of All Souls', and by the time that he had attained his twenty-fourth year his name had gone over Europe, and he was considered as one of that band of eminent men, whose discoveries were raising the fame of English science. Seldom, indeed, has the promise of youth been so well redeemed as in the case of Wren.

In August, 1657, he was appointed Professor of Astronomy at Gresham College; three years later, Savilian Professor at Oxford, and received the degree of D. C. L. in September 1661. It was after delivering his lecture on Astronomy at Gresham College, on the 28th November, 1660, that the foundation of a scientific society was discussed, and the great share he took in contributing to the success and reputation of the young Society is well known. Indeed, the archives of the Society bear the amplest testimony to his knowledge and industry, as exhibited in his commentaries on almost every subject connected with science and art. His inventions and discoveries alone are said to amount to fifty-three.

It is not very clear at what period he first commenced studying architecture: it was probably more owing to his general scientific reputation, that he was appointed in 1661 assistant to Sir John Denham, the surveyor-general, and was commissioned, in 1663, to survey and report upon St. Paul's cathedral, with a view to its restoration, or rather, the entire rebuild-

---

vations," published in the *Philosophical Transactions* for 1668, says, *Circa finem anni* 1656, *aut circiter, Mathematicus ille insignissimus, D. D. C. Wren, primus infusionem variorum liquorum in massam sanguineam viventium animalium excogitavit et Oxonii peregit.*" p. 678.

In the first Charter Wren is styled Doctor of Medicine.

ing of the body of the fabric, so as to reconcile it with the Corinthian colonnade of Inigo Jones.

Meanwhile, his knowledge of architecture was increasing, and to improve it still more, he went to Paris in 1665, to study the principal buildings in that city.

On his return, the great Fire decided the long-debated question, whether there should be a new Cathedral. What was calamitous in itself at the time, happened most opportunely for Wren. On the 2nd July, 1668, he was officially informed that the Archbishop of Canterbury, and the Bishops of London and Oxford, had resolved to have a Cathedral worthy of the reputation of the city, and the nation. It has been justly observed, that but for the Fire, Wren might have trifled away his genius, patching the old Cathedral, and perhaps adding a new wing to Whitehall.

It is well known that Wren's original design for the Cathedral contemplated one order instead of two, and no side oratories, or aisles. The first stone of the present edifice was laid June 21, 1675; the choir was opened for divine service in December 1697, and the last stone on the summit of the lantern was placed by the architect's son, Christopher, in 1710[12]. It is worthy of mention, that Wren's salary as architect was only 200*l.* a year, for which he drew all the designs, and was also at the expense of drawings and models of every part, the daily overseeing of the works, framing the estimates and contracts, and auditing the bills[13], not to mention, as the Duchess of

---

[12] St. Peter's had twelve architects, and occupied one hundred and forty-five years in building.

[13] A MS. in the Harleian collection states, that Wren entered

 into

Marlborough says in a letter, "his being dragged up in a basket three or four times a-week to the top of the building, at great hazard." His pay for rebuilding the churches in the City was only 100*l*. a year; it is however related, that on the completion of the beautiful church of St. Stephen, Walbrook, the parishioners presented his wife with twenty guineas!

His plan for rebuilding the city, had it been adopted, would have made it, as was said at the time, "the wonder of the world;" but there were too many individual interests to be combated, and the project did not go beyond paper.

One work, which would have probably not a little augmented his fame, was a design for a magnificent mausoleum to the memory of Charles I., yet though Parliament voted 70,000*l*. for the purpose in 1678, the design was abandoned, and the money applied more in accordance with the personal tastes of Charles II.

Wren was thwarted in his ideas for the monument commemorative of the Fire, which he designed in such a manner, that the shaft was to be adorned with gilt flames issuing from the loop-holes, but as no such pattern was to be found in the 'five orders,' the design was set aside for the pillar now on Fish-street Hill.

Though, as we have seen, Wren's abilities and acquirements would have adorned, as they would have fitted him for, almost any profession, his life was mainly occupied by architectural pursuits: in fact, from the time that he was appointed architect to the city and St. Paul's, he had little time for any

---

into the details connected with the great work, as examining accounts, agreeing for prices of workmanship, materials, &c.

other occupation. Mr. Cockerell, in his *Tribute to the Memory of Wren*, enumerates forty public buildings erected by that great man.

It is surprising, therefore, to find his name occurring so frequently in the Journal and Council-books of the Royal Society, presiding at the Meetings, and proposing and taking part in experiments. He also sat for some years in Parliament. According to the manuscript list of the pedigrees of knights, preserved in the British Museum, Wren was knighted at Whitehall on the 12th November, 1673. He was twice married; first to the daughter of Sir Thomas Coghill, by whom he had one son, Christopher; and afterwards, to a daughter of Lord Fitzwilliam, Baron of Lifford in Ireland, by whom he had a son and a daughter. After a long life spent, as his epitaph says, *non sibi, sed bono publico*, he retired in peace from the world to his home at Hampton Court, where the five remaining years of his existence were passed in repose, interrupted only by occasionally going to London, to superintend the repairs of Westminster Abbey, his only remaining public employment; for, to the disgrace of the government of that day, he had been deprived of his office of surveyor in 1718.

His death was as placid as the whole tenour of his existence. On the 25th of February, 1722—3, his servant found him dead in his chair. He had slumbered softly to wake in eternity. An attempt was made to compensate the denial of earthly honours in his latter days by a splendid funeral. His remains were laid within his own cathedral, surmounted by the well-known and sublimely eloquent legend, *Si Monumentum quœris, circumspice.*

Many great men have shed lustre upon the

chair of the Royal Society: few to a greater degree than Sir Christopher Wren. A remarkably fine portrait of him by Sir P. Lely, with St. Paul's in the back-ground, is in the meeting-room in Somerset House.

In 1681, Dr. Grew published his curious book, under the patronage of the Society, and Daniel Colwall, Esq., who defrayed the expense of the engravings, entitled, *Musæum Regalis Societatis; or, A Catalogue and Description of the Natural and Artificial Rarities belonging to the Royal Society, and preserved at Gresham College; whereunto is subjoyned the Comparative Anatomy of the Stomack and Guts.* The work is a folio, comprising 435 pages, and 31 sheets of plates. It is dedicated to Daniel Colwall, Esq., founder of the Museum. In the dedication, Dr. Grew hopes that the Royal Society "may always wear this Catalogue, as the miniature of Mr. Colwall's abundant respects, near their hearts;" and in another place, he says, "Besides the particular regard you had to the Royal Society itself, which seeming (in the opinion of some) to look a little pale, you intended, hereby, to put some fresh blood into their cheeks; pouring out your box of oyntment, not in order to their burial, but their resurrection."

The Catalogue is extremely curious, on account of the very quaint titles given to objects of natural history (now recognised by very different names), and the descriptions attached to them.

The Museum contained at this period several thousand specimens of zoological subjects and foreign curiosities: among the contributors, amounting collectively to 83, are the names of Prince Rupert, Duke of Norfolk, Boyle, Evelyn, Hooke, Pepys, Lord

Brouncker, and the East India, and Royal African Companies.

In the early part of 1682, Sir Christopher Wren, acting on the authority given him by the Council, sold Chelsea College and the surrounding lands to the King, for the sum of 1300*l.*[14] Small as this sum now appears, and was even at that period, for such an estate, yet the Council were so well pleased, that they voted their thanks to their President, "for the service rendered to the Society in thus disposing of a property which was a source of continual annoyance and trouble to them[15]." A long discussion arose concerning the best means of profitably investing the amount, which ended by a resolution to place it in the hands of the East India Company.

Thus strengthened in funds, the Council turned their attention to the expediency of putting a stop to the too indiscriminate and easy admission of Fellows, and with this view, drew up the following statute, which, after being fully debated, was put to the vote, and passed on the 5th Aug. 1682:—

"Every person that would propose a Candidate, shall first give in his name to some of the Council,

---

[14] We read in the Council-Minutes, that the College and lands "might have been well disposed of, but for the annoyance of Prince Rupert's glass-house which adjoined it." Sir Jonas Moore wrote to the Prince, at the request of the Council, urging him to "consider the Society, on account of the mischief that his glass-house was doing to the College," and this letter was supported by similar remonstrances from Sir R. Southwell and Mr. Pepys.

[15] The Royal Hospital was built upon the site of the College. The first stone was laid by the King, on the 16th February, 1681—2, and Sir C. Wren was appointed commissioner and architect in 1683.

that so in the next Council it may be discoursed, *vivâ voce,* whether the person is known to be so qualified, as in probability to be useful to the Society. And if the Council return no other answer, but that they desire farther time to be acquainted with the gentleman proposed, the proposer is to take that for an answer: and if they are well assured that the Candidate may be useful to the Society, then the Candidate shall be proposed at the next Meeting of the Society, and ballotted for, according to the Statute in that behalf; and shall immediately sign the usual bond, and pay his admission-money upon his admission."

At the same time it was resolved, "That no person shall be capable of being chosen into the Council, who hath not, at or before the 10th day of November preceding the election, accounted with the Treasurer, and paid his dues to the Michaelmas before: and in order thereunto, the names of those who have not paid till the Michaelmas preceding, shall not be inserted in the printed lists for the use of the Society, at the election day."

On the 13th December, 1682, it was "Ordered, that Dr. Grew take upon him the care of the Repository, under the name of *Præfectus Musei Regalis Societatis,* &c., and that he make a short Catalogue of the Raritys, with a method for the ready finding them out: as also a Catalogue of the Benefactors, and the particulars given by them. That he enter into a book all such things as shall be given hereafter, with the name of the donor, and from time to time observe what may be necessary for the preservation and augmentation of the said Repository, and make a report thereof to the Councill. And that he bring in to the usual Meetings of the Society such descriptions of naturall things there contained, as have not yet been published in his book." The Doctor readily accepted the office.

At the anniversary in this year, Wren retired from the Presidency, and Evelyn was solicited to allow himself to be put in nomination as his successor, but declined; he says in his *Diary*, under the date of November 30, 1682, "I was exceedingly indanger'd and importun'd to stand the election (for President of the Royal Society), having so many voices; but by favour of my friends, and regard of my remote dwelling, and now frequent infirmities, I desir'd their suffrages might be transferr'd to Sir John Hoskyns, one of the Masters of Chancery, a most learned virtuoso, as well as lawyer, who accordingly was elected[16]."

Sir John Hoskyns, of Harewood, in the county of Hereford, Bart., was grandson of Judge Hoskyns, a noted poet and critic in the reign of James I., and son of Sir Bennet Hoskyns, by his first wife Anna Bingley, daughter of Sir John Bingley[17]. He was born in 1633, and received his early education from his mother, who taught him Latin; "meaning," as the manuscript memoir of him states, "as 'tis supposed, that she familiarised that language to him in such a manner, as made the acquiring it the more easy. He was sent afterwards to Dinedor School, and from thence to Westminster, where he was scholar under Dr. Busby, from whom, 'tis most remarkable, he never received a blow."

He entered as student of the Temple, and although

---

[16] Vol. I. p. 512.

[17] Man. Brit. Mus. Sloane Collec. 4222.

called to the bar, did not practise. He was, however, eminent for his legal attainments, and esteemed for his invincible integrity in the discharge of official duties. He was appointed a Master in Chancery, which office he held until a year before his decease.

But Sir John Hoskyns was much better known to the world as a philosopher than a lawyer, and especially in the latter part of his life, when he devoted a considerable portion of his time to scientific pursuits and experiments. His general knowledge is said to have been very considerable, and to have been imparted freely and cheerfully to all who applied to him for information. Granger observes, "There was nothing at all promising in his appearance; he was hard favoured, affected plainness in his garb, walked the streets with a cudgel in his hand, and an old hat over his eyes. He was often observed to be in a reverie, but when his spirits were elevated over a bottle, he was remarkable for his presence of mind and quickness of apprehension, and became the agreeable and instructive companion. He was an excellent Master in Chancery, and a man of an irreproachable character. He was more inclined to the study of the new philosophy, than to follow the law, and is best known to the world as a virtuoso[18]." Le Neve says that "Hoskyns was much noted for his general knowledge, and vigorous searching after natural philosophy[19]."

Sir John Hoskyns was one of those elected by the Council on the 20th May, 1663, by virtue of the power given them by the Charter. He acted as Vice-

---

[18] *Biog. Hist. of England*, Vols. II. and III. pp. 267 and 539.
[19] *Mon. Angli.* p. 202.

President during the Presidency of Sir C. Wren, and we find him frequently occupying the chair, and attending the Council-Meetings. He married a daughter of Sir Gabriel Lowe, of Gloucestershire, by whom he had several children.

He represented the county of Hereford in Parliament for many years, and died on the 12th September, 1705.

In January, 1683, the publication of the *Philosophical Transactions* was resumed, and the 143rd Number published under the editorship of Dr. Robert Plot, who had been chosen Secretary in place of Hooke[20]. An explanatory Preface commences this number:—

"Although the writing of these *Transactions* is not to be looked upon as the business of the Royal Society, yet, in regard they are a specimen of many things which lie before them, contain a great variety of useful matter, are a convenient Register for the bringing in and preserving many experiments, which, not enough for a book, would else be lost, and have proved a very good ferment for the setting men of uncommon thoughts in all parts a-work; and because, moreover, the want of them for these four last years, wherein they have been discontinued, is much complained of; that the said Society may not seem now to condemn a work they have formerly encouraged, or to neglect the just expectations of learned and ingenious men, they have

---

[20] The Council-Minutes record, that it was "ordered the Treasurer should buy sixty coppys of the *Transactions*, at the current price, for the use of the Society." Vol. II. p. 24. The publication of the *Transactions* ceased again for a temporary period in 1687.

therefore thought fit to take care for the revival hereof, that they may be published once every month, or at such times whereof forenotice shall be given at the end of these and the following *Transactions.* Neither is it doubted, but that those who desire to be accommodated herewith, will most earnestly endeavour themselves, or by others, to supply and keep up that stock of experiments, and other philosophical matters, which will be necessary hereunto; with this assurance given them, that whatever they shall be pleased to communicate, shall be disposed of with all fidelity."

During the short Presidency of Sir John Hoskyns, who filled the chair at every Meeting, much activity prevailed, and numerous experiments in natural philosophy were performed. Those of Halley on magnetism are extremely curious. His object in making them was to illustrate his favourite theory of four magnetic poles[21]. At one of the Meetings, a great number of curiosities from China were presented by Captain Knox, which formed a valuable addition to the Museum. The description of some of these "rarities," is not a little amusing. Thus we have "A true dolphin's skin caught by the captain, and stuffed, being very different from the porpess, which is commonly here called and esteemed to be the dolphin."

"A sprig of a shrub called *Ki-vong*, of strange virtue; for being put into water, it driveth the crabs from it; and being put to the mouth of the holes where they have burrowed themselves, they immediately run out, or are killed in their holes. And it is for that end used by the natives to catch them."

---

[21] See *Phil. Trans.*, Vol. XII. p. 208.

The whole list shows how little was known of foreign countries and their productions at that period.

On November 30, 1683, Sir John Hoskyns resigned, and was succeeded in the chair by Sir Cyril Wyche. This gentleman was the second son of Sir Peter Wyche, and was born at Constantinople, whilst his father was ambassador from the English government to the Porte[22]. He was named after Cyril, the Patriarch of Constantinople, distinguished for his literary attainments, who is said to have sent to England for a fount of types, to set up a Greek press at Constantinople, where a Greek translation of Jewell's *Apology* was printed. A short time after he was murdered by the Jesuits[23].

Sir Peter Wyche, the eldest son, and brother of the subject of this memoir, was one of the original Fellows of the Royal Society, and translated from a Portuguese Manuscript a short account of the river Nile, at the desire of the Society[24]. His attention was frequently directed to the extension of geographical knowledge. His brother Cyril, who was also one of the early members of the Society, studied at Christ Church, Oxford, and became M.A. and LL.D. in 1665. He sat frequently as parliamentary representative for Kellington, in Cornwall; held the office of Secretary for Ireland, and took so active a part in the affairs of that country, as to be appointed one of

---

22 He was a great favourite of Sultan Amurath IV., who styles him in his letters to King Charles, "the famous lord, whose end be happy." During his ambassadorship he contrived to get the duty taken off English cloth.

23 *Fasti Oxon.*, Vol. XI. p. 163.

24 Published in 1669.

the Lords Justices in 1693. In 1692 he married the niece of Evelyn, who calls him "a man of perfect integrity, and a noble and learned gentleman." He purchased the estate of Poyning's Manor, and other lands at Hockwold, in Norfolk, where he died December 29, 1707.

In April 1684, Dr. Denis Papin, who had frequently exhibited experiments before the Society, and was known as the inventor of the "Bone-Digester[25]," was appointed temporary Curator, with a salary of 30*l.* per annum; in consideration of which he was required to produce an experiment at each Meeting of the Society. Evelyn, in his *Diary*, gives an amusing

---

[25] Papin published, under the patronage of the Royal Society, an account of this Machine, which he entitled, *A New Digester, or Engine for softening Bones, containing the description of its make and use in these particulars, viz. Cookery, Voyages at Sea, Confectionary, making of Drinks, Chemistry, and Dyeing, with an account of the price a good big Engine will cost, and of the profit it will afford.* 4to. London, 1681. In 1687 he published a further description of the Machine with improvements, which were occasioned, as he states, by the King having commanded him to make a Digester for his Laboratory at Whitehall. The invention appears to have excited a considerable degree of interest. In the Preface to the last-mentioned work, Papin says; "I will let people see the Machines try'd once a week, in Blackfriars, in Water Lane, at Mr. Boissonet's, over against the Blew Boot, every Monday at three of the clock in the afternoon; but to avoid confusion and crouding in of unknown people, those that will do me the honour to come, are desired to bring along with them a recommendation from any of the Members of the Royal Society." Subsequently, he adapted the piston of the common sucking-pump to a steam-machine, making it work in the cylinder, and applying steam as the agent to raise it. It is a curious fact, that, although Papin invented the safety-valve, he did not apply it to his steam-machine. We shall have occasion to allude to his communications on the steam-engine in a subsequent part of this work.

account of a supper prepared by Papin's Digesters, to which he was invited with other Fellows.

"Went this afternoone with severall of the R. S. to a supper, which was all dress'd, both fish and flesh, in *M. Papin's Digesters,* by which the hardest bones of beef itselfe, and mutton, were made as soft as cheese, without water or other liquor, and with lesse than eight ounces of coales, producing an incredible quantity of gravy; and, for close of all, a jelly made of the bones of beef, the best for clearness and good relish, and the most delicious that I have ever seen or tasted. We eat pike and other fish-bones, and all without impediment: but nothing exceeded the pigeons, which tasted just as if bak'd in a pie: all these being stewed in their own juice, without any addition of water, save what swam about the Digester, as *in balneo;* the natural juice of all these provisions acting on the grosser substances, reduc'd the hardest bones to tendernesse. This philosophical supper caus'd much mirth amongst us, and exceedingly pleas'd all the company. I sent a glasse of the jelley to my wife, to the reproach of all that the ladies ever made of the best hartshorn[26]."

At one of the Meetings of the Society, Dr. Papin exhibited several descriptions of jellies, made by his pneumatic engine, which he declared to have been prepared at one-third of the usual cost[27]. From the period of Papin's engagement as Curator until 1687, when he was appointed Professor of Mathematics in the University of Marburg, he was very zealous in providing "*entertainments*" for the Society. These principally consisted in Mechanical, Pneumatic, and

[26] Vol. I. p. 509.

[27] See Register-book, Vol. VI. p. 208.

Hydrostatic experiments, an account of which will be found in the *Philosophical Transactions*, and Register and Journal-books[28].

His duties at Marburg did not prevent his occasionally sending philosophical communications to the Society, and to the *Acta Eruditorum* of Leipsic. It was in the latter work that the celebrated addition to his paper, on the use of gunpowder, communicated to the Royal Society in 1687, first appeared, in which he proposes to use steam as a moving power[29]. It is thus noticed in the *Philosophical Transactions*, No. 226.

"A method of draining mines where you have not the conveniency of a river to play the engine; and, having touched upon the inconveniency of making a vacuum by gunpowder, the author proposes the alternately turning a small surface of water into vapour, by fire applied to the bottom of the cylinder that contains it, which vapour forces up the plug in the cylinder to a considerable height, and which, as the vapour condenses as the water cools when taken from the fire, descends again by the air's pressure, and is applied to raise the water out of the mine[30]."

This invention is highly creditable to Papin, and though much remained to perfect the engine, yet the philosophical principle is pointed out. The enormous strength required for his Digesters, and the means to which he was obliged to resort, for the purpose of confining the covers, must have early shewn him

---

[28] A complete account of these and other inventions by Papin, will be found in his *Recueil de diverses pièces touchant quelques nouvelles Machines*. 8vo. Cassel, 1695.

[29] 1690, p. 410.

[30] P. 482.

what a powerful agent he was using. It was while making many experiments for Boyle, that he discovered if vapour be prevented from rising, the water becomes hotter than the usual boiling point. This led to the invention of his Digester[31].

In 1684, the Society lost another of its original and valuable members, in the person of Dr. Croone[32], who died of a fever, on the 12th October, in that year. Besides contributing some important Papers to the Society on Anatomy, he left a plan of two lectureships which he designed to found; one, to be read before the College of Physicians, with a sermon to be preached at the church of St. Mary-le-Bow; the other, to be delivered yearly before the Royal Society, upon the nature and laws of muscular motion. But, as his Will contained no provision whatever for the endowment of these lectures, his widow, who subsequently married Sir Edwin Sadleir, Bart., carried out his intention by devising in her Will, dated Sept. 25, 1701, that "one-fifth of the clear rent of the King's Head Tavern, in or near Old Fish-street, London, at the corner of Lambeth Hill[33], be vested in the Royal Society, for the support of a lecture and illustrative experiment, for the advancement of natural knowledge on local motion, or (conditionally) on such other subject as, in the opinion of the President for the time being, should be most useful in promoting the objects for which the Royal Society

---

31 Papin presented one of his Digesters to the Society, with a very interesting letter descriptive of it.

32 So his name is signed in the Charter-book, but it is still in his will *Croune*. Printed books have it *Cron*, *Croun*, *Croone* and *Crone*.

33 This property was let on a lease of 99 years in 1786 for 15*l*. per annum; consequently the Society's share amounts to 3*l*.

was instituted[34]." Although prospectively, I may remark here, that the College of Physicians made a proposition to the Society, in 1739, to "get rid of Lady Sadleir's donation;" but the Council were of opinion that the transfer could not be legally made. The first lecture was read in 1738 by Dr. Stuart, the subject being "on the Motion of the Heart;" and it was the custom, for many years afterwards, for the authors to read their papers to the Society. It has been well observed, in the report above quoted, that "this lecture, as founded by the relict of the first Register to the Royal Society in 1660, before its incorporation, while it consisted only of the few, but illustrious, names of Brouncker, Boyle, Wilkins, Petty, Wren, Evelyn, Wallis, Oldenburg, and other successors of Lord Bacon, who met weekly in Gresham College, may be regarded as being coeval with the Society's first formation and establishment; and cannot fail at the present day to be considered as one of its most valuable foundations."

It will not be uninteresting to notice here, the impetus which the study of Geology received in 1683 and 1684, by the labours of Dr. Martin Lister, and the gift from Lord Clarendon, of a "large parcell of oares from New England." Dr. Lister's Paper is extremely curious. It is entitled, *An Ingenious Proposal for a new sort of Maps of Countrys, together with Tables of Sands and Clays.* The Paper which was laid before the Society in March, 1683—4, is published in the 14th volume of the *Transactions.*

The author commences: "We shall be better able to judge of the make of the earth, and of many phe-

---

[34] *Report on Medals of Royal Society*, p. 43.

nomena belonging thereto, when we have well and duely examined it, as far as human art can possibly reach, beginning from the outside downwards. As for the most inward and central parts thereof, I think we shall never be able to confute Gilbert's opinion, who will, not without reason, have it altogether *iron*[35]."

The annexed extract of a letter from Mr. Aston to Dr. Plot at Oxford, gives some further account of Dr. Lister's ideas. It is dated London, March 13, 1683—4.

"I received from Mr. Lister two schemes of the lands and clays found in England, made by himself about twenty years since. He mentioned besides the great advantage of a map of the earths peculiar to some places and counties; he considers the sands and clays as two of the coats of the earth; the sand, probably, the uppermost coat, (for some reasons he gives), whence it comes to be washt to the body of rivers and the sea-shore. By this opinion I perceive may be given an account of sand-beds, too often attributed to the sea[36]."

Dr. Lister's scheme for a Map of England, distinguishing the soils and their boundaries by colours, has certainly the merit of priority. "He was the first," says Mr. Lyell in his *Geology*, "who was aware of the continuity over large districts, of the principal groups of strata in the British series, and who proposed the construction of regular geological maps[37]."

---

35 *De Magni.* Lib. I. cap. 17. *Tellus in interioribus partibus magneticam homogenicam naturam habet.*

36 Ashmolean MSS. No. 1813.

37 Vol. I. p. 45, third edition.

At the anniversary in 1684, Sir Cyril Wyche resigned the Presidency, and Samuel Pepys, Esq. was elected his successor. During the last few years, as we have seen, the Society were extremely active and diligent in procuring information on all matters relating to philosophy. The Meetings were continued with great regularity, and we have a description of one, in the following extract from a letter written by Sir Thomas Molyneux, Bart., F.R.S., to his brother Mr. William Molyneux at Dublin[38]. The letter is dated May 26, 1683.

"The 23rd being Wednesday, I was at Gresham College; there I saw the *Bibliotheca Korfolcktiana*[39]; afterwards I went to the Repository, and viewed the rarities of that place, which do very much increase, there being new additions daily made. The Royal Society meeting together whilst I was in the house, by the favour of Mr. Haak and Dr. Green I was admitted to sit among them; the ceremony observed at their meeting is this: The President, one Sir John Hoskins, sits in a chair at the upper end of a table, with a cushion before him; the Secretary, Mr. Aston, a very ingenious man, at the side, on his left hand; he reads the heads one after another to be debated and discoursed of at the present Meeting; as also, whatever letters, experiments, or informations, have been sent in since their last Meeting; of all which, as they are read, the Fellows which sit round the room, spake their sentiments, and give their opinions, if they think fitting; and of the chief matters discussed at this time was the cause of the inundation of the Nile. When this is

[38] Published in *Dub. Univ. Mag.* 1841.

[39] Sir Thomas means *Norfolciana*.

over, if any of the company have made experiments, or have had particular information concerning any thing worthy the notice of the Society, they then make it known. At this convention, it happened that young Mr. Halley, the astronomer, made some magnetical experiments, in order for the laying down of an hypothesis to solve the variation of the needle in several parts of the world: they thanked him for them, and desired him to set down his notions in writing, which I believe you will have in some of their following *Transactions*.

"His hypothesis runs all upon the supposition, that there are four magnetical poles in the four quarters of the earth, which he proves by observations made by himself and others in long voyages. I had a great deal of discourse with him. He asked me for you, and whether you went on in your astronomical studies; he told me that he had heard some of the R. S. say, that wood turned into stone by Lough Neagh had a magnetical virtue: this I would have you try, and let me know your success in your next. At this meeting I had the opportunity of seeing several noted men, as Mr. Evelyn, Mr. Hooke, Mr. Isaac Newton, Dr. Tyson, Dr. Slare, &c. Round the room was hung the pictures of these men: Lord Brouncker, Bishop Wilkins, Gunter the mathematician, Signor Malpigi, sent by himself from Italy, and Mr. Hobbs. Over the chimney stood their own Arms."

In another letter, Sir T. Molyneux gives some account of distinguished Fellows of the Society, whose acquaintance he had made. "Dr. Green," he says, "is a very civil, obliging person; Hooke, the most ill-natured, conceited man in the world, hated and despised by most of the Royal Society, pretending to have all other inventions, when once discovered by their authors to the world; Dr. Tyson, a most understanding anatomist; Dr. Croone, and Dr. Slare, both extraordinary, civil, and ingenious men; the first a very exact observer of the weather, in whose study I saw several thermometers, hugroscopes, and baruscopes." Sir T. Molyneux appears to have been a keen observer; his letters are full of interesting matter, collected during his travels in England and on the Continent. But we must revert to the Society and their new President, a memoir of whom will be found in the next Chapter.

---

## CHAPTER XI.

Memoir of Samuel Pepys—Establishment of the Dublin Philosophical Society—Their Rules—Auxiliary to Royal Society—Mr. Aston resigns the Secretaryship—New Office created—Appointment of Halley as Clerk—His Duties—Attempt to establish a Philosophical Society at Cambridge—Newton's Letter on the subject—Death of Charles II.—His indifference to the Society—Sends receipt for curing Hydrophobia—Manuscript of *Principia* presented to Society—Halley's Letter respecting it—Council order it to be printed—Halley undertakes its publication—Correspondence with Newton—Facsimile of Title-page—Pepys resigns—Lord Carbery chosen President—Memoir of him—Hooke proposes to deliver a weekly Lecture—The Society in debt—Obliged to pay for Apartments in Gresham College—Professors let their Rooms—Scientific Business—Lord Carbery resigns—Lord Pembroke elected President.

---

1680—90.

SAMUEL PEPYS was born on the 23rd February, 1631—2, of a family which, he honestly acknowledges, "had never been very considerable." His father, John Pepys, was a citizen of London, where he followed the trade of a tailor. It appears, by his *Diary*, that Pepys passed his early days in or near the metropolis, and was educated at St. Paul's School; he remained there till 1650, early in which year his name occurs as a sizar on the books of Trinity College, Cambridge. Before he took up his residence at that University, he had removed to Magdalen College. There is no evidence to show how long he remained at Cambridge, nor what were his academical pursuits. In October, 1655, when only twenty-three years of age, he married Elizabeth St. Michel, a girl of fifteen, whose father is described as having been of a good family, and whose mother was a descendant of the Cliffords of Cumberland. The young couple were kindly noticed by Sir Edward Montagu, afterwards Earl of Sandwich, who received them into his house; to this gentleman Pepys was indebted for his subsequent advancement. In 1658 he accompanied Sir Edward in his expedition to the Sound, and, on his return, became a clerk in the Exchequer. About this period he commenced his celebrated and interesting *Diary*. In 1660 he was appointed Clerk of the Acts of the Navy, which commenced his connexion with a great national establishment, to which his diligence and acuteness were afterwards of the highest service. It is recorded, that when the metropolis was nearly deserted on account of the Plague, the whole management of the navy devolved on him, and he remained at his post, regardless of the danger which surrounded him. In a letter to Sir W. Coventry at this period, he observes, "The sickness in general thickens round us. You, Sir, took your turn at the sword; I must not, therefore, grudge to take mine at the pestilence." During the awful Fire in London he also rendered most essential service.

The Duke of York being Lord High Admiral, Pepys was by degrees drawn into close personal connexion with him, and as he enjoyed his good opinion, had also the misfortune to experience some part of the calumnies with which the Admiral was loaded during the time of the Popish Plot. The absence of evidence did not prevent his being thrown into the Tower (May, 1679), on the charge of instigating and abetting, and he was for a time removed from the Navy Board. His liberation took place in February following, and soon after he was fortunate in attracting the favourable notice of the King, who made him Secretary to the Admiralty. He filled this office during the remainder of Charles II.'s reign, and the whole of that of his successor.

Upon the accession of William and Mary, Pepys lost his official employments, and the electors of Harwich, unmindful of his having served them in Parliament, refused to return him to the convention. He retired consequently into private life, and was desirous of passing the remainder of his days in the enjoyment of scientific and literary society, for which his various acquirements peculiarly qualified him. But his enemies, actuated by malice, caused him to be committed to the Gatehouse in 1690, on pretence of his being affected to King James; he was soon permitted, however, on account of ill health, to return to his own house, where he resided until 1700. In that year his physicians persuaded him to retire, for the sake of change of air and repose, to the seat of his old friend and servant, William Hewer, at Clapham, where he expired, after a lingering illness, May 26, 1703.

Pepys was a very remarkable man. Of his official life it could not be said,

*Initia magistratuum nostrorum meliora et firma finis inclinat;*

for the same zeal and energy which marked his entrance into office was conspicuous throughout the whole of his career. In fact, his skill and experience in naval affairs could not be dispensed with by Government; and for many years the whole management of the Admiralty was borne by him.

Yet, amidst all his official business, surrounded by political intrigues and court dissipation, he contrived to find time for scientific and literary pursuits.

He was one of the earliest Fellows of the Royal Society, having been elected in 1663, from which period, until within a very few years of his death, his name frequently occurs in the Council and Journal-books.

At the anniversary in 1667, he says, in his *Diary*, "I was near being chosen of the Council, but am glad I was not, for I could not have attended, though above all things I could wish it; and do take it as a mighty respect, having been named." Shortly after, he was elected into the Council, on which he often served, and, as we have seen, was chosen President in 1684, on account of his high literary attainments, and probably his court-influence. "After he resigned the Presidency, he was in the habit of entertaining the most distinguished Fellows, on Saturday evenings, at his house in York Buildings, where they assembled for the discussion of literary subjects, and the encouragement of the liberal arts. To the dissolution of these meetings, occasioned by the increasing infirmities of their founder, Evelyn adverts in his letters, in terms of the strongest regret; nor could a person of his enlightened mind fail to derive the most heartfelt gratification from witnessing so many of his contemporaries eagerly devoting the small portion of their lives that remained to the cultivation of science, and the acquirement of useful knowledge[1]."

Pepys was a munificent patron of literature. To Magdalen College, Cambridge, he left an invaluable collection of naval memoirs. "These," says Lord Braybrooke, "he had obtained at immense cost, for the general history of the *Navalia* of England, which he had promised to the public; but age and ill health

[1] Life, prefixed to Pepy's *Diary*, p. 44.

intervening, he was deprived of the vigour and opportunities requisite for completing the work." Besides these memoirs he left to the same college a large collection of prints and ancient English poetry, amongst which are five large folio volumes of English ballads.

He contributed sixty plates to Willughby's *Historia Piscium*[2], and presented the Royal Society with 50*l.* the year after he was elected President. Indeed, nearly all his leisure time appears to have been occupied by scientific and literary pursuits. He felt the force of the words,

*Otium sine literis mors est, et hominis vivi sepultura;*

and, had his health[3] been better, he would, in all probability, have been a more diligent labourer.

A fine portrait of him, by Kneller, is in the Meeting-room of the Society.

In 1684, a Society, consisting of about twenty individuals, was established in Dublin[4], whose objects

---

[2] Thus alluded to in the work:—*Amplissimus Vir D. Samuel Pepys, Societatis Regiæ Præses, ingenuarum Artium et Eruditorum fautor et patronus eximius, qui operi illustrando exornandoque Icones plurimas ad Tabulas usque sexaginta, privatis impensis et proprio ære sculptas, raro magnificentiæ exemplo largitus est.* The book is dedicated to Pepys.

[3] He appears never to have recovered from the disease of stone, for which he underwent an operation in early life; he was in the habit of celebrating the anniversary of this event with religious gratitude to Providence.

[4] It appears by the "Life of Sir Thomas Molyneux," in the *Dublin University Magazine*, before quoted, that his brother William originated this Society. In a letter written by the latter to his brother, dated Dublin, Oct. 30, 1683, N.S., he says, "I have also here promoted the rudiments of a Society for which I have drawn up Rules, and called it *Conventio Philosophica.* About half-a-score or a dozen of us have met about twelve or fifteen times, and we

were similar to those of the Royal Society. Indeed, it may be regarded as an auxiliary of the latter, to which it regularly transmitted copies of its Minutes and Philosophical Communications; these were read at the ordinary Meetings of the Royal Society, and are preserved amongst its archives.

The following *Advertisements*, as they are called, were drawn up for the Society by Sir William Petty, who was elected President[5]:—

"1. That they chiefly apply themselves to the making of experiments, and prefer the same to the best discourses, letters, and books, they can make or read, even concerning experiments.

"2. That they doe not contemne and neglect common, triviall, and cheap experiments and observations.

"3. That they provide themselves with rules of number, weight, and measure; not only how to measure the plus and minus of the qualitys and schemes of matter; but to provide themselves with scales and tables, whereby to measure and compute such qualitys and schemes in their exact proportions.

"4. That they divide and analyze complicate matters into their integrall parts, and compute the proportions which one part bears to another.

"5. That they be ready with instruments and other

we have very regular discourses, concerning philosophical, medical, and mathematical matters. Our Convention is regulated by one chief, who is chosen by the votes of the rest, and is called *Arbiter Conventionis*, at present Dr. Willughby, (the name president being yet a little too great for us). What this may come to, I know not; but we have hopes of bringing it to a more settled Society."

[5] He had been frequently on the Council of the Royal Society: at this time he was devoting considerable attention to the improvement of his estates in Ireland, and the promotion of trade.

apparatus, to make such observations as doe rarely offer themselves, and doe depend upon taking opportunities.

"6. That they provide themselves with Correspondents in severall places, to make such observations as doe depend upon the comparison of many experiments, and not upon single and solitary remarks.

"7. That they be ready to entertain strangers and persons of quality with great and surprizing experiments of wonder and ostentation.

"8. That they carefully compute their ability to defray the charge of ordinary experiments fforty times per annum, out of their weekly contributions, and to procure the assistance of Benefactors for what shall be extraordinary, and not pester the Society with useless or troublesome Members for the lucre of their pecuniary contribution.

"9. That whoever makes experiments at the publick charge, doe first ask leave for the same.

"10. That persons (tho' not of the Society) may be assisted by the Society to make experiments at their charge, upon leave granted.

"11. That for want of experiments, there shall be a review and rehearsall of experiments formerly made[6]."

It was the custom, as we learn, of some of the members, including Sir William Petty, the Bishop of Ferns, and Dr. Willughby, to meet every Sunday-evening to discourse upon theology; and, in the words of Dr. Huntington's letter to Dr. Plot, "endeavour to

---

[6] Sir W. Petty also drew up "a Catalogue of mean, vulgar, cheap, and simple experiments" for the Dublin Society—it is printed in the 168th number of the *Phil. Trans.;* he always took great interest in the Society, up to the period of his death (which occurred in 1687).

establish religion, and confute atheism, by reason, evidence, and demonstration."

The Royal Society hailed the establishment of the Dublin Society with great pleasure, and directed their Secretary to write, tendering their assistance towards promoting the objects of the new Association[7].

Several letters passed between Mr. William Molyneux and Halley, in which the philosophical labours of the two Societies were detailed. In a communication from the former to the latter, written in 1686, he says:—

"I thank you exceedingly for your philosophical communications, and your kind promise of the continuance of them. I wish I may in any wise be able to make you suitable returns, but that I must despair of yet; I promise you, however, that nothing shall happen here worth your notice, which I shall not timely communicate. You may have heard of a girl in this town, strangely overgrown with excrescences, vastly numerous and very large; my next shall bring you the sketches of her, as well as my rude hand can draw them."

In the latter part of 1685 a change was made in the official constitution of the Society, occasioned by Mr. Aston's resigning his office of Secretary, to which he had been re-elected on the 30th November in the same year. By a letter from Halley to Mr. William Molyneux, dated London, March 27, 1686, it appears that Mr. Aston threw up the Secretaryship in so sudden and violent a manner, that the Council resolved

---

7 About the close of 1684 the Dublin Society numbered 33 Members. The Council of the Royal Society passed a resolution, that all Members of the Dublin Society, who were Fellows of the Royal Society, should only pay half the usual amount of subscription.

not to run the risk of being similarly treated on any future occasion, and determined on having an officer more immediately under their command. Halley's letter will better explain the circumstances of the case:—

"The history of our affairs is briefly this. On St. Andrew's day last, being our anniversary day of election, Mr. Pepys was continued President, Mr. Aston, Secretary, and Dr. Tancred Robinson chosen in the room of Mr. Musgrave; every body seemed satisfied, and no discontent appear'd anywhere, when on a sudden Mr. Aston, as I suppose willing to gain better terms of reward from the Society than formerly, on December 9th, in Council, declared that he would not serve them as Secretary; and therefore desired them to provide some other to supply that office; and that after such a passionate manner, that I fear he has lost several of his friends by it. The Council, resolved not to be so served for the future, thought it expedient to have only Honorary Secretaries, and a Clerk or Amanuensis, upon whom the whole burthen of the business should lie; and to give him a fixt salary, so as to make it worth his while; and he to be accountable to the Secretaries for the performance of his office; and on January 27th last, they chose me for their under-officer, with a promise of a salary of fifty pounds per annum at least[8]."

Halley was not elected without opposition. The candidates for the new office were, Dr. Papin, Dr. Sloane, Mr. Salisbury, and Mr. Halley. The number of Fellows present amounted to 38. Dr. Sloane had 10 votes, Dr. Papin 8, Mr. Salisbury 4, and Mr. Halley 16; but the majority of the members present being requisite to an election, the ballot was repeated,

[8] Supp. to Letter-books, Vol. IV. p. 330.

when Dr. Sloane had 9 votes, Dr. Papin 6, and Mr. Halley 23. Previous to Halley's election, the following resolutions were agreed to by the Council, respecting the new office :—

"1. If a Fellow of the Society be chosen into the office of Clerk, he shall, before his admission to his office, resign his Fellowship.

"2. If any person other than a Fellow shall be chosen Clerk, he shall be incapable of being chosen a Fellow while he holdeth the office of Clerk.

"3. That he shall have no other employment.

"4. That he shall constantly lodge in the College where the Society meeteth.

"5. That he shall be a single man without children.

"6. That he shall obey all orders from the President, Council, or Secretaries.

"7. That he shall be master of the English, French, and Latin tongues.

"8. That he shall be able to write a fair and legible hand.

"9. That he shall be completely seen in the Mathematics and Experimental Philosophy.

"10. That all letters of philosophical correspondence shall be signed by one of the Secretaries, and not by the Clerk.

"11. That the Clerk shall be accountable to the Council for the performance of his office, as it shall be from time to time appointed to him.

"12. That his salary for copying, entering, and the performance of all other parts of his office, shall be after the rate of 50*l.* per annum at least; he being found as above, and performing his duty to the satisfaction of the Council.

"THE DUTY OF THE CLERK.

"1. He shall take the Minutes of the Society in a book, and not on loose papers.

"2. He shall draw up the Minutes at large against the next Meeting.

"3. He shall enter the Minutes, after they have been read at the board, in the Journal-books.

"4. He shall draw up all letters, and bring them to be signed by the Secretaries.

"5. He shall index the books of the Society.

"6. He shall keep a Catalogue of all gifts to the Society."

Sir John Hoskyns and Dr. Thomas Gale were elected Secretaries in place of Mr. Aston and Dr. Robinson, and it was "Ordered, that Mr. Aston be presented with a gratuity of £60, and Mr. Musgrave, who had held the office of second Secretary from 1684 to 1685, with a piece of plate of 60 oz., with the thanks of the Society, and their Arms upon it."

An attempt was made, during this year, to establish a Philosophical Society at Cambridge, which was to co-operate with the Royal Society. The annexed letter, from Sir Isaac Newton, will best explain the cause of the plan being abandoned. It is addressed to Mr. Aston.

"*Cambridge, Feb.* 23, 1684—5.

"THE designe of a Philosophical Meeting here, Mr. Paget, when last with us, pusht forward, and I concurred with him, and engaged Dr. More to be of it; and others were spoke to partly by me, partly by Mr. Charles Montague; but that which chiefly dasht the business, was the want of persons willing to try experiments, he whom we chiefly rely'd on refusing to concern himself in that kind. And more what to add further about this business I know not; but only this, that I should be very ready to concur with any persons for promoting such a designe, so far as I can do it without engaging the loss of my own time in those things[9]."

This affords evidence that the spirit of philosophic inquiry had been awakened by the labours of the energetic band at Gresham College; who held on their course undaunted by difficulties sufficient to damp the ardour of any one less warm in the noble cause in which they were engaged.

The year 1685 was marked by the death of Charles the Second, whose name will ever be honourably associated with the Royal Society, as its founder and earliest patron[10]. During the latter years of this Monarch's reign, court intrigues and pleasures so entirely engrossed his time and attention, that in all probability the Royal Society, in all its struggles to advance science, was scarcely ever thought of. The only record contained in the archives of the Society, of any communication between the Fellows and the King, during a period of several years previously to his decease, consists in his having ordered Sir Robert Gourdon, F.R.S., to send the Society a recipe to cure hydrophobia, invented by his physician Thomas Frasier[11].

---

[9] Letter-book, Vol. x. p. 28.

[10] "If," says Dr. Sprat, "the first Monarchs deserved a sacred remembrance for one natural or mechanical invention, your Majesty (alluding to Charles II.) will certainly obtain Immortal Fame, for having established a perpetual succession of Inventors." Ded. of *Hist. of Royal Society* to Charles II.

[11] This recipe is so curiously illustrative of the times, that I extract it from the Archives of the Society: "Agrimony roots,

Thus neglected by the Sovereign, and occupied in pursuits so totally at variance with those of the Court, it will not be very surprising that the decease of Charles the Second is not alluded to in the Council or Journal-books. The King died on the 6th of February, 1684—5, and the Society met as usual on the 9th of the same month; the Minutes contain no reference to the Monarch's death, and they are equally silent respecting any endeavours to gain the patronage of his successor, James the Second.

At the ordinary Meeting on the 28th April, 1686, Dr. Vincent presented the Society with the MS. of the first book of Newton's immortal work, entitled, *Philosophiæ Naturalis Principia Mathematica,* which the illustrious author dedicated to the Society. It was "ordered that a letter of thanks be written to Mr. Newton, and that the printing of his book be referred to the consideration of the Council; and that in the mean time, the book be put into the hands of Mr. Halley, to make a report thereof to the Council." On the 19th May following, the Society resolved, that "Mr. Newton's *Philosophiæ Naturalis Principia Mathematica,* be printed forthwith in quarto, in a fair letter; and that a letter be written to him to

primrose roots, dragon roots, single peony roots, the leaves of box, of each a handfull; the starr of the earth two handfulls, the black of crab's claw prepared, Venice treacle, of each one ounce: all these are to be bruised together, and boyled in a gallon of milk, till the half be boyled away; then put it into a bottle unstrained, and give of it about three or four spoonfulls at a time, three mornings together, before new or full-moon." Sir Hans Sloane sent a specimen of the plant called "Starre of the Earth" to Ray, who pronounced it to be the *Sesamoides Salamanticum magnum,* of Clusius, and that it was found in abundance upon Newmarket-heath. See Ray's *Phil. etters*, p. 209.

signify the Society's resolution, and to desire his opinion as to the volume, cuts," &c. Halley, accordingly, wrote to Newton[12].

"*London, May* 22, 1686.

"SIR,

"YOUR incomparable treatise intitled, *Philosophiæ Naturalis Principia Mathematica,* was by Dr. Vincent presented to the Royal Society on the 28th past; and they were so very sensible of the great honour you have done them by your dedication, that they immediately ordered you their most hearty thanks, and that the Council should be summoned to consider about the printing thereof. But by reason of the President's attendance upon the King, and the absence of our Vice-Presidents, whom the good weather has drawn out of town, there has not since been any authentic Council to resolve what to do in the matter; so that on Wednesday last, the Society in their meeting judging that so excellent a work ought not to have its publication any longer delayed, resolved to print it at their own charge in a large quarto of a fair letter; and that this their resolution should be signified to you, and your opinion thereon be desired, that so it might be gone about with all speed. I am instructed to look after the printing of it, and will take care that it shall be performed as well as possible. Only I would first have your directions in what you shall think necessary for the embellishing thereof, and particularly whether you think it not better, that the schemes should be inlarged, which is the opinion of some here; but what you signify as your desire shall be punctually observed.

"There is one thing more that I ought to inform

---

[12] Supp. to Letter-books, Vol. IV. p. 340.

you of, viz. that Mr. Hooke has some pretensions upon the invention of the rule of decrease of gravity being reciprocally as the squares of the distances from the center. He says, you had the notion from him, though he owns the demonstration of the curves generated thereby, to be wholly your own. How much of this is so, you know best; as likewise what you have to do in this matter. Only Mr. Hooke seems to expect you should make some mention of him in the Preface, which, it is possible, you may see reason to prefix. I must beg pardon that it is I, that send you this ungrateful account; but I thought it my duty to let you know it, that so you might act accordingly, being in myself fully satisfied that nothing but the greatest candour imaginable is to be expected from a person who has of all men the least need to borrow reputation[13].

"I am, &c.,

"E. HALLEY."

The Council met on the 2nd June, when it was ordered that "Mr. Newton's book be printed, and that Mr. Halley undertake the business of looking after it, and printing it at his own charge; which he engaged to do." The latter part of this resolution, which is at variance with the decision of the Society at their Meeting on the 19th May, is only to be explained by the fact, that the Council,—who were much better informed of the state of the Society's finances than the Society generally,—were aware that the publication of Willughby's *De Historia Piscium* had exhausted

[13] It is not the province of this work to enter into the claims of Hooke. The reader who is anxious to do so, is referred to the article Hooke, *General Dictionary*, Vol. VII.; to Brewster's *Life of Newton;* Whewell's *History of the Inductive Sciences*, Vol. II.; and Rigaud's *Essay on the First Publication of the Principia.*

their finances to such an extent, that the salaries even of their officers were in arrears[14]. Bearing this in mind, it is difficult to conceive how the Council could have entered with prudence on any fresh printing expenses during this year. Mr. Rigaud, in his *Essay on the First Publication of the Principia*, most justly remarks, that "under these circumstances, it is hardly possible to form a sufficient estimate of the immense obligation which the world owes in this respect to Halley, without whose great zeal, able management, unwearied perseverance, scientific attainments, and disinterested generosity, the *Principia* might never have been published[15]." When Newton was apprised of Hooke's claims, he conceived the intention of suppressing the third book of the *Principia*, as the following extract of a letter to Halley, dated Cambridge, June 20, 1686, explains:—"The proof you sent me I like very well. I designed the whole to consist of three books; the second was finished last summer,

---

[14] It is recorded in the Minutes of Council, that the arrears of salary due to Hooke and Halley were resolved to be paid by copies of Willughby's work. Halley appears to have assented to this unusual proposition, but Hooke wisely "desired six months' time to consider of the acceptance of such payment."

The publication of the *Historia Piscium*, in an edition of 500 copies, cost the Society 400*l*. It is worthy of remark, as illustrative of the small sale scientific books met with in England at this period, that a considerable time after the publication of Willughby's work, Halley was ordered by the Council to endeavour to effect a sale of several copies with a bookseller at Amsterdam, as appears in a letter from Halley requesting Boyle, then at Rotterdam, to do all in his power to give publicity to the book. When the Society resolved on Halley's undertaking to measure a degree of the earth, it was voted that "he be given 50*l*., or fifty books of fishes."

[15] P. 35.

being short, and only wants transcribing, and drawing the cuts fairly. Some new Propositions I have since thought on, which I can as well let alone. The third wants the theory of Comets. In autumn last, I spent two months in calculations to no purpose, for want of a good method, which made me afterwards return to the first book, and enlarge it with divers Propositions, some relating to comets, others to other things, found out last winter. The third, I now design to suppress. Philosophy is such an impertinently litigious lady, that a man had as good be engaged in lawsuits, as have to do with her. I found it so formerly, and now I am no sooner come near her again, but she gives me warning. The two first books without the third, will not so well bear the title of *Philosophiæ Naturalis Principia Mathematica*, and therefore I have altered it to this, *De Motu Corporum Libri duo:* but, upon second thoughts, I retain the former title. 'Twill help the sale of the book, which I ought not to diminish now 'tis yours[16]. The articles are, with the largest, to be called by that name; if you please, you may change the word to *Sections*, though it be not material. In the first page I have struck out the words *uti posthac docebitur*, as referring to the third book[17]."

---

[16] At this period, London publishers were extremely averse to undertake the printing of mathematical books. Collins, in a letter to Newton says, "our Latin booksellers have no vent for mathematical works; and so, when such a copy is offered, instead of rewarding the author, they rather expect a dowry with the treatise." It is recorded that the Royal Society gave 5*l.* with the copy of Horrox's *Opera Posthuma*, to encourage a bookseller to print it.

[17] Archives: Royal Society.

In consequence of Halley's arguments and intreaties, Newton was eventually prevailed on not to suppress the third book, which, under the title of *De Systemate Mundi*, was presented to the Society on the 6th April, 1687. One of Halley's arguments to induce Newton to give this book to the world, was "the application of your mathematical doctrine to the theory of Comets, and several curious experiments, which, as I guess by what you write, ought to compose it, will undoubtedly render it acceptable to those who will call themselves Philosophers without mathematics, which are much the greater number[18]." It will be remembered, that Newton himself talks in the *Principia* of having drawn it up *methodo populari*, and such was probably his primary intention; but, as Professor Rigaud remarks, "the value of the change was an ample compensation for the diminished number of readers, who, in consequence, would be likely to study it."

The entire work was probably finished at the close of 1686; for Pemberton says, "this treatise was composed by Newton in the space of a year and a half[19]." At a Meeting of the Council on the 30th June, 1686, the President was ordered to "license Mr. Newton's book, entitled *Philosophiæ Naturalis Principia Mathematica*, and dedicated to the Society;" and the *imprimatur*, signed by Pepys, is dated the 5th of July following. In the Minutes of the Oxford Philosophical Society, is recorded, under the date of May 20, 1687:—"Mr. President (Dr. Wallis) was pleased to communicate a letter from Mr. Halley,

---

18 Halley to Newton, in Archives: Royal Society.

19 View of Newton's *Phil.*

which gives an account of Mr. Newton's book, *De Systemate Mundi*, now in the press[20]." The *Principia* was published about the middle of 1687, Halley prefixing to it a set of Latin hexameters in praise of the illustrious author. When it is remembered that the volume contains 64 sheets, and above 100 diagrams cut in wood, besides an engraving on copper, it is wonderful that Halley should have performed his laborious task so well, and in so short a space of time. Indeed his zeal outstripped that of Newton, for, in a letter from the latter to Halley, preserved in the archives of the Royal Society, he says, "Pray, take your own time; if you meet with any thing which you think needs either correcting, or further explaining, be pleased to signify it to me;" and adds, "I wish the printer be careful to mend all you note." It will not be out of place to state that the number of copies printed of the first edition is not known. Professor Rigaud supposes it to have been small, and adduces as a proof, that in 1692, when the reputation of the work was established, Huyghens, who was anxious for a second edition, was of opinion that "200 *exemplaires suffiroient*."

According to Sir William Browne, the price of a copy of the first edition did not exceed twelve shillings[21]. The second edition was so soon exhausted by purchases in England, that it was reprinted by a company of booksellers at Amsterdam. The third edition appeared in 1726, under the care of Dr. Henry Pemberton.

---

[20] The title-page of the *Principia* attests that it was published *jussu* (but not *sumptibus*) *Societatis Regiæ*.

[21] Nichols's *Lit. Anec.*, Vol. III. p. 322.

The *Principia* contains the dedication to the Royal Society; a brief preface; verses by Halley in honour of Newton; definitions, axioms; a short book on unresisted motion, a second on resisted motion, and a third on the system of the universe. Halley's verses were somewhat altered by Bentley in the second edition, but the original readings were very nearly restored in the third. Newton wrote a short preface for each of the editions, and Cotes one of considerable length for the second. The dates of the Newtonian prefaces are, May 8, 1686; and March 28, 1713.

The manuscript of this immortal work, entirely written by Newton's own hand, is in admirable preservation, and is justly esteemed the most precious scientific treasure in the possession of the Royal Society. A fac-simile of the title is annexed, which, as will be observed, was first written *De Motu*, and subsequently altered to *Philosophiæ Naturalis Principia Mathematica.*

At the anniversary in 1686, Pepys resigned the office of President, and was succeeded by John, Earl of Carbery.

This nobleman was descended from one of the most considerable families in Wales; powerful—according to tradition—before the Norman Conquest, and remarkable in later times for many worthy persons, conspicuous for their talents and their high employments, and connected with literary history at several points.

The first nobleman was John, son of Walter Vaughan, of the Golden Grove, in Caermarthenshire, who was created Lord Vaughan of Mullingar, by James I., and afterwards by Charles I., Earl Carbery. His son Richard, the second Earl, having adhered to

# Philosophiæ Naturalis principia Mathematica

## Definitiones

Def. I.

the Royal cause in the time of the civil war, was rewarded by Charles I. with an English peerage, under the title of Baron Vaughan of Emlyn; and, after the Restoration, he was constituted Lord President of the Principality.

John, his son by his second wife, and third Earl of Carbery, was Governor of Jamaica for some years; he passed the greater part of his life in retirement, and but very little of him appears to be known. Dedications are treacherous evidence, and must be received with caution. We are, however, told by Beaumont, who dedicated his *Treatise on Spirits and Apparitions* to Lord Carbery, that his "Lordship's great genius to a contemplative life, which raises human nature to an excellency above itself, and highly influences the economy of this world, has naturally induced him to dedicate his book to his Lordship." This work was published in 1705, eight years before Lord Carbery's decease, which took place in his house at Chelsea, January 16, 1712—13, at the age of 73. His Countess was daughter of George Savile, Marquis of Halifax. With him the title became extinct, as he left only a daughter, who married the Marquis of Winchester. There are two engraved portraits of Lord Carbery extant, after paintings by Kneller.

In the early part of 1687, the Society received a proposition from Hooke, offering to deliver a discourse, illustrated by one or two experiments, weekly, provided his salary was increased to 100*l.* per annum. The Council-minutes record that the proposition "was much debated, and it was concluded that Mr. Hooke should have 50*l.* a year from the Society, and their lawful assistance and recommendations towards the recovery of the 50*l.* a year, which Sir John Cutler stands obliged to pay him during his life; and that in consideration thereof, Mr. Hooke should at every Meeting produce one or two new experiments, together with a discourse concerning them in writing, to be left with the Societie, and that the said experiments should proceed in a naturall method." It does not appear that Hooke was satisfied with this determination of the Council, for although he continued to make scientific communications to the Society, yet there is no record of his delivering an experimental lecture weekly.

The debts of the Society weighed so heavily upon their finances, that in all probability the Council were unable to accede to Hooke's proposition. At a Meeting in 1687, they record in their Minutes, "It appearing to this Councill that the debts of the Societie are such, that they cannot otherwise be satisfied, it is ordered that their stock in the East India Company be sold."

Circumstances, however, occurred, which fortunately prevented this sacrifice, though the Society still continued to suffer considerable pecuniary embarassment. Their difficulties were increased in 1688, by the necessity to rent apartments in Gresham College. At a Council, held on the 20th June, 1688, it was "Ordered, that Mr. Perry and E. Halley do make a conclusive bargain with Mr. Wells, the Divinity Professour, for his lodgings in Gresham College;" and again at the next Meeting on the 11th July, "the Council were pleased to signifie their consent to the agreement made with Mr. Wells for his lodgings, viz. that they will give him 22*l.* per ann., to enter at Michaelmas next." It was also resolved to give Hooke 10*l.* a year for the use of his rooms; which sum he received to the period of his decease.

In a scarce volume, entitled, *Extracts from the Records of the City of London, with other documents respecting the Royal Exchange and the Gresham Trusts*, it appears that a Sub-committee of the Joint Gresham Committee was appointed in 1697, "to enquire into the manner in which the Professors' lodgings were used and occupied;" when it was found that the majority, instead of living in their rooms, as was intended by Sir T. Gresham, had let them to other parties. "The Physick lodgings," says the Report, "have been let by Dr. Woodward and his predecessor to one Mr. Styles, a merchant, for ten years or more; and the said Mr. Styles, his two nieces and two servants, are now in the said lodgings, and Dr. Woodward has converted his kitchen into lodging rooms for his own use, but he seldom lodges in the College[22]."

During the period that the Earl of Carbery occupied the chair, a great number of valuable inventions were brought forward, and experiments made by Hooke, Halley, and Papin. The latter frequently exhibited his *Pneumatic Tube*, which propelled a leaden ball of two ounces with considerable force. An account of this instrument will be found in the 16th Volume of the *Transactions*, under the title *Shooting by the Rarefaction of the Air*. Several pendulum and magnetical experiments were made; and a large telescope was erected in the quadrangle of the College, which was much used by Hooke.

Dr. Sloane frequently sent communications to the

---

[22] P. 123.

Society on botanical subjects, accompanied by specimens of the plants[23]. In 1688, a very interesting set of maps was laid before the Society by Dr. Cox, showing the lakes in North America, which had been surveyed for the first time by Englishmen, who, as was said, strongly recommended the immediate establishment of hunting and trading companies, which, they were certain, would reap immediate and great profit, as the quantity of beavers and other animals, that they had seen, was immense. A Committee was specially appointed to take this subject into consideration.

At the Anniversary in 1689, the Earl of Carbery retired from the chair, and the Earl of Pembroke was chosen President.

---

[23] At one of the Meetings he exhibited "Irish sea-weed called *Dulesk*, which he stated the Irish who were afflicted by scurvy were in the habit of chewing."

## CHAPTER XII.

Memoir of Lord Pembroke—Anxiety to continue the *Transactions*—Evelyn again solicited to become President—Declines—Election of Sir R. Southwell—Memoir of him—Advertisement to *Transactions*—Uninterrupted publication since 1691—Death of Boyle—Eulogium on him—Leaves his Minerals to Society—His great respect for the Society—He deposits sealed packets—Huyghens's Aërial Telescope-glasses—Sir R. Southwell retires from the Chair—Charles Montague elected—Memoir of him—Dr. Woodward's Geological Works—His Scientific Labours in the Society—Accused of insulting Sir Hans Sloane—Expelled the Council—Institutes legal proceedings—Is defeated—Resignation of Mr. Montague—Election of Lord Somers as President.

---

1685—1700.

THOMAS, eighth Earl of Pembroke, and fifth of Montgomery, succeeded his only brother Philip, who died without male issue, in 1683. He enjoyed the honours of the family for nearly fifty years, and held various high offices in the State, being appointed President of the Council; in 1708 Lord High Admiral; and, on the demise of Queen Anne, one of the Lord's Justices, until the arrival of George I. from Hanover, at whose coronation he carried the sword called *Courtana*. He was also a Knight of the Garter. He was more devoted to the arts and archæological pursuits than to natural science. He formed the celebrated cabinet of Coins and Medals, and collected the Marbles at Wilton. Both in his political character, and relation to the intellectual progress of the country, he was one of the most eminent and valuable persons of his age, and a distinguished member of the house of Herbert, to whom literature, from its dawn in England, owes so many obligations. Lord Pembroke was three times married: in 1684 to Margaret, only daughter and heir of Sir Robert Sawyer, of High Cleer in the county of Southampton, Attorney-General in the time of Charles II. By this lady he had twelve children. His second wife was Barbara, daughter of Sir Thomas Slingsby of Scriven, Yorkshire, by whom he had one daughter; and his third, Mary, sister of Viscount Howe. He was elected a Fellow of the Society May 13, 1685, chosen into the Council February 16, 1686—7, and appointed a Vice-President April 13, 1687. He seldom attended the Meetings, but made communications on mechanical subjects to the Society, which created considerable interest. He died on the 22nd January, 1732—3. The present Earl of Pembroke and Caernarvon is lineally descended from this nobleman.

That Lord Pembroke gave little attention to the Society is fairly deducible from the fact, that his name does not appear as presiding, on any one occasion, at the Council or ordinary Meetings. His place was generally occupied by Sir John Hoskyns, one of the Vice-Presidents, or by Sir Cyril Wyche. The most important event in 1689 is the effort made to republish the *Transactions*, which were frequently asked for by the public. At a Meeting of the Council, on the 22nd October, 1690, it was "Ordered, that Mr. Hooke have the postage of all letters of Philosophical Correspondence allowed him, on condition that he publish *Transactions*, or *Collections*, as formerly, and that in consideration thereof the Society will take off 60 books." Hooke agreed to this proposition, but nothing further was done in the matter until Sir Robert Southwell became President.

On the retirement of Lord Pembroke from office, a general wish once more prevailed to elect Evelyn to the chair. In his *Diary* he writes, that he was "chosen President in the first instance, but desired to decline it, and with great difficulty devolv'd the election on Sir R. Southwell, Secretary of State to King William, in Ireland."

Sir Robert Southwell was descended from a very ancient and distinguished family, who took their name from the town of Suelle, or Southwell, in Nottinghamshire, the chief place of their residence, from the reign of Henry III. to that of Henry VI. The descendants of the elder branch passed over to Kinsale in Ireland, which town was held by Sir Robert Southwell, as well as the estate of King's Weston, in Gloucestershire.

The subject of this memoir was born at Kinsale, in 1635. He was educated at Queen's College, Oxford, and afterwards entered at Lincoln's Inn. On the completion of his law studies he travelled on the Continent. On the 27th September, 1664, he was sworn one of the clerks of the Privy Council; and on the 20th November, 1665, received the honour of knighthood[1]. He was employed on several diplomatic missions: first as Envoy to mediate a peace between Spain and Portugal, in which he was successful. In 1672 he was sent as Envoy Extraordinary to Portugal, and afterwards, in the same capacity, to Flanders, and to the Elector of Brandenburg at Berlin, visiting the Prince of Orange on his way. After his return in 1681, he retired from public business to his seat at King's Weston. The accession of King William brought him again into the world. His Majesty ap-

---

[1] Wood, *Athen. Oxon.*, Vol. II. p. 879.

pointed him principal Secretary of State for Ireland, to which country Sir Robert accompanied the King, in the expedition of 1690. He served in three parliaments, and died at King's Weston in 1702, aged 66. He was buried in the beautiful church of Henbury. His abilities, improved by a critical knowledge in polite literature, qualified him eminently for the various offices which he held. His letters are written in a most masterly style, exhibiting great thought, perspicuity, and penetration. He contributed several Papers to the *Transactions*, mostly on physiological and chemical subjects; and appears to have been more attached to literature and science than to politics. His name stands honorably associated with Dampier, who he patronised, and to whom he even advanced money to equip his ships. In a letter to Mr. Sidney, dated from King's Weston, published in the *Diary and Correspondence of Charles II.*, Sir Robert says, "I am neither in office, nor dare desire it; but am resolved to pass my life between Virgil's *Georgics* and Mr. Evelyn *On Trees.*" Evelyn calls him "a sober, wise, and virtuous gentleman;" and one of his biographers says, "It is not then to be wondered, if, when his course was ended, and he was called to receive the reward of his piety, his death was greatly lamented, his memory held in the highest veneration, and his example esteemed by all his friends and relations as a most complete pattern for imitation."

Such was the individual who presided over the Society for five years. A portrait of him, by Kneller, is preserved in the Meeting-room.

Soon after the election of Sir R. Southwell, we find him urging the expediency of resuming the publication of the *Transactions*.

At a meeting of the Council on the 28th January, 1690—1, at which fifteen Members were present, it was "Resolved, that there shall be *Transactions* printed, and that the Society will consider of the means for effectually doing it. And Dr. Tyson, Dr. Slare, Mr. Waller, and Mr. Hooke, were desired to be assistants to E. Halley, in compiling and drawing up the *Transactions*." It was further "Ordered, that Mr. Boyle be desired to continue his designe of communicating his small tracts, to be published in the *Transactions*."

In consequence of this decision, measures were taken to collect Papers for publication; and in February, the 192nd number, for January and February, 1690—1, was published under the editorship of Halley. The following advertisement is prefixed.

"The publication of these *Transactions* having for some time past been suspended, chiefly by reason that the unsettled posture of publick affairs did divert the thoughts of the curious towards matters of more immediate concern, than are Physical or Mathematical enquiries, such as for the most part are the subjects we treat of, with exclusion to many others wherewith the forein journalists usually supply their monthly tracts: These are now to advertise that, for the future, the *Royal Society* has commanded them to be published as formerly, and, if possible, monthly. And all lovers of so good a work are desired to contribute their discoveries in art or nature, addressing them, as formerly, to Mr. H. Hunt, at Gresham College, and they shall be inserted herein, according as the Authors shall direct."

The succeeding numbers of the *Transactions* were published under the ostensible editorship of Mr.

Richard Waller, Secretary to the Society; and on the 15th February, 1692—3, the Council-minutes state it was "Resolved, that Dr. Plot shall print the *Transactions*, and that, for his encouragement therein, he have the 60 books agreed by the bookseller to be allowed for the copy, and that the Society will make up to the Dr. what the value of the said books shall fall short of 40*l.* per annum, and they will take the said sixty books as for money, and allow him 12 of each sort to present to his chief correspondents: to which the Dr. agreed." Mr Halley, however, still continued to afford considerable assistance, in supplying Papers, and occasionally superintending the publication of the *Transactions*, until he departed on his voyage to the southern hemisphere, in 1698. At a meeting of the Council on the 7th December, 1692, he undertook to "furnish *de proprio* five sheets in twenty."

The Preface to the *Transactions* for 1693 bears so much upon the scientific progress and labours of the Society, and, at the same time, so curiously exemplifies the general state of science at that period, as to give it general interest:—

PREFACE TO THE XVII. VOLUME OF TRANSACTIONS.

"So many and so large steps having been made towards the discovery of Nature by the indefatigable industry of this last Age, it may seem as if the subject were almost exhausted, and Nature herself wearied with the Courtships of so many Pretenders: But if, on the other side, we consider the vast, not to say boundless, extent of the Universe, and that the discovery of one Phænomenon leads to, as well as entices to, the search after another: together with how easie a thing it is even to impose upon ourselves groundless Opinions, instead of Real Knowledge; we must own the work at least great enough for the Age of the World, and sooner doubt our own Resolutions and abilities, than fear the failure of fit Subjects to entertain our Thoughts. Real Knowledge is a nice thing; and as no man can be said to be master of that which he cannot teach to another, so neither can the mind itself, at least as to physical matters, be allowed to apprehend that, whereof it has not in some sense a Mechanical Conception; for this knowledge entering wholly by the senses, whose objects only are bodies, whereof their Organs have the perception, but from the Magnitude, Figure, Situation, and Motion of them, which are all mechanically to be considered, or we come short of a Satisfactory Information, it follows, that Number, Weight, and Measure, must be applied to analize the Problems of Nature by which they were compounded.

"This has been the Employment of the Experimenting part of mankind, and the design of that Glorious Institution of the Royal Society, whose youthful vigour carried them warmly on in the pursuit of Nature, then at a farther distance off, catch't and grasp'd the Proteus thro' all its changes. And, since Publications of this Nature have been thought no small advancement to that great Design, because it collects and preserves several small Tracts, which otherwise might possibly be lost, the Publisher has yielded to the sollicitation of some friends to undertake this Work, with an engagement to the learned, of communicating (as constantly as hath ever been at any time practised) whatever of a Philosophical Nature shall come to his hands, clearing the Royal Society (which is no way concerned therein) from all the miscarriages he may possibly commit; and promises himself he shall never fail of materials from the ingenious, since he proposes neither the mean end of private advantage, nor thinks himself capable of the BASENESS, to stifle any person's discovery till another may pretend to it; being resolved immediately to insert in the next Transaction what ingenious Communications shall be so desired, that the true Author may not be defrauded of his due Merit and Glory."

From this period (1691) to the present, the publication of the *Transactions* has continued uninterruptedly year by year. It may have been owing to the additional impetus given by the re-appearance of the *Transactions*, that the Society acquired a considerable accession of Members during the years immediately following this event. Prior to 1691, the average number of Fellows elected annually was only 5, but in subsequent years the average rose to 18. The consequence of this increase was a corresponding augmentation of funds, amounting, on the 30th November, 1697, to 250*l.* in the stock of the East India Company, and 800*l.* in that of the African Company.

The year 1691 was marked by a melancholy event, the death of Boyle, which occurred on the 31st of December. By his decease the Society lost a most sincere friend, one who ever manifested an interest in its welfare, not surpassed, if indeed equalled, by any other member.

Boyle died at the age of 65; his tomb is in St. Martin's Church, Westminster. In addition to his expenditure to promote science, his charities were so munificent as to exceed 1000*l.* a year. Of his merits as an inquirer into nature, Dr. Boerhaave, after declaring Lord Bacon to be the father of experimental philosophy, observes: "Mr. Boyle, the ornament of his age and country, succeeded to the genius and inquiries of the great Chancellor Verulam. Which of Mr. Boyle's writings shall I recommend? All of them. To him we owe the secrets of fire, air, water, animals, vegetables, fossils; so that from his works may be deduced the whole system of natural knowledge[2]." This may now seem extravagant praise, but, taking all circumstances into consideration, it may be regarded as a just tribute to extraordinary merit and indefatigable perseverance.

The following is an extract from his Will, made a few months before his death:

"Item: to the Royal and learned Society, for the advancement of experimental knowledge, wont to meet at *Gresham College*, I give and bequeath all my raw and unprepared minerals, as ores, marchasites, earths, stones, (excepting jewels), &c., to be kept amongst their collections of the like kind, as a testimony of my great respect for the illustrious Society and design, wishing them also a most happy success in their laudable attempts to discover the true nature of the works of God, and praying that they, and all other searchers into physical truths, may cordially refer their attainments to the glory of the Author of Nature, and the benefit of mankind."

Only a few months prior to his decease, he proposed to the Council, "that a proper person might be found out to discover plagiarys, and to assert inven-

[2] Dr. Johnson says, "It is well known how much of our philosophy is derived from Boyle's discoveries, yet very few have read the detail of his experiments. His name is indeed reverenced, but his works are neglected; we are contented to know, that he conquered his opponents, without inquiring what cavils were produced against him, or by what proofs they were confuted." *Rambler*, No. 106.

tions to their proper authors, whereupon," adds the Minutes, "it was put to the question by ballot, whether the Council were of opinion that it would be a useful work if such a person could be found, and it was carried in the affirmative."

Some sealed packets, which Boyle had deposited with the Society, were opened in February, 1691—2. They contained Papers on chemical subjects, and bore the dates of 1680, 1683, and 1684[3].

A portrait of the philosopher, by Kerseboom, was presented to the Society by Boyle's executors, a short time after his decease: it is now in the Meeting-room.

Among the donations presented to the Society in 1691, must be mentioned the celebrated object-glass of 122 feet focal length, made by Huyghens, for an "aërial telescope[4]." Hooke was entrusted with this glass, with the view of constructing an apparatus for its use, and in the mean time, Halley was ordered to "view the scaffolding of St. Paul's Church, to see if that might not conveniently serve for the present, to erect the object-glass thereon, for viewing such of the celestial objects as now present themselves[5]." Much difficulty attended the mounting of these large telescopes. "To think," says Captain Smyth, "of a hundred, or a couple of hundred, feet for a refractor! Auzout's was three hundred feet long, and therefore

---

[3] The earliest of these communications is printed in the *Phil. Trans.* for Jan. 1683.

[4] Two other object-glasses of Huyghens' were afterwards presented to the Society, (one of 170 feet focal length), by Sir Isaac Newton, and the other (of 210 feet, with two eye-glasses, by Scarlet), by the Rev. Gilbert Burnet, F.R.S. in 1724.

[5] Journal-book, Vol. IX. p. 87. The 35th volume of the *Phil. Trans.* contains an account of Huyghens' Aërial Telescope.

useless; another one was contemplated, and all but completely constructed, of nearly double that length. I was so puzzled to know how they contrived to get the eye and object-glasses of these unwieldy machines *married*, or brought parallel to each other for perfect vision, and so desirous of comparing the performance of one of them, that I was about to ask the Royal Society's permission to erect the aërial 123-foot telescope in their possession. The trouble, however, promised to be so much greater than the object appeared to justify, that I laid the project aside,—not wishing to furnish a parallel to the case of the worthy captain, whose veracity having been doubted as to the length of a West Indian cabbage-tree, took the trouble to bring one from Barbados, to prove his assertion[6]."

At the anniversary in 1693, Evelyn, according to his *Diary*, was again requested, or, as he styles it, "importun'd," to accept the office of President: he however persisted in declining the honour, and Sir Robert Southwell continued in office until 1695, when Mr. Charles Montague, afterwards Earl of Halifax, was elected President.

Charles Montague, Earl of Halifax, was the fourth son of George Montague, of Harton in Northamptonshire. He was born on the 16th April, 1661, and gave early evidence of an active genius, so much so that, according to authorities, he "was the admiration of all who came near him." At fourteen he was sent to Westminster, when his facility of versifying attracted the notice of Dr. Busby; and, being elected King's scholar, he was removed, in 1682, to Trinity College,

---

[6] *Cycle of Celestial Objects*, Vol. I. p. 370.

Cambridge, where he made great progress in his studies. Here he formed the acquaintance of Newton, and co-operated with him in attempting to establish a philosophical Society in that town. In 1684 he wrote a poem upon the death of Charles II., which led to his being noticed by the Earl of Dorset, who invited the author to London. The circumstances connected with his advancement in the metropolis, are curious.—It is recorded, that in conjunction with Matthew Prior, he published a parody, entitled, *The Country Mouse and the City Mouse.* His patron, Lord Dorset, introduced him to King William, with the words, "'May it please your Majesty, I have brought a mouse to have the honour of kissing your hand;' at which the King smiled, and being told the reason of Mr. Montague's being so called from the parody above-mentioned, he replied, with an air of gaiety, 'You will do well to put me in the way of making a man of him,' and ordered him an immediate pension of 500*l.* per annum out of the privy purse, till an opportunity should offer of giving him an appointment[7]."

This soon occurred: after displaying high abilities in parliament, he was made one of the Commissioners of the Treasury, and subsequently Chancellor of the Exchequer[8], in which office he accomplished the great work of re-coining all the current money of the kingdom. In this he was assisted by Newton, Locke, and

---

[7] *Memoirs of the Life of Charles Montague*, p. 17.

[8] Prior was greatly discontented at being less fortunate than his friend. He angrily expostulates:

"My friend Charles Montague's preferr'd,<br>
Nor would I have it long observ'd,<br>
That one mouse eats, while t'other's starved."

Halley[9]; and it was on this occasion that he obtained for Newton the office of Warden of the Mint, which the philosopher held until subsequently promoted to the higher appointment of Master of that establishment.

"This," says Sir David Brewster, "must have been peculiarly gratifying to the Royal Society, and it was probably from a feeling of gratitude to Mr. Montague, as much as from a regard to his talents, that this able statesman was elected President of that learned body, on the 30th November, 1695."

In 1698 he was appointed first commissioner of the treasury, and, in 1700, created Baron Halifax in the county of York. In 1701 and 1702 he was impeached by the House of Commons, for breach of trust; but he was entirely exculpated by the House of Lords, who dismissed the impeachment. His great attachment to Newton's niece, Mrs. Catherine Barton, widow of Colonel Barton, is well known. The lady was young and beautiful. It is not explained why he did not marry her instead of the Countess of Manchester; he, however, left her a large portion of his fortune. As may be imagined, she was not exempted from severe and unkind criticisms and censures.

On the death of Queen Anne, Lord Halifax was appointed one of the regents; and on the accession of George I. was created Earl of Halifax, and First Commissioner of the Treasury, with a grant to his nephew of the first auditorship of the Exchequer.

He died rather suddenly on the 19th May, 1715,

---

[9] Halley was appointed Master of the Mint at Chester, where a considerable amount of money was re-coined.

in the 54th year of his age, and was interred in Westminster Abbey.

The Earl of Halifax was, as Pope says, "fed with dedications;" but we must remember the age in which he lived, and the extensive patronage that he enjoyed. Besides several political pamphlets, he published a volume of poems, which, however, have not shared the immortality of his verses on the toasting-glasses of the Kit-Kat Club, of which he was an original member[10].

Lord Halifax was a most liberal patron of genius, and, apart from his literary attainments, his name will ever be intimately and honourably associated with that of Newton.

The author of the *Principia* must have felt an especial pleasure to have addressed to him, whilst President of the Royal Society, his solution of the celebrated problems, proposed by John Bernouilli.

A remarkable circumstance in the history of the Society, during the presidency of Mr. Montague, was the publication of Dr. Woodward's *Essay towards a*

---

[10] It was customary at this once-celebrated Club to inscribe on the glasses, the name of the lady who was the toast for the year. "When she is regularly chosen," says the *Tatler* (No. 24), "her name is written with a diamond on the drinking-glasses. The hieroglyphic of the diamond is to shew her, that her value is imaginary; and that of the glass to acquaint her, that her condition is frail, and depends on the hand which holds her." The 'Duchess of Richmond' was an enthusiastic toast in the time of Lord Halifax. To her he inscribed the following lines, which, according to custom, were cut on the glasses:—

> "Of two fair Richmonds, different ages boast;
> Theirs was the first, and ours the brighter toast;
> The adorer's offering proves whose most divine,
> They sacrificed in water, we in wine."

*Natural History of the Earth*, printed in 1695, and reviewed at considerable length in the *Transactions* for that year. His theory attracted a great deal of attention, and gained him considerable reputation. "Among the contemporaries of Hooke and Ray," says Mr. Lyell in his *Geology*, Woodward, a Professor of Medicine, had acquired the most extensive information respecting the geological structure of the crust of the earth[11];" and Dr. Whewell, in his *History of the Inductive Sciences*, observes, that "one of the most remarkable occurrences in the progress of descriptive Geology in England, was the formation of a geological museum by William Woodward, as early as 1695. This collection, formed with great labour, systematically arranged, and carefully catalogued, he bequeathed to the University of Cambridge; founding and endowing at the same time a professorship of the study of Geology. The Woodwardian Museum still subsists, a monument of the sagacity with which its author so early saw the importance of such a collection[12]."

Dr. Woodward was only 30 years of age when his book was published; his life is one among many examples of the triumph of abilities and application over difficulties. Of humble origin, he was placed apprentice to a linen-draper in London, but this situation not according with his philosophical turn of mind, he soon left it, and devoted himself to science. "He was so fortunate as to attract the attention of Dr. Peter Barwick, an eminent physician, who finding him of a very promising genius, took him

[11] Vol. I. p. 53.

[12] Second edition, Vol. III. p. 542.

under his tuition in his own family. In this situation he began to apply himself to philosophy, anatomy, and physic, until he was invited by Sir Ralph Dutton to his seat at Sherborne, in Gloucestershire[13]." Here it was that he began those observations and collections relating to the present state of our globe, which laid the foundation for his discourses afterwards on that subject, concerning which he has given the following account. "The country about Sherborne, and the neighbouring parts of Gloucestershire, to which I made frequent excursions, abounding with stone, and there being quarries of this kind open almost everywhere, I began to visit these in order to inform myself of the nature, the situation, and the condition of the stone. In making these observations, I soon found there was incorporated with the sand of most of the stone thereabouts, great plenty and variety of sea-shells, with other marine productions. I took notice of the like lying loose on the fields in the ploughed lands so thick, that I have scarcely observed pebbles or flints more frequent and numerous on the ploughed lands of those countries that most abound in them. This was a speculation new to me, and what I judged of so great moment, that I resolved to pursue it through the other remoter parts of the kingdom; where I afterwards made observations upon all sorts of fossils, collected such as I thought remarkable, and sent them up to London[14]."

In 1692, Dr. Woodward was appointed Professor of Physic at Gresham College, and the following year was elected a Fellow of the Royal Society. He took

[13] Ward's *Lives of the Gresham Professors*, p. 284.

[14] Preface to *Catalogue of English Fossils*.

a very active part in its pursuits, and particularly in those relating to Geology, or, as this science was then termed, "the Natural History of the Earth." The Journal-books of that period show that Geology was frequently discussed at the ordinary Meetings.

Dr. Woodward was often placed on the Council of the Royal Society. In 1710, when associated among others with Sir Hans Sloane, he made use of expressions in reference to the knight, which were considered so insulting, that he was required by the other members to make an apology. This he refused to do; and consequently, after solemn deliberation, he was expelled the Council. He brought an action at law against that body, with the view of being reinstated in his place, but was unsuccessful. The history of this quarrel is detailed at considerable length in the Council-minutes, and is remarkable as being the first recorded in the annals of the Society[15].

In 1693 Dr. Sloane was elected Secretary in the place of Dr. Gale, who withdrew from office. Mr. Montague continued to occupy the chair until the Anniversary in 1698, when he resigned; and Lord Somers, then Lord Chancellor, was unanimously chosen President.

---

[15] Woodward fought a duel with Dr. Mead, under the gate of Gresham College. Woodward's foot slipped, and he fell. "Take your life!" exclaimed Mead. "Any thing but your physic," replied Woodward. The quarrel arose from a difference of opinion on medical subjects.

# CHAPTER XIII.

Memoir of Lord Somers—Committee appointed to wait upon him—Society receive valuable present from the East India Company—Halley sails on a Scientific Expedition—Mr. Jones sent by the Society on an Expedition of Discovery—Resolution not to give opinions in Scientific Controversies—The *Transactioneer*—Dr. Woodward disowns the Work—Favour shown to the Academy of Sciences—Letter of M. Geoffroy—Zeal of Sir Hans Sloane—Savery exhibits his Steam-engine—Presents Drawing of it to Society—Receives a Certificate—Performance of the Engine—Death of Hooke—His interest in the Society—His design of endowing the Society—His Wealth—Proposal to rebuild Gresham College—Wren furnishes plan of rooms for the Society—Scheme abandoned—The Society resolve on building or buying a House—Lord Somers resigns—Sir Isaac Newton elected President.

---

1695—1705.

THE election of Lord Somers to the office of President, reflects great lustre upon the Royal Society, "He was a man," says Lord Campbell, in his *Lives of the Lord Chancellors*, "eminent as a lawyer, a statesman, and a man of letters,—the whole of whose public career and character I can conscientiously praise—and whose private life, embellished by many virtues, could not have been liable to any grave imputation, since it has received the unqualified approbation of Addison."

The family to which Lord Somers belonged, had long been proprietors of a small estate in the parish of Severn Stoke, in the county of Gloucester; and of the site of a dissolved nunnery, called the "White Ladies," a short distance from Worcester.

The Chancellor's father, John Somers, was bred to

the law, which he practised with great success. In 1649 he married Catherine Ceavern, of a respectable family in Shropshire. Her first child was a daughter, Elizabeth, afterwards Lady Jekyll. Her second, the future Chancellor, who was born in 1651.

The first notice of the boy is exceedingly curious, and is stated, in Cooksey's *Life and Character of Lord Somers*, to be perfectly well authenticated. It is to the effect, that when walking with one of his aunts, under whose care he was placed at the time, "a beautiful roost-cock flew upon his curly head, and while perched there, crowed three times very loudly[1]." Such an occurrence was of course viewed as an omen of his future greatness.

At the proper season he was placed at the College-school at Worcester, where, under Dr. Bright, an eminent classical scholar, he was thoroughly grounded in Greek and Latin. He afterwards went to a private academy at Walsall, and to another in Shropshire. In Seward's *Anecdotes*, it is stated that "though he was the brightest boy in the College-school, instead of joining his young companions in their boyish amusements, he was seen walking and musing alone, not so much as looking on while they were at play[2]."

Lord Campbell informs us, that when only sixteen years of age he was matriculated, and admitted of Trinity College, Oxford; a few years after, he occupied a desk in his father's law-office at White Ladies. The drudgery of an attorney's office was far from agreeable to him, and he eagerly seized every opportunity to exchange the study of parchments for the more congenial pursuits of literature.

---

[1] Cooksey, p. 10. [2] Vol. II. p. 114.

Parliamentary business having led Sir Francis Winnington, afterwards Solicitor-General, to White Ladies, he became acquainted with young Somers, and perceiving his merit, recommended that he should study for the bar, pointing out that Littleton and other Worcestershire men had risen to be judges. With considerable difficulty old Somers yielded his consent to this change, and on the 24th May, 1669, his son went to London, and entered as a student of the Middle Temple.

He forthwith commenced the study of law under Sir Francis Winnington, spending his vacations at White Ladies, where he made the acquaintance of the young Earl of Shrewsbury, which early ripened into a friendship that lasted through life. He returned to London with the Earl, who introduced him to Dryden, and other distinguished men of letters, and also to several noblemen. His manners soon acquired that "most exquisite taste of politeness[3]," for which he was afterwards so distinguished. It appears that his new acquaintances made him painfully aware of his defective education, which, says Lord Campbell, "must have arisen either from a very short stay at the University, or from idleness while resident there." He returned to his College for the purpose of acquiring a sound education, keeping his terms at the Middle Temple during his residence at the University. On the 5th May, 1676, he was called to the Bar, but did not begin to practise until 1681, when, says Lord Campbell, he was "a ripe and good scholar as well as lawyer; and, regard being had to his acquaintance with modern languages and literature, perhaps the

---

[3] *Freeholder*, No. 39.

most accomplished man that ever rose to high eminence in the profession of the law in England." His progress was extremely rapid; so much so, that in a very few years his professional profits amounted to 700*l.* a year, a very large sum for those times[4].

It would greatly exceed our limits to follow Lord Somers at length through his brilliant professional career. The nature of this work is much more closely connected with his literary than forensic life. I shall, therefore, only glance at his various official promotions. At the age of thirty-seven he was elected to the Convention Parliament, as representative of Worcester; appointed Solicitor-General, and knighted, upon the accession of William and Mary. On the 2nd May, 1692, he was promoted to be Attorney-General, and it was while holding this office that he acted as counsel for the plaintiff in the case of the Duke of Norfolk *v.* Germaine, the first instance on record in England of an action to recover damages for criminal conversation with the plaintiff's wife[5]. On the 23rd March, 1693, the Great Seal was placed in his hands as Lord Keeper, and he was at the same time sworn of their Majesties' Most Honourable Privy Council. Evelyn thus alludes to the promotion of Lord Somers, in his *Diary:* "The Attorney-General Somers made Lord Keeper,—a young lawyer of extraordinary merit." He presided in the Court of Chancery for seven years, winning the applause of all

---

[4] Cooksey's *Life*, p. 15.

[5] The damages were laid at 100,000*l.* The jury found a verdict for the plaintiff, with 100 marks damages. The court, it appears, reprimanded them severely for giving so small and scandalous a fine.

parties for his learning, impartiality, and courtesy. Only one of his decrees has been discovered to have been reversed. During the period that he remained in office, the administration of affairs at home was chiefly intrusted to him[6]. He had been frequently offered a peerage, but steadily refused it until 1697, when being appointed Lord Chancellor, he was at the same time created Baron Somers of Evesham, in the county of Worcester. To support the dignity, the King granted to him and his heirs the manors of Reigate and Howleigh in Surrey, and 2100*l.* a year out of the fee-farm rents of the crown. "Lord Somers had now reached his highest pitch of worldly prosperity. He was not only the favourite of the King, but he could influence a decided majority in both houses of Parliament, and his general popularity was such, that the High Church party expressed a wish that he were theirs. The Tory fox-hunters could say nothing against him, except that he was 'a vile Whig;' the merchants celebrated him as the only Lord Chancellor who had ever known any thing of trade or finance; the lawyers were proud of him as shedding new glory on their order; and so much was he praised for his taste in literature, and his patronage of literary men, that all works of any merit in verse or prose were inscribed to him[7]."

Lord Campbell adds, that "he assisted Montague in the appointment of Newton as Warden of the Mint, and, on his recommendation, Locke was nominated a Lord of Trade, to carry into effect the sound commercial principles which this great philosopher had propounded in his writings."

---

[6] Campbell, Vol. IV. p. 119. [7] Ibid., p. 133.

His income enabled him to encourage rising merit, and to assist literary men, among whom Addison stands prominent. Tickell, in his *Life of Addison*, relates that "he was in his twenty-eighth year, when his inclination to see France and Italy was encouraged by the great Lord Chancellor Somers, one of that kind of patriots who think it no waste of the public treasure to purchase politeness to their country. The poem upon one of King William's campaigns, addressed to his Lordship, was received with great humanity[8], and occasioned a message from him to the author, to desire his acquaintance. He soon after obtained, by his interest, a yearly pension of three hundred pounds from the Crown, to support him in his travels."

Addison's *Remarks on Italy* were also dedicated to Lord Somers. It appears to be the fate of all great and good men to undergo severe trials; and the life of Lord Somers was no exception to the law. It was, indeed, hardly to be supposed, that the Lord Chancellor of those days could hold his high office without provoking enemies. On the 17th April, 1700, the King yielded to popular clamour, and deprived Lord Somers of the great seal. Oldmixon gives the following account of this transaction, on the authority of a gentleman who had it from Lord Somers. "The

---

[8] The dedication thus commences:

"If yet your thoughts are loose from state affairs,  
Nor feel the burden of a nation's cares,  
If yet your time and actions are your own,  
Receive the present of a muse unknown;  
A muse that in adventurous numbers sings,  
The rout of armies and the fall of kings;  
Britain advanc'd, and Europe's peace restored,  
By Somers' counsels, and by Nassau's sword."

King some time before the prorogation, which was April 11, had given his Lordship a hint of the necessity he should be under to part with him, in order to accommodate matters with those in opposition to the measures of the administration. His Lordship, upon this, told his Majesty, that he knew very well what his enemies aimed at, by their abusing and persecuting him as they had lately done. That the Great Seal was his greatest crime, and if he quitted it he should be forgiven; but knowing what ill use would be made of it, if it were put into their hands, he was resolved, with his Majesty's permission, to keep it in defiance of their malice, and to stand all the trials they should put upon him with the support of his innocence, and the hopes of his being serviceable to his Majesty. That he feared them not; but if he would be as firm to his friends, as they would be to him, they should be able to carry whatever points he had in view for the public welfare in the new Parliament. The King shook his head as a sign of his diffidence, and only said, *It must be so*[9]."

Evelyn writes in allusion to the Chancellor's fall: "The Seal was taken from Lord Somers, though he had been acquitted by a great majority of votes for what was charged against him in the House of Commons. It is certain that this Chancellor was a most excellent lawyer, very learned in all polite literature, a superior pen, master of a handsome style, and of easy conversation."

Party feeling ran too high to be satisfied with the mere dismissal of Lord Somers from office. On the 1st April, 1701, Simon Harcourt, the great Tory law-

---

[9] Oldmixon's *Hist.*, p. 208.

yer, appeared at the bar of the House of Lords, "in the name of the House of Commons, and all the Commons of England, and impeached John, Lord Somers, of high crimes and misdemeanours[10]."

The ex-chancellor was most honourably acquitted; but although powerful attempts were made to restore him to office, they all proved fruitless. Indeed, such was the prejudice of Queen Anne against him, that he was excluded from the Privy Council, and even from the Commission of the Peace.

It was during this period that his duties as President of the Royal Society were peculiarly grateful to him. An ardent lover of literature and science, he had, as a Fellow, long been in the habit of frequenting the Meetings. As President, he now regularly attended, presiding over the Council and ordinary Meetings, and doing all in his power to extend the reputation and usefulness of the Society. Lord Campbell says, that at this time "he presents the *beau idéal* of an ex-chancellor,—active in his place in Parliament when he could serve the state, and devoting his leisure to philosophy and literature." He was an original member of the Kit-Cat Club, where he was in the habit of meeting literary men[11]. Horace Walpole, in his *Life of Vertue, the Engraver*, states that Lord

[10] The Commons carried the following resolution by a majority of 198 to 188:—"That John, Lord Somers, by advising His Majesty, in 1698, to the treaty for partition of the Spanish monarchy, whereby large territories of the King of Spain's dominions were to be delivered up to France, was guilty of a high crime and misdemeanour."

[11] "At the distance of many years, Swift, notwithstanding the hardness of his nature, retained a tender recollection of the pleasant literary reunions at the houses of Pope, Somers, and Montague." Campbell, Vol. IV.

Somers was the first to bring him into notice, by employing him to engrave the portrait of Archbishop Tillotson; and it is well known that he made a superb collection of paintings, engravings, medals, and books, with *objets de vertu*. "If," says Addison, "he delivered his opinion upon a piece of poetry, a statue, or a picture, there was something so just and delicate in his observations, as naturally produced pleasure and assent in those who heard him[12]."

He carried on a correspondence with several distinguished men of letters in foreign countries, and was master of seven languages, without having been out of England.

"Besides his collection of printed tracts (which were twice published, first in 1748, and again in 1809, under the superintendence of Sir W. Scott), he left behind him an immense mass of MSS., partly his own composition, partly others. These came into the possession of the Hardwicke family, who were allied to him by marriage, and being deposited in the chambers of the Hon. Charles Yorke, in Lincoln's Inn, were there nearly all destroyed by an accidental fire. Mr. Yorke collected a few of the papers saved, which he bound in a folio volume. From this a selection was given in the *Miscellaneous State Papers*, published in 1778, by the second Earl of Hardwicke[13]."

So great a man could not remain many years neglected. In 1708 he was appointed Lord President of the Council, which appears to have given general satisfaction, and was so fortunate as even to have partly overcome the Queen's prejudice against him.

---

[12] *Freeholder*, No. 39.

[13] Campbell, Vol. IV. p. 221.

He retained this office for two years, when he was dismissed with his Whig colleagues. He now entered into violent opposition against the Tories, which he maintained until the Queen's death. On the arrival of George I. in England, he was sworn of the Privy Council, had a seat in the Cabinet assigned to him, and would have received the Great Seal, had his health permitted him to undergo the labours of office. But this not being the case, an additional pension of 2000*l.* a-year was settled upon him. His infirmities, from which he had been long suffering, continued to increase, and he expired at his villa in Hertfordshire, on the 26th April, 1716, not old in years, but in worldly greatness and glory. He never married, though he paid his addresses to Miss Anne Bawdon, daughter of Sir John Bawdon, a wealthy alderman of London. This happened when he was Solicitor-General. The lady, it is stated, reciprocated his affection, but her sordid father conceived the offer of a rich Turkey merchant to be more advantageous than that of the future Chancellor, and peremptorily forbade Lord Somers' attention. The latter keenly felt the disappointment, which had the effect of causing him to forego all matrimonial considerations, and to cultivate literature and science more assiduously. Remembering that Newton was the successor of Lord Somers, as President of the Royal Society, we must not regret that the latter occupied the chair for the short period of five years; otherwise we might justly regret that the Royal Society was not presided over by so great and good a man as Lord Somers for a much longer time. It is probable that he would have retained this high office, had not the brilliancy of Newton's name pointed out the distinguished philosopher as the proper person to occupy the chair of the great, and at that time only scientific, body in the kingdom. Lord Campbell expressly states, that Lord Somers "gracefully resigned the Presidency to Sir Isaac Newton," whom he, no doubt, conceived was the fittest individual to succeed him.

It would be easy to multiply extracts from the works of statesmen, historians, and poets, all alike loud and eloquent in their praise of Lord Somers. Addison says, "He had worn himself out in his application to such studies as made him useful or ornamental to the world[14], in concerting schemes for the welfare of his country, and in prosecuting such measures as were necessary for making those schemes effectual. His great humanity appeared in the minutest circumstances of his conversation. You found it in the benevolence of his aspect, the complacency of his behaviour, and the tone of his voice[15]. His great application to the severer studies of the law had not infected his temper with any thing positive or litigious. He did not know what it was to wrangle on indifferent points, to triumph in the superiority of his understanding, or to be supercilious on the side of truth. He joined the greatest delicacy of good breeding to the greatest strength of reason. This great man was not more conspicuous as a patriot and a statesman, than as a person of universal knowledge and learning. As by dividing his time between

---

[14] His motto was, *Prodesse quam conspici.*

[15] De Foe said of him, in his *Jure Divino :*

"Somers, by nature great, and born to rise,
In counsel wary, and in conduct wise;
His judgment steady, and his genius strong,
And all men own the music of his tongue."

the public scenes of business, and the private retirements of life, he took care to keep up both the great and good man; so by the same means he accomplished himself not only in the knowledge of men and things, but in the skill of the most refined arts and sciences[16]." A fine portrait of Lord Somers, by Sir G. Kneller, is suspended in the Meeting-room of the Society. It was presented by Sir Joseph Jekyll.

The Society were highly sensible of the honour conferred on them by Lord Somers assuming the Presidency. A Committee was immediately appointed to wait on his Lordship, and acquaint him with the duties of his new office. Under the date of Dec. 7, Evelyn records :—"Being one of the Council of the Royal Society, I was named to be of the Committee to wait upon our new President, the Lord Chancellor. Our Secretary Dr. Sloane, and Sir R. Southwell, last Vice-President, carrying our book of Statutes, the Office of the President being read, his Lordship subscribed his name, and took the oaths according to our Statutes, as a corporation for the improvement of natural knowledge. Then his Lordship made a short compliment concerning the honour the Society had done him, and how ready he would be to promote so noble a designe, and come himself among us as often as his attendance on the public would permit, and so we took our leave[17]."

In 1698 the Society received a very valuable present from the East India Company, the nature of which will be understood by the following letter.

---

[16] *Freeholder*, No. 39.

[17] *Diary*, Vol. II. p. 61.

"*Gresham College, Nov.* 23, 1698.

"GENTLEMEN,

"THE Royal Society in a full assembly this day commanded me to return you their most humble and hearty thanks for your late valuable and acceptable present of plants, drugs, &c., of the East Indies, with the account of the uses of them in those parts. Their Society being a body instituted for the advancement of naturall knowledge, they will not be wanting in their duty and endeavours to bestow a gift so generously and properly bestowed to the general use of mankind, and, more particularly, that of their own country. They will also take care that your most Honorable Company, whose prosperity they truly wish, shall have all publick acknowledgments due for such a favour[18]. These things I was ordered to write to you in their name, and remain,

"Your, &c. HANS SLOANE.

"*The Hon. East India Company.*"

In this year the Society were deprived of the valuable services of Halley, who sailed on a scientific expedition, in the *Paramour*, fitted out by Government, to lay down the longitudes and latitudes of his Majesty's settlements in America, and to endeavour to verify his theory of the variation of the compass, which he had published in 1683. He set out on the 24th November, but on crossing the line his men grew sickly, and his lieutenant having mutinied, he returned home in June, 1699. On the 16th August, he laid before the Society the variations of the needle, which he had observed during his voyage; and showed that

[18] A description of these plants is printed in the 22nd volume of the *Transactions*.

Brazil was erroneously placed in the maps. It is also stated in the Journal-book, that he submitted some barnacles to the Society, which he observed to be of quick growth. Having succeeded in getting his lieutenant tried and cashiered, he sailed again in September, 1699, in the same ship, accompanied by another of less size, of which he had also the command. He now traversed the Atlantic from one hemisphere to the other, and having made observations at St. Helena, Brazil, Cape Verd, Barbados, Madeira, the Canaries, the Coast of Barbary, and in many other latitudes, returned to England in 1700, and the next year published a general chart, showing at one view the variation of the compass in all those places[19]. This will be more particularly mentioned, with reference to the recent labours of the Society, in connexion with terrestrial magnetism.

About the same time that Halley sailed on his first voyage, Mr. Jezreel Jones, who had acted as clerk to the Society for two years, set out under their patronage on an expedition of discovery into the interior of Africa. The Council appear to have entertained a very high opinion of Mr. Jones; they voted him 100*l.* towards his journey, regretting their ina-

---

[19] Respecting Halley's voyages, Humboldt says: "Never before, I believe, did any government equip a naval expedition for an object, which, whilst its attainment promised considerable advantages for practical navigation, yet so properly deserves to be entitled scientific or physico-mathematical. Halley, as soon as he returned from his voyages, hazarded the conjecture that the Aurora Borealis is a magnetic phenomenon. Faraday's brilliant discovery of the evolution of light by magnetism, has raised this hypothesis, enounced in 1714, to the rank of an experimental certainty." *Cosmos,* Vol. II. p. 333.

bility to defray all the expenses of his expedition[20]. The sum of 20*l.* was voted at the same time to Mr. Vernon, to assist him in exploring the Canaries, and the Council-minutes add, "to let him know that they expect of him, besides natural history collections, to take care to answer such queries as should be recommended to him from the Society[21]."

In 1700, Dr. Bidloo published a work, which he dedicated to the Society, entitled *Gulielmus Cowper criminis literarii citatus coram Tribunali Nobiliss. Ampliss. Societatis Britanno Regiæ*, in which he brought forward several serious charges against Mr. Cowper, a surgeon. With a copy of the work he addressed a letter to Dr. Sloane, requesting him to urge the Council to adjudicate between him and Mr. Cowper. Dr. Sloane did so, and was ordered by the Council to acquaint Dr. Bidloo, "that the Society are not erected for determining controversies, but promoting naturall and experimentall knowledge, which they will do in him or anybody else[22]." This transaction is interesting, as showing the early determination of the Council not to mix themselves in quarrels between scientific men, whether Fellows of the Royal Society or not.

Soon afterwards a pamphlet, called the *Transactioneer*, made its appearance, the tendency of which was to cast ridicule on the Society and its *Transac-*

---

[20] Council-minutes, Vol. II. p. 111. Mr. Jones communicated some of his observations and discoveries to the Society. See *Phil. Trans.*, Vol. XXI. p. 248.

[21] Council-minutes, Vol. II. p. 101.

[22] The reader is referred to Dr. Sloane's letter to Dr. Bidloo, for further information on this subject. Letter-book, Vol. XII. p. 210.

*tions*. It is so frequently alluded to in the Minutes of Council, with the view of taking legal proceedings against the author, that I became anxious to see the work. The library of the Royal Society does not possess a copy of this scurrilous publication. One is preserved in the British Museum: it is entitled *The Transactioneer, with some of his Philosophical Fancies, in two dialogues.* The preface, which consists mainly of an attack upon the Secretary, commences: "By the following dialogues it is apparent that by industry alone a man may get so much reputation, almost in any profession, as shall be sufficient to amuse the world, though he has neither parts nor learning to supply it." But the whole publication is of so low and ridiculous a nature that it is surprising the Council should have thought it worth their while to notice it[23]. They appear, however, to have used every exertion to discover the author's name; and it seems, by the following letter, the original of which exists in the British Museum, that Dr. Woodward was suspected to be concerned in the publication, but he warmly disowns it[24].

"It cannot be any news to you, y^t there was a while ago published a pamphlet, entitled y^e *Transactioneer*,

---

[23] This will appear from these heads, taken at random from the table of contents:

Eggs in the cauda of a barnacle.
Four sorts of lady's bugs.
A buck in a snake's belly.
A shower of whitings.
A shower of butter to dress them with.

[24] Dr. Johnson says: "Dr. William King, born in 1663, a man of shallowness, wrote the *Transactioneer* in 1700, in which he satirized the Royal Society, at least Sir H. Sloane." *Works*, Vol. x. p. 32.

which calls Dr. Sloane's management of the *Philos. Transactions* in question: and I am sorry to find two or three Members of $y^e$ Society, and my particular friends, ill-treated in it: the writer of it is but meanly qualified for what he undertakes; tho' whether there was not occasion given, may be worth your consideration. This I'm sure, the world has been now for some time past very loud upon $y^t$ subject: and there were those who laid $y^e$ charges so much wrong, $y^t$ I have but too often occasion to vindicate even the Society it self, and $y^t$ in publick company too. 'Twas a regard I justly ow'd $y^t$ excellent Body, and shall be always ready to pay it, as well as to contribute all that is in my power towards carrying on the design of it. At present, Sir, 'tis some dissatisfaction to me to ly under a necessity of laying a complaint before you. The matter is this: Dr. Sloane and his friend Mr. Pettiver cause it to be spread abroad that I am the author, or at least concerned in writing $y^e$ aforesaid pamphlet. They do not directly charge me with it: that is not their way, but they do the thing as effectually by insinuating in their clubbs and meetings, from whence all the rumour comes, $y^t$ $y^e$ world ascribes the pamphlet to me. At other times they assert that it was wrote by a Member or Members of $y^e$ Society. I cannot but believe they know $y^e$ true Author all $y^e$ while: at least they know I utterly disown it. What reason they can have to suspect me of any such practises, they best know. For my own part, I'm sure I'm far from them: and will never say or write any thing $y^t$ I will not own and justify. I am so far from writing the *Transactioneer*, or having any hand in it, $y^t$ I can averr it in great truth I do not so much as know who wrote it. At the same time $y^t$ they intitle me to it, they assert $y^t$ $y^e$ pamphlet is very foolish and very malicious. 'Tis certainly not fit $y^t$ without any grounds I should thus publickly be brought under so very heinous a charge: and I submit it to you and $y^e$ Council to direct in what manner I shall have reparation and justice done me. They are pleased likewise to assign Mr. Harris a share in the work; which I have acquainted him with, and I do not make any question but he will vindicate himself[25].

"I am

"Your obedient humble Servant,

"J. WOODWARD."

In 1699 the Academy of Sciences at Paris received a new charter, which gave the members considerable powers, and at the same time advanced and rewarded science. The fact is worthy of mention, as marking the different manner in which the great learned Societies of England and France were treated by their respective Sovereigns. In the latter country science was thus early fostered and rewarded, while in England the Royal Society were left to struggle with poverty. The letter of M. Geoffroy (an Academician) to Dr. Sloane, giving an account of the new organization of the Academy, is sufficiently interesting to be preserved here. It is dated Paris, March 7, 1699.

"I shall here give you an account of the great splendor that the Académie des Sciences has received by the regulations, increase, encouragement, and orders, M. L'Abbé Bignon has obtain'd to it from the King. That Academy is now composed of ten honorary academicians, which are chosen learned and eminent gentlemen; of eight strangers, associates, each of which are distinguished by his learning;

[25] Sloane MSS. 4441.

twenty pensioners, fellows; twenty éleves; twelve French associates. Between the honorary Academicians, two are elected every year, one for President, the other for Vice-president. Twenty pensioners have every year 1500 French livres, and after the death of one pensioner, the Académie will propose to the King three persons, associates or éleves, or sometimes others, and His Majesty will call one of the three for pensioner[26]." The letter concludes with a list of the Academicians, and it is gratifying to find the name of Newton as a foreign Member[27].

During the Secretaryship of Sir Hans Sloane, the scientific correspondence of the Society received considerable impulse. The Letter-books contain copies of a great number of letters which he addressed to persons, at home and abroad, requesting communications on subjects relating to the objects of the Society. In one of these we read:—

"The Royall Society are resolved to prosecute vigorously the whole designs of their institution, and accordingly they desire you will be pleased to give them

---

[26] Letter-book, Vol. xii. p. 91.

[27] Dr. Lister, in his *Journey to Paris,* in giving an account of the Académie des Sciences, says, "I was informed by some of them, that they have this great advantage to encourage them in the pursuit of Natural Philosophy, that if any of the Members shall give in a bill of charges of any experiments which he shall have made; or shall desire the impression of any book, and bring in the charges of graving required for such book, the president allowing it and signing it, the money is forthwith reimbursed by the King. Thus, if M. Merrie for example, shall require live tortoises for the making good the experiments about the heart, they shall be brought him, as many as he pleases, at the King's charge." p. 79. Such royal patronage, it must be confessed, was wholly unknown to the English philosophers.

an account of what you meet with or hear of, that is curious in nature, or any ways tending to the advancement of naturall knowledge, or usefull arts. They in return will always be glad to serve you in any thing in their power."

The result of this correspondence is apparent in the large number of communications made to the Society at the ordinary Meetings, and recorded in the Journal and Register-books. Several of these are of great interest; and it is not a little curious to trace in them the germ of discoveries in science which, in a more perfected state, have effected such extraordinary changes in the condition of mankind. Of such a nature was Savery's condensing steam-engine, a model of which was exhibited to the Society on the 14th June, 1699. In the Minutes of that date we find, that "Mr. Savery entertained the Society with shewing his engine to raise water by the force of fire. He was thanked for shewing the experiment, which succeeded according to expectation, and was approved of." Savery presented the Society with a drawing of his engine[28], accompanied by a description, which was printed in the 21st volume of the *Transactions*. At his request, the Society gave him a certificate, that the engine exhibited before them "succeeded according to expectation, and to their satisfaction." It will be remembered, that although the Marquis of Worcester undoubtedly invented a steam-engine "to drive up water by fire," as specified in his *Century of Inventions*, and which Cosmo de' Medici, Grand Duke

[28] This is preserved among the Society's collection of prints and drawings. It is entitled, *An Engine for Raising Water by Fire*, by Thomas Savery.

of Tuscany, describes in his *Diary* as having seen in operation at Vauxhall in 1656[29], yet we are indebted to Savery for the introduction of a vacuum, which enabled his engine to perform double the work of that invented by the Marquis of Worcester.

The certificate granted to Savery by the Society, was the means of his procuring a patent from the Crown for the manufacture of steam-engines. It is recorded, in Switzer's *System of Hydrostatics*, published in 1729, that "the first time a steam-engine play'd was in a potter's house at Lambeth, where, though it was a small engine, yet it (the water) forced its way through the roof, and struck up the tiles, in a manner that surprised all the spectators[30]."

The Museum was also indebted to Sir Hans Sloane for many curiosities sent from abroad. At a Meeting in 1700, a live crocodile and some opossums were exhibited, which afforded considerable interest to the Fellows.

In 1703 the Society lost another of their ablest men,

---

29 "His Highness, that he might not lose the day uselessly, went again after dinner to the other side of the city, extending his excursion as far as Whitehall beyond the Palace of the Archbishop of Canterbury, to see an hydraulic machine, invented by my Lord Somerset, Marquis of Worcester. It raises water more than forty geometrical feet by the power of one man only; and in a very short space of time will draw up four vessels of water through a tube or channel not more than a span in width; on which account it is considered to be of greater service to the public than the other machine near Somerset-house." p. 325.

30 The engineer may be interested to learn, that a few years after Savery's engine was exhibited, a Paper was read before the Society, entitled, *A description of an Engine to raise water by the help of Quicksilver*, invented by the late Mr. Joshua Hoskins, and made practicable and useful by J. Desaguliers. See *Phil. Trans.*, No. 370.

by the death of Robert Hooke, who died on the 3rd March in the above year. He had been very infirm for several months previous to his decease, so much so, indeed, that one of his biographers speaks of him as "living a dying life." Yet, although in this unhappy state, the Royal Society still occupied his thoughts[31], for we find Halley giving an account of his marine barometer, which Hooke personally was unable to do; and on the 24th February, only eight days before his decease, there is an entry in the Journal-book of his applying for Cheyney's *Fluxionum Methodus Inversa*, which had been shortly before presented to the Society.

"He was," says his biographer Waller, "of an active, restless, indefatigable genius, even almost to the last; and always slept little, to his death, seldom going to sleep till two, three, or four o'clock in the morning, and seldomer to bed, oftener continuing his studies all night, and taking a short nap in the day. His temper was melancholy, mistrustful, and jealous, which more increased upon him with his years." This failing was probably aggravated by the long chancery-suit in which he was involved, to recover the salary accorded to him by Sir John Cutler, which terminated eventually in his favour, on the 18th July, 1696. How heavily this affair weighed upon his spirits, appears, not only from his repeated applications to the Council, to grant him a certificate of his having delivered the required lectures[32], but also

---

[31] His *Vindication of the Royal Society* abundantly testifies the interest which he took in all its affairs.

[32] The following is a copy of the certificate. The original is preserved in the British Museum among Hooke's Papers:

"We

from the entry in his *Diary*, recording the happy issue of his law-suit, which runs thus: "D.O.M.S.H. L.G.I.S.S.A.[33] I was born on this day of July, 1635, and God has given me a new birth: may I never forget his mercies to me; whilst he gives me breath may I praise him!"

Waller states that he, as well as others, often heard Hooke declare, that "he had a great project in his head, to devote the greatest part of his estate to promote the design of the Royal Society." "With this view," Waller pursues, "he proposed building a handsome fabric for the Society's use, with a library,

---

"We, the President and Councell of the Royall Society of London for improving naturall knowledge, doe hereby certify all whom it may concern, that Robert Hooke, Dr. of Physick, and one of the Fellows of our Society, was for his great learning in naturall and mathematicall sciences, his diligence and readiness of invention in things of art, made choice of and employed as Director and Curator of Experiments to be made at the publique Meetings of the said Society, and was by them paid eighty pounds a year for the said employment, untill Sir John Cutler, Kt. and Bart., (both for the said Dr.'s encouragement, and also to express his respects to the said Society,) did by bond in the penalty of 1000*l*. secure to the said Dr. an annuity of 50*l*. to be paid him half-yearly, during his the Dr.'s life; whereupon, so much of the said allowance was abated, and only 30*l*. yearly was afterwards paid by the said Society, and the same was accepted by the said Dr., in consideration of the aforesaid annuity settled on him by Sir John Cutler. And we doe further certify, that the said Dr. Hooke hath ever since the said annuity was settled, constantly continued to read lectures and to discourse on such subjects as were recommended to him by the said Society. All which he hath performed to their great satisfaction, and to the full of his undertaking. In testimony whereof we have hereunto caused our common seal to be affixed."

[33] This is read *Deo Opt. Max. summus Honor, Laus, Gloria, in secula seculorum. Amen.*

repository, laboratory, and other conveniences for making experiments, and to found and endow a perpetual physico-mechanick lecture, of the nature of what he himself read. But though he was often solicited by his friends, to put his designes down in writing, and make his will as to the disposal of his estate to his own liking in the time of his health; and after, when himself and all thought his end drew near, yet he could never be prevail'd with to perfect it, still procrastinating it, till at last this great design prov'd an airy phantom and vanish'd into nothing. Thus he died at last, without any will and testament that could be found." "It is indeed," adds his biographer, "a melancholy reflection, that while so many rich and great men leave considerable sums for founding hospitals, and the like pious uses, few since Sir Thomas Gresham should do anything of this kind for the promoting of learning, which, no doubt, would be as much for the good of the nation, and glory of God, as the other of relieving the poor."

It may appear singular that Hooke, who, when an officer of the Royal Society, was almost clamorous for an increase of salary, and, at a time when the Society's funds were extremely low, prayed that he might be allowed six months' time to consider the resolution of Council respecting the payment of his salary in books[34], should talk so largely of building a handsome edifice for the Society. That he had the means to do so was, however, certain; for a large iron chest was found after his death, "locked down with a

---

[34] It will be remembered, that for some years previously to his death he received 10*l.* per annum from the Society for the use of his lodgings in Gresham College.

key in it, and a date of the time, showing it to have been so shut up for above thirty years, in which were contained many thousand pounds in gold and silver[35]." It is right to add, that Hooke was extremely parsimonious, scarcely allowing himself the common necessaries of life.

His errors and frailties were alike forgotten over his grave, to which he was attended by all the members of the Royal Society in London at the time of his decease, and who unanimously lamented him as one of their greatest ornaments, and promoters of science. His energy was truly astonishing, and although this fact is most amply confirmed by his posthumous works, we must examine the Journal and Register-books of the Royal Society to become fully aware of the labours of this great philosopher. They are a wonderful monument of his mathematical and mechanical genius; for there is hardly a page, during many years, in which his name does not appear, in connection with new inventions.

In 1701 the Trustees of Gresham College having obtained the consent of all the Professors with the exception of Hooke, brought a Bill into Parliament for rebuilding the College, upon the plea "that it had become old and ruinous, and the repairs thereof very expensive; but the said College standing upon a con-

---

[35] Waller's *Life*, p. 13. It was as one of the city surveyors that Hooke acquired the greater part of his estate. Dr. Waller says: "He might by this place acquire a considerable estate, every particular person being in haste to have his concerns expedited; so that, as I have been inform'd, he had no rest early and late from persons soliciting to have their grounds set out, which, without any fraud or injustice, deserv'd a due recompense in so fatiguing an employ."

siderable quantity of ground, and great part of it lying waste, good improvement might be made by rebuilding it." The Bill passed through the Commons, but at the second reading in the House of Lords it was thrown out upon the petition of Hooke, who had opposed it from the beginning in the strongest manner. Shortly after Hooke's death, the Trustees, who were exceedingly annoyed at the loss of their Bill, again brought it forward. Before introducing it into the House of Commons, they acquainted the Royal Society that they were desirous of "accommodating them with conveniences for their meetings, repository, and library." The Council "Ordered their humble thanks to be given to the Committee by the President, and desired Sir Christopher Wren that he would please to take the trouble of viewing the design and project, and consider what accommodations the Society wanted, and to resolve by changing or purchasing ground fit for their affairs, to add to what the Committees offer for their accommodation[36]."

Wren accordingly examined the design, and drew up a document entitled, *Proposals for building a House for the Royal Society*:—

"It is proposed as absolutely necessary for the continuing the Royall Society at Gresham Colledge, that they should have a place so seated in the said ground, that the coaches of the Members, (some of which are of very great quality) may have easy access, and that the building consist of these necessary parts.

"1. A good cellar under ground, so high above it as to have good lights for the use of an elaboratory, and housekeeper.

---

[36] Council-minutes, Vol. II. p. 120.

"2. The story above may have a fair room and a large closet.

"3. A place for a repository over them.

"4. A place for the Library over the repository.

"5. A place covered with lead for observing the heavens.

"6. A good stair-case from bottom to top.

"7. A reasonable area behind it, to give light to the back-rooms.

"All which may be comprised in a space of ground 40 foot in front, and 60 foot deep[37]."

In consequence, however, of the insertion of a clause in the new Bill, "that the Trustees should be obliged and required to build these houses, hall, and almshouses, for the lecturers and almsfolks, within five years from the passing of this Act, upon the penalty of two thousand pounds," exception was taken, and the bill was rejected by the Commons on its first reading[38]. All prospect of procuring better accommodation being thus at an end, the Council considered the expediency of removing from the College. At their Meeting, held on the 21 April, 1703, it was "Resolved, that the Society should purchase a place of abode for themselves; and it was ordered that a Committee, consisting of Mr. Isted, Mr. Hill, Dr. Tyson, Sir John Hoskyns, Dr. Sloane, and the Treasurer, should consider of a place to build on, or buy, and lay their thoughts before the Society."

---

[37] The original of this document is preserved in the Archives of the Royal Society.

[38] Stow's *London*, Second Appendix, Vol. IV. p. 22, edit. 1720. *Commons' Journals*, Vol. XIV. p. 426. It would appear by the petition that the Mercers' Company were considerably embarrassed.

From this period until the Society removed, the Council-minutes make frequent mention of the labours of the Committee, who reported on various localities and houses, which appeared suitable for the purposes of the Society. Amongst these was a house in Whitehall, and ground for building in the Savoy, and near St. James's Park. The Duke of Bedford also offered the Society "an estate of inheritance, or a lease of ground for 61 years;" but as these proposals did not meet the approbation of the Council, the Society meanwhile continued to occupy their apartments in Gresham College.

At the Anniversary in 1703 Lord Somers retired from the Presidency, and Sir Isaac Newton was elected to this high office, which even in those early days of the Society was regarded as conferring great honour and distinction upon the individual selected to fill it. In this case, the election was alike honourable to the Society and to Newton. It is not a little remarkable that he was chosen into the Council for the first time, and elected President, on the same day. The cause of his not having been called earlier to the Councils of the Society, arose probably from the jealousy of Hooke, which betrayed itself in so melancholy a manner for some years previous to his decease, that it is hardly possible to conceive how Newton could have sat at the same board with him. It is well known that Newton decided not to publish his *Optics* during the lifetime of Hooke; tolerably conclusive evidence of his wish not to expose himself to the attacks of the irritable philosopher[39], of whom Biot said, in the

[39] In the Preface to the *Optics*, written a short time after Hooke's

words which had before been applied by D'Alembert to Fontaine, *Hooke est mort;—c'était un homme de génie, et un mauvais homme; la Société y gagne plus que la géométrie n'y perd!*

---

Hooke's death, Newton says: "To avoid being engaged in disputes about these matters, I have hitherto delayed the printing."

MANOR-HOUSE, WOOLSTHORPE,

THE BIRTH-PLACE OF SIR ISAAC NEWTON;

*Showing the Solar Dials which he made when a boy.*

# CHAPTER XIV.

Memoir of Sir Isaac Newton—His constant attendance at the Meetings—Presents his *Optics* to the Society—Prince George of Denmark elected—Requested by the Society to print Flamsteed's *Observations*—He consents to defray the Expense—Committee appointed to superintend the Publication—Flamsteed's Dissatisfaction—Painful Dispute—He burns the *Historia Cœlestis*—Prints a more perfect edition at his own expense—Newton's Propositions for Financial Improvements—Papin's Proposal to construct Steam-vessel—Edinburgh Philosophical Society—Death of Sir G. Copley—His Bequest—Devoted at first to Experiments—Gold Medal afterwards adopted—Awarded to Dr. Franklin—Mercers' Company give notice of their intention to withhold Apartments—Petition to the Queen for Land in Westminster—Application to Trustees of Cotton Library—Purchase of Dr. Brown's House in Crane Court—Objections by some of the Fellows—Proceedings of Council with respect to the Removal—First Meeting in Crane Court—Regret of the Gresham Professors on the Departure of the Society.

---

## 1700—10.

IT has devolved on me, in the course of this work, to notice the principal discoveries of Newton, which, as already shown, were communicated to the Royal Society. To these, therefore, we need only slightly allude, the object of this memoir being to give such an outline of the life of their author, as may refresh the memory of the reader, who will see that, amidst occupations of great care and responsibility, which would alone have engrossed all the time and thoughts of an ordinary mind, the greatest of all philosophers found opportunities to attend not only to onerous public duties, but also to the affairs and interests of the Royal Society, over which he presided for twenty-five years.

The mind of Newton has been happily and popularly compared by Professor De Morgan, to "a person who is superior to others in every kind of athletic exercise; who can outrun his competitors with a greater weight than any one of them can lift standing[1]." Upwards of a century has passed away, and the name of Newton still remains amongst us, "an object of unqualified wonder," a name to be pronounced with reverence.

Isaac Newton was born at Woolsthorpe, near Grantham, in Lincolnshire, on the 25th December, 1642, exactly one year after the death of Galileo. He was a puny and weakly infant, giving little promise of a vigorous maturity or prolonged life. According to his statement, made at the Herald's Office, his great grandfather's father was John Newton, of Westby, in Lincolnshire.

His father, Isaac Newton, married the daughter of James Ayscough, of Market Overton, in Rutlandshire. He died before the birth of his son, the only issue of the marriage. Mrs. Newton contracted a second alliance with the Rev. Barnabas Smith, rector of North Witham, and confided her son, who was then about three years old, to the care of his maternal grandmother, by whom he was sent to a day-school, and at the age of twelve to the public school at Grantham. He lodged in this town at the house of Mr. Clarke, an apothecary, where he met Miss Storey, daughter of Dr. Storey of Buckminster, and for whom he formed a friendship, which grew into a warmer feel-

---

[1] *Life of Newton*, p. 106.

ing. Their union was only prevented by their poverty. Miss Storey was afterwards twice married. The decease of Mrs. Newton's second husband in 1656, caused her to return to Woolsthorpe, and, in that year, her son was taken from school, where he had risen to the highest form, and distinguished himself by his mechanical contrivances, and brought home to assist in the management of the farm.

It was soon seen that agricultural pursuits were utterly opposed to his literary and studious habits, which increasing years further confirmed. Mrs. Smith, in consequence, wisely resolved on sending him back to Grantham school, where he remained for some time actively engaged in academical studies. By the advice of his maternal uncle, the Reverend W. Ayscough, he was sent to Cambridge[2], and admitted on the 5th June, 1660, into Trinity College.

According to the College-books, he was sub-sizar in 1661, scholar in 1664, Bachelor of Arts in 1665, Junior Fellow in 1667, Master of Arts and Senior Fellow in 1668, and Lucasian Professor of Mathematics in 1669. The chair of the latter had been filled by Dr. Barrow, one of the greatest mathematicians of his age, by whom it was resigned in favour of Newton.

Some of the new Professor's discoveries had been made anteriorly to this period. In 1663—4, he discovered the celebrated Binomial Theorem. "In 1665

[2] Biot, in his *Life of Newton*, relates that his uncle Ayscough found him under a hedge absorbed in the solution of a mathematical problem. A writer in the *Gentleman's Magazine*, who states that when a boy he received great kindness from Newton, gives this story, with the unimportant difference of the scene of Newton's labours being a hay-loft instead of a hedge. See Vol. for 1772.

he arrived at his discoveries in series, and substantially at his method of fluxions:" and in 1666 he made his great discovery of the unequal refrangibility of the rays of light.

In the last of these years he retired to Woolsthorpe, on account of the plague which was then raging; and there he began to reflect more particularly upon the nature of the force by which bodies at the earth's surface are drawn towards its centre; and to conjecture that the same force might possibly extend to the moon, with a sufficient intensity to counteract the centrifugal force of that satellite, and retain it in its orbit about the earth.

He returned to Cambridge when the plague subsided, but did not pursue his hypothesis further for several years. "The subject," says Professor De Morgan, "was not resumed till 1679; not, as commonly stated, because he then first became acquainted with Picard's measure of the earth (we think Professor Rigaud has shown this), but because leisure then served, and some discussions on a kindred spirit at the Royal Society had awakened his attention to the question. In 1679 he repeated the trial with Picard's measure of the earth; and it is said, that when he saw that the desired agreement was likely to appear, he became so nervous that he could not continue the calculation, but was obliged to intrust it to a friend. From that moment the great discovery must be dated; the connexion of his speculations on motion with the actual phenomena of the universe was established[3]."

The truth of this splendid discovery of Gravitation

---

[3] *Life*, p. 86.

has been lately confirmed in a wonderful manner. A new planet, by Newton's laws, has been computed with unerring accuracy into visible existence. "There is nothing," observes Mr. Airy, "in the whole history of astronomy, I had almost said in the whole history of science, comparable to this[4]."

Pursuing the chronological order of events, we find that in 1671 Newton sent his reflecting telescope to the Royal Society. In 1671—2 he was elected a Fellow, and during the succeeding years made several communications to the Society. In 1686 the manuscript of the *Principia* was presented to the Society, and published in 1687. In 1688 he was returned to Parliament, as one of the representatives of his University. He was again returned in 1703, but lost his election in 1705. He does not appear to have taken any conspicuous part in the debates of the House. In 1692 occurs the curious episode in his history, which led to the report and belief abroad that he had become insane.

The beautiful, but traditionary story of his dog Diamond having overturned a lighted taper upon his desk, and set fire to several precious manuscripts, and the calmness of Newton under the circumstances, is well known. "But," observes Mr. De Morgan, "the truth, as appears by a private *Diary* of his acquaintance, Mr. de la Pryme, recently discovered, is, that in February 1692 he left a light burning when he went to chapel, which, by unknown means, destroyed his papers, and among them a large work on *Optics*, containing the experiments and researches of twenty years. When Mr. Newton came from chapel, and

---

[4] *Proceedings of Astronomical Society*, Vol. VII. p. 121.

had seen what was done, everybody thought that he would have run mad; he was so troubled thereat, that he was not himself for a month after. Such phrases reported, gave rise to a memorandum in the *Diary* of the celebrated Huyghens (the first foreigner who understood and accepted the theory of Gravitation), stating that he had been told that Newton had become insane either from study, or from the loss of his laboratory and manuscripts by fire; that remedies had been applied, by means of which he had so far recovered, as to be then beginning again to understand his own *Principia*. That Newton was in ill health in 1692 and 1693 is known, but his letters to Dr. Bentley on the Deity, written during that period, are proof that he had not lost his mind[5]."

In 1695 Newton was appointed Warden, and in 1699 Master, of the Mint. For these offices he was indebted to Charles Montague, afterwards Lord Halifax.

"*J'avais cru, dans ma jeunesse,*" says Voltaire, "*que Newton avait fait sa fortune par son extrême mérite. Je m'étais imaginé que la cour, et la ville de Londres l'avaient nommé par acclamation grand maître des monnaies du royaume. Point du tout. Isaac Newton avait une nièce assez aimable nommée Madame Conduit, elle plut beaucoup au grand trésorier Halifax. Le calcul infinitésimal et la gravitation ne lui auraient servi de rien sans une jolie nièce*[6]."

The reader will scarcely arrive at Voltaire's flippant conclusion, which is not in keeping with his general remarks upon Newton and England, contained in the very volume from which the preceding passage is taken.

Honours were now rapidly showered upon the

---

[5] *Life*, p. 84. [6] *Dict. Phil.*, Tom. IV. p. 61.

great philosopher, his claims to which were fully acknowledged.

In 1699 he was elected a Foreign Member of the French Academy. In 1703 he was chosen President of the Royal Society, an office to which he was annually re-elected during the remaining twenty-five years of his life. On the 16th April, 1705, he was knighted at Cambridge by Queen Anne, who always treated him with marked esteem. In 1709 he entrusted to Roger Cotes the preparation of the second edition of the *Principia*, which appeared in 1713.

On the Accession of George I. to the throne, Newton became an object of interest at Court, and was honoured by the friendship of the Princess of Wales, who corresponded with Leibnitz. At her request he drew up a Paper containing his ideas on chronology, which was first printed surreptitiously at Paris. In 1722 he became subject to a disorder in the urinary organs, accompanied by cough and gout. His health declined gradually, and he died on the 20th March, 1727, in the eighty-fifth year of his age.

His body was interred in Westminster Abbey, and was honoured by a magnificent funeral. In 1731, a monument to his memory, designed by Kent and sculptured by Rysbrack, was erected in the Abbey; and in the same year, a medal was struck in his honour, bearing on one side his head, with the motto *Felix cognoscere causas*, and on the reverse, a personification of science displaying the solar system. In 1755 the beautiful statue of Newton, from the chisel of Roubilliac, was erected in Trinity College Chapel, Cambridge.

The Royal Society possess several most interesting relics of Sir Isaac Newton, which are duly noticed in a subsequent part of this work.

In contemplating the life of this great man, our feelings must be those of profound admiration of his high qualities, and gratitude to Almighty Providence, who has permitted so good a being, endowed with the highest mental faculties, to have lived before us, an example and model for all succeeding ages. His faults and imperfections were so few, as to be justly compared to the spots which feebly obscure the summer sun's noon-tide brilliancy. His intellectual capacity seems almost superhuman. "Even," says Dr. Whewell, "with his transcendent powers, to do what he did, was almost irreconcileable with the common conditions of human life, and required the utmost devotion of thought, energy of effort, and steadiness of will,—the strongest character, as well as the highest endowments, which belong to man[7]."

And yet, how humbly he thought of himself, and of his stupendous labours, best appears in the beautiful words which he is reported to have uttered a short time previous to his death. "I know not what I may appear to the world, but to myself I seem to have been only like a boy playing on the sea-shore, and diverting myself in now and then finding a smoother pebble or a prettier shell than ordinary, whilst the great ocean of truth lay all undiscovered before me."

Numerous and ready were the pens and voices which mourned his decease and chronicled his glory. The lines of Thomson in the poem on his death are amongst the most beautiful:

> "Say, ye who best can tell, ye happy few,
> Who saw him in the softest lights of life,
> All unwithheld, indulging to his friends

---

[7] *History of the Inductive Sciences*, Vol. II. p. 193.

> The vast unborrowed treasures of his mind.
> Oh, speak the wondrous man! how mild, how calm,
> How greatly humble, how divinely good,
> How firm established on eternal Truth!
> Fervent in doing well, with every nerve.
> Still pressing on, forgetful of the past,
> And panting for perfection; far above
> Those little cares and visionary joys,
> That so perplex the fond impassioned heart
> Of ever-cheated, ever-trusting man."

Such was Sir Isaac Newton, and well may the Royal Society reflect with pride and satisfaction that he was their President for a quarter of a century.

From the period of Newton's election until within a few weeks of his decease, the Journal and Council-books show that he presided at almost every meeting of the Fellows: so anxious indeed was he to do this, that a short time after his election, finding that his duties at the Mint interfered with his attendance at the Royal Society, he caused the day of Meeting to be changed from Wednesday to Thursday, in order that he might devote his undivided time on the latter day to the Society.

On the 16th February, 1703—4, the Journal-book records, that "Mr. Newton presented his *Opticks* to the Society. Mr. Halley was desired to peruse it, and give an abstract of it." This work contains his researches on the *inflexion* of light, made long before this period, but which he could not be prevailed on to publish during the lifetime of Hooke. The *Optics* was first published in English, and afterwards translated into Latin by Dr. Samuel Clarke. Newton was so much pleased with this translation, that he presented Dr. Clarke with 500*l*., as a testimony of his acknowledgments: many editions of the work itself, and of the translation, rapidly succeeded each other, both in England and on the Continent.

At the Anniversary in 1704, Prince George of Denmark was elected a Fellow of the Society. The Journal-book states, that "the Society were extremely pleased with the honour the Prince did them, in suffering them to choose him a member," and the Council desired "the President and Secretary to wait on the Prince with the Statute-book, to have the honour of his subscription."

At the period of the Prince's election, the Society were very anxious to publish Flamsteed's *Observations*, and were only prevented by the expense, which was estimated at 863*l*. Under these circumstances, they brought the matter before Prince George, praying his favourable consideration. The result was the following letter from Mr. Clarke to the President, written by order of the Prince.

"*December* 11, 1704.

"SIR,

"THE Prince has perused the estimate of the intended *Historia Cœlestis Britannica*, which you presented him. His Royal Highness is persuaded of Mr. Flamsteed's fitness for a work of this nature, and being unwilling that the *Observations*, designed for the benefit of navigation, and encouraged so well in the beginning, should want any necessary assistance to bring them to perfection, he has been pleased to command me to desire yourself, Mr. Roberts, Sir C. Wren, Dr. Gregory, Dr. Arbuthnot, and others of your Society as you think proper, and will share the trouble with you, to inspect Mr. Flamsteed's papers, and consider what is fit for the press; and when His Royal Highness knows your opinions, you may be sure he will do any thing that will conduce to the making them of use to the public.

"I am, &c.,

"GEO. CLARKE."

It appears by Flamsteed's history of his own life, that some friend of his had "acquainted the Prince with his performances," and in another place he mentions the "Prince's inclination to make him easier in his work." His Royal Highness was consequently not wholly unprepared for the application made to him by the Society.

In conformity with the Prince's wish, a Committee consisting of the persons mentioned in Mr. Clarke's letter examined Flamsteed's manuscripts, and reported that "all the Observations, together with the two Catalogues of the Fixed Stars in Latin, are proper to come abroad." "This set of Observations," they add, "we repute the fullest and completest that has ever yet been made; and as it tends to the perfection of astronomy and navigation, so, if it should be lost, the loss would be irreparable; and we have no prospect that a work so expensive will ever see the light, unless your Highness will please to be at the charge of publishing it." The report bears the date of 23rd January, 1704—5, and soon after the *Observations* were sent to press.

It would be foreign to the object of this work, to enter into the unhappy contentions which accompanied the printing of the *Historia Cœlestis.* It is painful at all times to witness the failings of poor humanity, and doubly so, when they are exhibited by philosophers, whom we wish to regard as bright examples worthy of all imitation. The rupture between Newton and Flamsteed, attended by suspicions which Mr. Baily has well designated as "unaccountable, unwarrantable, and extremely revolting," is a melancholy instance, that even giants in intellect are not free from the failings of their less gifted brethren[8]. It is sufficient to state here, that a quarrel arose between Newton and Flamsteed, respecting the printing of the latter's *Observations*, which, putting aside all private considerations, had the effect of delaying the publication of the *Historia Cœlestis* for several years, and rendering it incorrect when published[9].

---

[8] The reader desirous of examining into some of the particulars of this unhappy dispute, and the proceedings of the Royal Society relative to it, is referred to Baily's *Account of Flamsteed.* In a curious and interesting communication, made by Maskelyne to the Society, upon the Greenwich Observatory and the *Observations*, inserted in 25th volume of the Journal-book, he says: "Mr. Flamsteed the first Astronomer Royal's tenaciousness in witholding his observations, not only from the public, but even from his learned friends, a long while deprived Newton of the means of establishing his theory of the moon; and might have done so longer if that great man, in conjunction with other learned ornaments of this Society, had not obtained an order from superior authority, for selecting and publishing such of the observations made at the Royal Observatory as they should think proper." This was written when the facts were fresh in men's minds.

[9] The *Historia Cœlestis* was published in 1712, but in a form so distasteful to Flamsteed, that he collected as many copies as he could of the edition, and burnt them. A copy is preserved in the Library of the Royal Society, which was presented by Sir Isaac Newton. Another exists in the Bodleian Library, in which is written the following memorandum by Sir R. Walpole: "*Exemplar hoc Historiæ Cœlestis quod in thesauraria Regia adservabatur, et cum paucis aliis vitaverat* iram *et* ignem Flamstedianum, *Bibliotheca Bodleiana debet honorabili admodum viro* R. Walpole, 1725." In 1725 Flamsteed published at his own expense a correct edition of the *Historia Cœlestis,* in three folio volumes. "The referees," says Mr. Baily, "instead of printing his observations in detail, as practised at the present day, selected such only as tended to show the place of the moon or a planet when they passed the meridian: rejecting all the other observations of the stars, and the means of verifying and correcting the catalogue, as totally useless. This might be

In 1706 Sir Isaac Newton brought the following propositions before the Council, which were approved and passed into laws :—

"That every person, newly elected a Member of the Royal Society, do, before his admission, pay his admission-money, and give a bond to pay his weekly contribution, excepting Foreigners.

"That no person be capable of being a Member of the Council, who hath not given bond to pay his weekly contributions, or who hath not paid them till the Quarter-day then last past."

At a subsequent Meeting of Council, it was "Ordered that the Professors of Gresham College may be admitted Fellows of the Royal Society without paying admission-money, or giving bonds to pay their weekly contribution[10]; and a few months later, the Officers of the Society were exonerated from the payment of their subscription. These laws proved highly beneficial to the Society, and were the means of causing the subscriptions to be paid much more regularly.

---

pardonable where it was an object to save expense, but ought not to have been adopted where no such excuse was to be pleaded. Fortunately for the science, Flamsteed viewed the subject in a more comprehensive manner; and, to show his decided and fixed opinion upon this matter, he burnt the spurious edition, published at the expense of government, as soon as he got it into his possession, and *at his own cost* printed a correct transcript of all his observations." *Life,* p. 98.

I may add, that in 1798, Miss Caroline Herschel compiled a catalogue of stars, taken from Flamsteed's observations, and not inserted in the *British Catalogue,* with a collection of errata in the second volume of the *Historia Cœlestis.* This volume was printed at the expense of the Royal Society.

[10] The Professors were very grateful for this liberality on the part of the Council.

On the 11th February, 1708, Dr. Papin submitted a proposition of great interest to the Society, "concerning a new-invented boat to be rowed by oars, moved with heat." At the two ensuing meetings the matter was again brought forward, accompanied by letters of recommendation from Leibnitz.

Papin's proposal is thus recorded:—

"It is certain that it is a thing of great consequence to be able to apply the force of fire for to save the labour of men; so that the Parliament of England granted, some years ago, a patent to Esquire Savery, for an Engine he had invented for that purpose; and his Highness Charles, Landgrave of Hesse, hath also caused several costly experiments to be made for the same designe. But the thing may be done several ways, and the Machine tryed at Cassell differs from the other in several particulars, which may afford a great difference in the quantity of the effect. It will be good, therefore, to find out clearly what can be done best in that matter, that those which will work about it may surely know the best way they are to choose. I am fully persuaded, that Esquire Savery is so well minded for the publick good, that he will desire as much as any body that this may be done.

"I do therefore offer, with all dutifull respect, to make here an Engine, after the same manner that has been practised at Cassell, and to fit it so that it may be applyed for the *moving of ships*. This Engine may be tryed for an hour or more, together with some one made after the Saveryan method. The quantity of the effect should be computed both by the quantity of water driven out of each Machine, and by the height the said water could ascend to. And to know the said height, we should use the method advised by the Illustrious President; viz. to try to throw bullets by the said Engine, with the inclination of forty-five degrees, and reckon that the said height is half the horizontal distance to which the bullets will be driven; and this would be the rule so well for one as for the other Machine.

"I wish I were in a condition to make the said Cassellian Engine at my own charges; but the state of my affairs doth not permit me to undertake it, unless the Royal Society be pleased to bear the expense of the Vessel called 'Retort,' in the description printed at Cassell; but after that I will lay out what is necessary for the rest, and I will be content to lose that expense, in case the contrivance of the Landgrave of Cassell doth not as much again as that of Esquire Savery; but in case the effect be such as I do promise it, I do humbly beg that my expense, time, and pains, may be paid, and I reckon this to amount to 15 pounds sterling. If the Royal Society be pleased to honour me with their commands upon such conditions, the first thing to be done is to let me see the place where the Machine must be set, and I will work for it with all possible diligence; and I hope the effect will yet be much greater than I have sayd[11]."

The Society appear to have had some difficulty in arriving at any decision on the subject, which was eventually referred to the President, who reported: "if the pump proposed by Dr. Papin can spout out 400lbs of water every other second minute with the swiftness of 128 Paris-feet in a second, it will spout it up 100 yards high, or cast it to the distance of 200 yards upon level ground, and do this 30 times in a

[11] Register, Vol. IX. p. 108.

minute. Whether this can be done, is to be known only by experience; and if it can be done, I do not see but that such pump may be successfully applied to several uses, as the making artificial fountains, to the draining of water out of trenches, morasses, mines, &c., in difficult cases, and to the towing and *moving of ships and galleys*, by the recoil of the Engine and force of the stream duly applied. But the force and uses of the Engine must be learned gradually by trying the simplest and cheapest experiments first; and reasoning from those experiments." Neither the Journal nor Register-books contain any account of experiments having been made to test Papin's scheme, in consequence, probably, of the expense which would have attended them.

Papin, it is clear, conceived that steam might be employed to propel ships by paddles; for, as early as 1690, in a Paper, published in the *Acta Eruditorum*, he says, "without doubt, oars, fixed to an axis, could be most conveniently made to revolve by our tubes. It would only be necessary to furnish the piston-rod with teeth, which might act on a toothed wheel properly fitted to it, and which, being fixed on the axis to which the oars were attached, would communicate a rotary motion to it[12]."

These communications of Papin anticipate, by several years, the patent taken out by Jonathan Hulls[13] for the same object.

---

[12] P. 412.

[13] Hulls took out a patent in 1736, for "carrying vessels and ships out of, and into, any harbour, port, or river, against wind and tide, or in a calm." They were to be towed by a steam-boat furnished with a pair of wheels.

In 1705 the Society enlarged their scientific correspondence, by entering into communication with a small philosophical association at Edinburgh, whose objects appear in the annexed letter from Dr. Preston to Dr. Sloane :

"Our College in this City had almost dwindled into nothing, after the death of Sir Andrew Balfour, by their divisions; but now that they are come to a good understanding, they begin to make a greater figure; we have of late purchased a piece of ground within the City, where we design a house and physick-garden; but at present we content ourselves with a little convenient hall for meeting, a room for a library, and repository: our repository is but three or four weeks' standing, but it has increased considerably in that time, and we shall want your assistance now and then. We have also begun a *virtuoso* meeting every Monday, and discourse of philosophic matters: it begins to be pretty well frequented, and Dr. Gregory was pleased to give us his presence at one, before he took journey."

Several valuable papers were received from this new association, as also from the Dublin Society, whose meetings, after having been discontinued for some years, were resumed about this period. Mr. Molyneux's letter, communicating this pleasing intelligence, is worthy of being preserved here. It is addressed to Dr. Sloane :

"The kind acceptance your illustrious Royal Society was formerly pleased to show the weak endeavours of the Dublin Society, has made them again venture to beg the favour of your correspondence, now that their meetings are revived under the protection and favor of our excellent Lord Lieutenant. The enclosed are two very odd accounts of the effect of lightning, which we thought might prove acceptable to you. The undeserved honour our Society has lately done me, in commanding me to bear the office of their Secretary, I am never more sensible of my little ability to go through, than when I compare myself with him that holds that place with the illustrious Royal Society. Though we have done but little yet that may claim a right to the favour of your correspondence, yet give us leave to beg it, in hopes that possibly one day the good effects of your encouragement may make us not ungrateful children to that great body, whom we should be glad to call our favouring and protecting parent, and for whom we have the greatest respect."

In 1709 the Society lost one of their oldest Fellows, Sir Godfrey Copley, of Sprotborough, Yorkshire, Bart.

In his will, dated October 14, 1704, and proved in the Prerogative Court, April 11, 1709, he bequeathed to Sir Hans Sloane, Bart., and Abraham Hill, Esq., the sum of "One hundred pounds, in trust for the Royal Society of London for improving natural knowledge, to be laid out in experiments, or otherwise, for the benefit thereof, as they shall direct and appoint[14]."

There is, perhaps, no name of any Fellow of the Royal Society more generally known than that of Sir Godfrey Copley, which may be attributed to the

[14] The Council have been censured for throwing the medal arising from this bequest open to foreigners. It is however manifest by the terms of the bequest, that Sir G. Copley's great object was the encouragement and advancement of science either at home or abroad.

medal bearing his name, the most honourable in the power of the Society to bestow, originating in the bequest above mentioned. This medal, not unaptly termed, by Sir Humphry Davy, "the ancient olive-crown of the Royal Society," has been awarded, for upwards of a hundred years, to the authors of brilliant discoveries; and there is hardly a name eminent in science that does not appear as the recipient of this honourable testimonial of appreciated merit. Although Sir Godfrey Copley's legacy was received shortly after his decease, it was not until 1736 that the interest was applied to the purchase of a medal. Previously to this period it was given to Dr. Desaguliers, Curator to the Society, for various experiments made before them[15].

---

[15] John Theophilus Desaguliers was born in 1683, at Rochelle, from whence he was brought to England by his father while an infant, on the revocation of the edict of Nantes. He studied at Christ Church College, Oxford, where he took the degree of LL.D., and succeeded Dr. Keill in reading lectures on experimental philosophy. In 1712 he settled in London, where he introduced the reading of public lectures on such subjects; which he continued during the rest of his life, and frequently read lectures before the King and Royal Family. In 1714 he was elected a Fellow of the Society, but was excused from paying the subscription, on account of the number of experiments which he shewed at the Meetings. He was subsequently elected to the office of Curator, and communicated a vast number of curious and valuable Papers between the years 1714 and 1743, which are printed in the *Transactions*. Besides those numerous communications he published several works of his own, particularly his large *Course of Experimental Philosophy,* in two 4to volumes, being the substance of his public lectures, and abounding with descriptions of the most useful machines and philosophical instruments. He acted in the capacity of Curator to within a year of his decease, which occurred in 1744. It does not appear that he received a fixed salary, but was remunerated according to the number of experiments and communica-

 tions

At a Council-meeting on the 10th November, 1736, Mr. Folkes :—

"Proposed a thought to render Sir Godfrey Copley's Donation for an annual Experiment, more beneficial than it is at present; which was to convert the value of it into a Medal, or other honorary Prize, to be bestowed on the person whose experiment should be best approved: by which means he apprehended a laudable emulation might be excited among men of genius to try their invention, who in all probability may never be moved for the sake of lucre."

In consequence of this proposition, the Council "Resolved to strike a gold Medal of the value of five pounds, to bear the Arms of the Society[16]; and that the same should be given as a voluntary reward, or honorary favour, for the best experiment produced within the year, and bestowed in such a manner as to avoid any envy or disgust in rivalship." In 1736 the latter clause was repealed, and it was "Resolved, to award the medal to the author of the most important scientific discovery, or contribution to science, by experiment or otherwise." The awards were made on the nomination of Sir Hans Sloane and Mr. Hill, the trustees under the will of Sir Godfrey Copley, to the time of Mr. Hill's death, and afterwards on the nomination of Sir Hans Sloane alone, as surviving trustee, until his decease in 1753. The adjudication then devolved on the President and Council for the

tions which he made to the Society, sometimes receiving a donation of 10*l.* and occasionally 30*l.*, 40*l.*, or 50*l.*

[16] It was resolved that the weight of the Medal should be one ounce and two pennyweights of fine gold: a representation of it appears at the end of this Chapter.

time being; and it is gratifying to find that the first award, under these circumstances, was made to Dr. Franklin[17].

The time was now approaching for the Society to assume a more independent existence; or, in the words of their President, Sir Isaac Newton, "to have a being of their own." We have seen that for a long time, the tenure by which they held their apartments at Gresham College was very uncertain; and that steps had been taken to procure accommodation elsewhere. In 1705 the Council received a communication from the Mercers' Company, to the effect that the latter had come to a resolution "not to grant the Society any room at all[18]." This notice seems not to have been wholly unexpected; as some time before it was received, a Committee was appointed to superintend the removing of "the goods of the Society, in case any warning comes from the Committee of Gresham College[19]." It had, however, the effect of causing the Council to use every means to obtain a *locale* for the Society; one of their first acts was to petition the Queen for a grant of ground in Westminster. The Petition was as follows:—

---

[17] The Earl of Macclesfield in his address as President, delivered on this occasion, states, that when the Council originally determined the award of the Medal, they had kept steadily in view "the advancement of science and useful knowledge, the honor of the Society, and not confining the benefaction within the narrow limits of any particular country, much less of the Society itself." Journal-book, Vol. XXII. p. 411.

[18] Council-book, Vol. II. p. 129.

[19] Ibid., p. 124.

"*To the Queen's most Excellent Majesty.*

"The humble Petition of the President, Councill, and Fellows of the Royall Society of London:—

"SHEWETH,

"THAT your Petitioners are a Corporation founded by your Royall Uncle King Charles the Second, for improving Naturall Knowledge, and that for want of a place of their own to meet in, they have, by the favour of the Trustees and Professors of Sir Thomas Gresham, been allowed the use of rooms in Bishopgate-Street to meet in, and for their Library and Repository, and that the said Trustees and Professors are about pulling down and rebuilding the said College in a new form, which will afford your Petitioners no convenient accommodation. And that a seat nearer Westminster would be more convenient for persons of quality, and render our Meetings more numerous, and thereby conduce more to the improvement of naturall knowledge. Wherefore your Petitioners most humbly pray, that your Majesty would be graciously pleased to give them leave to petition the Parliament for a clause to be inserted into the Bill, now depending about disposing of the Mews, for empowering your Majesty to grant them a piece of ground not exceeding the quantity of fifty foot by sixty, to build upon in any part of the said Mews, where it can be best spared, or in any other place as your Majesty shall think fit[20]."

This failed to procure the boon requested. It was, however, hardly to be expected that Anne, who was not remarkable for her patronage of literature and science, would pay much attention to the prayer of a learned Society, particularly at a period when State-

20 *Commons' Journals*, Vol. XIV. p. 471.

affairs demanded, if they did not obtain, the greatest consideration.

The Council next applied to the Trustees of the Cotton Library, then kept in Cotton House, Westminster, for permission to meet in their apartments; but here again they were disappointed[21]. It would be tedious and useless to enumerate all the plans for building or renting houses, which were successively proposed, discussed, and abandoned. Some idea, however, may be formed of the labour attendant upon the search for suitable premises from the fact, that during a period of six years the Committee appointed to procure accommodation for the Society were actively engaged. At last, on the 8th September, 1710, the President (Sir I. Newton) summoned the Council, for the purpose of informing them that the house of the late Dr. Brown, in Crane Court, was to be sold, "and being in the middle of the town, and out of noise, might be a proper place to be purchased by the

---

[21] It was confidently expected that accommodation would be procured in this house. When examining the MSS. in the Bodleian Library, I found a letter from Sir H. Sloane to Dr. Charlott, Master of University College, dated April 26, 1707, in which he says: "Here are great designes on foot for uniting the Queen's Library, the Cotton, and Royal Society together. How soon they may be put in practice, time must discover." Burnet mentions that "Lord Halifax moved the House of Lords to petition the Queen, that the Cotton Library and the Queen's Library should be joined, and that the Royal Society, who had a very good library at Gresham College, would remove, and hold their assemblies there as soon as it was made convenient for them. This was a great design, which the Lord Halifax, who set it on foot, resolved to carry on till it was finished. It will set learning again on foot among us, and be a great honour to the Queen's reign." *Burnet's Own Times*, Vol. II. p. 441.

Society for their Meetings." Accordingly, "the President, Sir Christopher Wren, Mr. Hill, Mr. Pitfield, Dr. Sloane, Dr. Arbuthnot, Mr. Waller, Mr. Wren, Mr. Isted, Dr. Mead, Sir John Percival, and Dr. Cockburn, four of them to be a quorum, were ordered a Committee to take care of this matter." On the 20th September, the President reported to the Council, "that he, and several members of the Committee, had been to view the late Dr. Brown's house, and found it very convenient for the Society." "Then the question was put, whether they agreed to buy the interest of the late Dr. Brown's house at fourteen hundred and fifty pounds; this question was ballotted for, and twelve votes were for it, and one against it."

It was then resolved, "that the former Committee appointed the 8th September last, four of them to be a quorum, be appointed to contract for, and purchase the house of the late Dr. Brown, and adjoyning little house, for a sum not exceeding 1450*l.*, not including the repairs; and seeing the deeds and writings prepared to be perfected, and the house fitted up for the reception of the Society, with all necessary expedition."

On the 26th October following, the President reported, at a Meeting of Council, that "he had, with the Committee appointed for that purpose, agreed for 1450*l.* for the two houses in Crane Court, mentioned in the last Minutes."

It was thereupon "Ordered, that Mr. Pitfield pay 550*l.* to Mr. Brigstock and the Trustees of Dr. Brown, for the use of the Society towards the purchase of the said houses in Crane Court, and that Mr. Collier, who is ready to advance nine hundred pounds, pay the same to the said Mr. Brigstock, for the purchase of the said houses, and that he have a mortgage of the houses for the same, with interest at 6 per cent. This was ballotted for, and all were affirmatives[22]."

Thus was the Society at last in possession of a house of their own; and after the uncertainty that attended their occupancy of Gresham College, it might naturally be imagined that the Fellows generally would hail with pleasure the announcement of the Council having taken an house where the Society would be more independent. But, according to a pamphlet entitled, *An Account of the late Proceedings in the Council of the Royal Society, in order to remove from Gresham College into Crane Court in Fleet Street*, published in 1710[23], it would appear that this was not the case, but that, on the other hand, considerable dissatisfaction was felt by the Fellows at the change.

The writer, after observing that the Council had resolved on taking Mr. Brown's house, goes on to say, that "the President (Sir Isaac Newton) gave orders at night to summon as many Fellows as were in town, or could be found, to meet at Gresham College on the 1st Sept., previous to a Meeting of the Council, held on the same day." At this extraordinary Meeting he told them that they were without any *being* of their own; that their continuing in Gresham College was very precarious; that Dr. Brown's house had been proposed to them, and a Committee

[22] This Council consisted at its opening of thirteen members: upon Sir I. Newton's announcement, two, viz. Dr. Harris and Mr. Clavel, withdrew.

[23] I have only been able to meet with this pamphlet in the British Museum.

had view'd it, and that he thought it very convenient for the uses of the Society. He added, that he had called them thither, that he might hear what objections they had to offer against the proposal, that the Council might consider of them, and take their final resolutions accordingly. The profound silence that followed, sufficiently exprest a general surprise; till the President (after a little while) began the debate, and addressing himself to some particular members, ask'd their objections. They told him that the very *embryo* of the Society had been form'd in Gresham College, and that they kept their weekly Meetings in that place some time before they obtain'd the Royal Charter of Incorporation; that the Society had continued there almost ever since, even in their most flourishing condition; that they yet enjoy'd the same freedom and convenience as formerly, without the least disturbance or impediment, and therefore they hoped to hear the reasons that induced him, and a few others who appeared as zealous and earnest, to remove from thence. Till that question was debated and determin'd, it was out of season to enquire into the inconveniences of the house he had recommended. The President was not prepared (or perhaps not *instructed*) to enter upon that debate: but freely (though methinks not very civilly) reply'd, That he had *good reasons for their removing, which he did not think proper to be given there.* The acting Secretary, who has engross'd the whole management of the Society's affairs into his own hands, and despotically directs the President, as well as every other member, took upon him to relate a fact, which he thought would determine every vote. He told them that one of the Gresham Committee ask'd him (not long ago), Why the Royal Society did not remove from Gresham College, since the City had several times sent them warning to that purpose?" This does not appear to have satisfied the Members, who were of opinion that there were no grounds for removing; they likewise remonstrated, "that that season of the year, and the short notice he had given of this Meeting, made it very improper to determine an affair of so great *importance* to the *Society* at that time; and therefore they moved that the debate should be adjourn'd to St Andrew's day, or at least to some other extraordinary Meeting." This the President would not hear of; they then offered to give him their opinion either by ballotting, or voting *vivâ voce.*—But in vain; his scruples were unmoveable: so that some of the gentlemen, with warmth enough, ask'd him, To what purpose then he had call'd them thither? Upon which the Meeting broke up somewhat abruptly, and not only the Members of the Society, but most of those of the Council also, left the President with Dr. Sloane, Mr. Waller, and one or two more, to take such measures at the Council as they best lik'd." The writer then proceeds to contrast the large rooms in Gresham College with those in Dr. Brown's house, which he describes as small and inconvenient, and so dilapidated, as to require the large sum of 1800*l.* to put them into habitable condition. His description of their "new purchase," as he calls it, is so minute, as to warrant insertion.

"The approach to it, I confess, is very fair and handsome, through a long court: but then they have no other property in this than in the street before it; and in a heavy rain a man can hardly escape being thoroughly wet before he can pass through it.

The front of the house, towards the garden, is about 42 ft. long; but that towards Crane Court not above 30 foot. Upon the ground-floor there is a little hall, and a direct passage from the stairs into the garden, about 4 or 5 foot wide; and on each side of it, a little room about 15 ft. long, and 16 ft. broad. The stairs are easie which carry you up to the next floor. Here there is a room fronting the Court, directly over the hall, and of the same bigness. And towards the garden is the Meeting-room, which is 25½ foot long, and 16 foot broad. At the end of this room there is another (also fronting the garden) 12½ foot long, and 16 broad. The three rooms upon the next floor are of the same bigness with those I have last described. These are all that are as yet provided for the reception of the Society; except you will add the garrets, a platform of lead over them, and the usual cellars, &c. below, of which they have more and better at Gresham College. The garden is but 42 foot long, and 27 broad, and the coach-house and stables are 40 foot long, and 20 foot broad."

He then states, that on the Wednesday following effectual care was taken, at a Meeting of Council, to give the finishing stroke before they parted. Of the fourteen Fellows that then appear'd[24], there were two only who dissented." "The President was so elated with his success," adds the author, "that he presently summon'd another Council, to meet at the house in Crane Court; and at the same time gave notice that the ordinary Meetings would begin there on Wednesday, Nov. 8."

It is rather remarkable, that neither the Coun-

---

[24] There are only thirteen enumerated in the Council-book.

cil nor Journal-books contain any Minute to this effect.

An ordinary Meeting was held on the 8th November (the Minutes of which are duly entered), but whether at Crane Court or Gresham College, is not mentioned. There is no doubt, however, that on the above day the Royal Society met for the first time in Crane Court. Ward, in his preface to the *Lives of the Gresham Professors*, says, "The year 1710 proved very unfortunate to the College, by the removal of the Royal Society; who, having purchased the house of the late Dr. Brown, in Crane Court, Fleet Street, began their Meetings there on the 8th November that year. And not long after, their library and repository were also removed thither. Thus were these two learned bodies, both founded for the improvement of knowledge, and benefit of the public, at length separated, after they had continued together 50 years, except when necessarily parted for a time by reason of the Great Fire. While the Royal Society held their Meetings at Gresham College, such of the Professors who were Members of it were, in civility, excused from their annual payments, and felt little inconvenience from the want of a college-library; but after the books of the Society were removed, they became sensible of that disadvantage[25]."

In the *Account of the Rise, Foundation, Progress, and Present State of Gresham College*, published anonymously in 1707, we have evidence that the College was deriving at that period, and for many years previously, its chief renown from the Royal Society.

---

[25] *Lives of the Gresham Professors*, p. 18.

The writer says: "It must indeed be acknowledged, to the great honour of the Royal Society, that Gresham College has not been altogether barren and unfruitful; but that illustrious body has been the main, if not only occasion that this College has not by this time lost its name[26]." This, it will be seen, was written only three years before the change took place.

It is easy to understand that the removal of the Society from Gresham College, where, as an incorporated body, they drew their first breath, and which cradled them during half-a-century, was keenly felt by many of the Fellows, who were probably ignorant of the reasons that influenced the Council; but an acquaintance with all the circumstances of the case leads us to the conclusion, that the Council had no other alternative but to engage the most suitable accommodation for the Society that was to be found. It will be remembered, that they were apprised by the Mercers' Company that they had resolved not to grant the Society any room at all; and it appears, moreover, that the College was in a most dilapidated condition. In the petition of the Mercers' Company, praying for leave to introduce a Bill into Parliament, to enable them to pull down, and rebuild Gresham College, the latter is represented as being "in a most crazy state, not worth repairing, if even the Company had the means of doing so, which was not the case[27]." This was probably another motive for determining the Council to remove; and it must not be forgotten, that at the period of the change, and for some years previously, the Society were actually pay-

---

[26] P. 15.

[27] This petition may be seen at length in the *Journals of the Commons*, Vol. XIV. p. 426.

ing 32*l.* a year for the use of some of the Professors' apartments.

With regard to the question, whether Dr. Brown's house was more eligible for the purposes of the Society, than other houses which were to be rented or purchased, it is manifestly impossible to offer any opinion; but when we reflect that the Society occupied the building, as we shall see, for a period of 72 years, gathering fresh renown, and extending their reputation more and more, as time flowed on, we have strong presumptive evidence that the Council did not make an injudicious choice.

COPLEY MEDAL.

*Size of Original.*

## CHAPTER XV.

House occupied by the Society in Crane Court—Fellows subscribe to defray the expense of Repairs—Ode by a Frenchman to the Society—Society appointed Visitors of Royal Observatory, by Royal Warrant—Flamsteed's Vexation—His Conduct on the occasion—Visitors examine the Observatory—Recommend Repairs and new Instruments—Ordnance decline to undertake the work—Queen wishes the Society to take care of the Observatory—Appointment of celebrated Committee on the disputed Invention of Fluxions—Historical Account of the Dispute—Report of Committee—Society adopt the Report—Leibnitz dissatisfied—Appeals to the Society through Chamberlayne—Society confirm the Report of their Committee in favour of Newton—Remarkable Error of Writers on this Subject—Probable Origin of the Error—Foreign Ambassadors attend the Meetings—Experiments exhibited before them—Queen orders her Foreign Ministers to assist in promoting the Objects of the Society—Fossil Remains sent from America—Philosophical Society established at Spalding—Curious List of the Fellows published—Bequests to the Royal Society—Foreign Secrtary appointed—Opinion of Attorney-General—Petition to George I. for Licence to purchase or hold lands in Mortmain—King grants the prayer of the Society.

---

1710—25.

THE march of improvement, which has changed so much of old London, has not yet penetrated into the quiet court, made classical by the abode of the Royal Society for so many years.

The exterior of the house remains unaltered, though the interior has undergone some changes, with the view of adapting it to the wants of the Scottish Hospital, to which it is devoted. Happily, however, the room in which the Society met is in the same condition as when Newton occupied the presidential chair, and it is impossible to stand in that ancient apartment, a representation of which is annexed, without feeling the associations connected with those days stealing over their mind[1].

It is gratifying to find that several noblemen and gentlemen came forward with donations to assist in discharging the debt due to Mr. Collier. Amongst them was the Earl of Halifax, who gave 100*l.*; Newton, who contributed 120*l.*; and Mr. Richard Waller, 100*l.* towards building a repository, to contain the Society's Museum. This addition, together with the the necessary repairs, cost 800*l.* By Newton's order, the porter was clothed in a suitable gown[2], and provided with a staff, surmounted by the Arms of the Society in silver; and on the meeting-nights a lamp was hung out over the entrance to the court from Fleet Street.

The Fellows appear to have felt all the pleasures of independence, which they determined to enjoy; for we find, when the President stated that the Company of Mine Adventurers, who were established by Act of Parliament, wished to meet in the Society's house, paying a yearly rent for the convenience, it was unanimously resolved, that the Society "would continue as they were, and not let their house to the Mine Adventurers."

The Society's removal seems to have inspired a

---

[1] As I write (1847), the destruction of Crane Court has been resolved on, to make way for a new Record Office.

[2] At the period of my election to the office I hold, the Porter had abandoned the gown; but on my representation of previous custom, the Council resolved that he should be furnished for the future with livery.

French Poetaster, who wrote an ode in praise of the body when they moved to Crane Court. The only copy that I have seen of this poem is in the British Museum. It is entitled, *Crane Court, ou le Nouveau Temple d'Apollon à Londres: Ode à Messieurs de la Société Royale.* Some idea may be formed of it by the following extract:

"*Que vois-je? Quelle ardeur soudaine*
*M'agite, echauffe mes transports?*
*C'est icy la celèbre Athêne*
*Qu'élèvent les Dieux sur ces bords;*
*De loin un Temple se découvre:*
*Son superbe portique s'ouvre,*
*Couvert de lauriers verdissants,*
*Pour Apollon, et pour sa gloire.*
*Ce nouveau temple, à la memoire*
*Y rapelle un antique encens.*
*Pénétrons, Muse, en ces détours;*
*Minerve y rit avec les graces,*
*Les Muses avec les amours;*
*Clio, Calliope, Uranie,*
*Melpomène, Euterpe, Thalie,*
*Des lauriers en main tour a tour,*
*Jalouses mais sages rivales;*
*Ceignent de couronnes égales,*
*Cent rivaux qui forment leur cour.*"

Shortly after the removal, the Society acquired additional importance, by being appointed Visitors and Directors of the Royal Observatory at Greenwich. The Warrant to this effect was read before the Council on the 14th December, 1710, and runs as follows:—

"*To our Trusty and Well-beloved the President of the Royal Society for the time being;*

"ANNE R.

"TRUSTY and well-beloved, we greet you well. Whereas we have been given to understand that it would contribute very much to the improvement of

Astronomy and Navigation, if we should appoint constant visitors of our Royal Observatory at Greenwich, with sufficient powers for the due execution of that trust; we have therefore thought fit, in consideration of the great learning, experience, and other necessary qualifications of our Royal Society, to constitute and appoint, as we do by these presents constitute and appoint, you, the President, and in your absence the Vice-President, of our Royal Society for the time being, together with such others as the Council of our said Royal Society shall think fit to join with you, to be constant *Visitors* of our said Royal Observatory at Greenwich: authorising and requiring you to demand of our Astronomer and Keeper of our said Observatory, for the time being, to deliver to you, within six months after every year shall be elapsed, a true and fair copy of the annual observations he shall have made. And our further will and pleasure is, that you do likewise, from time to time, order and direct our said Astronomer and Keeper of our said Royal Observatory, to make such Astronomical Observations as you in your judgment shall think proper. And that you do survey and inspect our Instruments in our said Observatory; and as often as you shall find any of them defective, that you do inform the principal officers of our Ordnance thereof; that so the said instrument may be either exchanged or repaired. And so we bid you farewell. Given at our Court of St. James's, the 12th day of December, 1710, in the ninth year of our reign.

"By her Majesty's command,

"H. St. John."

A copy of this Warrant was sent to the Ordnance Office, with the subjoined letter:—

"GENTLEMEN, "*Whitehall*, 12 *Dec.* 1710.

"I SEND you enclosed, by the Queen's command, a copy of Her Majesty's letter to the Royal Society, appointing the President, and in his absence the Vice-President, together with such others as the Council of the said Royal Society shall think fit to join with them, to be constant Visitors of the Royal Observatory at Greenwich: and I am at the same time to signify Her Majesty's pleasure to you, that you do receive and take notice of such representations as the said Visitors shall think fit to make to your Board, concerning Her Majesty's instruments at any time remaining in the said Observatory: and that you order them to be repaired, erected, or changed, as there shall be occasion: and if any instruments be now there which do not belong to Her Majesty, you are to give necessary directions for purchasing the same. Her Majesty is likewise pleased to direct that you should have regard to any complaints the said visitors may make to you of the behaviour of Her Majesty's Astronomer and Keeper of the said Observatory, in the execution of his office.

"I am, Gentlemen, &c.,

"H. ST. JOHN."

On the same day that the Warrant was received, it was ordered by the Council, that "the President, Mr. Roberts, Dr. Arbuthnot, Dr. Halley, Dr. Mead, Mr. Hill, Sir Christopher Wren, Mr. Wren, and Dr. Sloane, be appointed a Committee to go to Greenwich, any three of them (of which the President, or Vice-President to be one) to be a quorum, and to report their opinion of the condition of the Observatory, and the instruments therein, and to take an inventory of the instruments."

This Warrant occasioned Flamsteed the greatest vexation. In his *Autobiography* he accuses Newton of having procured it, and states that he waited on Mr. Secretary St. John, and told him, "that he was injured, and would be hindered by this new constitution; that he wanted no new instruments, and that if he did, the Visitors were not skilful enough to contrive them[3]." All his arguments and aspersive remarks proved unavailing; for Mr. St. John, according to Flamsteed's own account, "seemed not to regard what he said, but answered haughtily, 'The Queen would be obeyed.'"

It must be admitted that Flamsteed's excited remarks would ill become any man, more especially a clergyman, which the Astronomer-Royal was; but his dislike to Newton appears to have increased with his infirmities, and we find him writing, in 1713, to his friend Mr. Sharp, "Sir I. Newton still continues his designs upon me, under pretence of taking care of the Observatory, and hinders me all he can: but, I thank God for it, hitherto without success[4]."

Notwithstanding the objections of the Astronomer-Royal, the newly-appointed Visitors, acting on the Queen's warrant, visited the Observatory, and directed Flamsteed to send copies of his observations to the Royal Society; they also addressed a letter to the Ordnance-office, representing the inefficiency of seve-

---

[3] Baily's *Account of Flamsteed*, p. 92.

[4] Some idea may be formed of Flamsteed's feelings towards Newton, by his saying in this letter, "I think his new *Principia* worse than the old." And in another letter, he declares that "he does not know whether the alterations and additions be worth 12*d*."

ral instruments[5], and the expediency of replacing them by new ones, adding, that "they would accompany able workmen to Greenwich, and shew them what is wanting to be done, and give the best advice they can for doing every thing after the best manner."

The Ordnance-office thus replied:—

"*Office of Ordnance*, 4 *Sep.* 1713.

"GENTLEMEN,

"WE received yours, and in answer acquaint you that we do not very well apprehend what is meant by repairing the instruments in the Royal Observatory at Greenwich; this office never having been at the charge thereof, or any other, except of repairing the house and paying Mr. Flamsteed's salary[6].

"W. BRIDGES.
"C. MUSGRAVE."

It will be seen from this how little interest in science was taken at that period by public departments of Government; and, consequently, how much is due to the Royal Society for cultivating and fostering it[7].

When Sir Isaac Newton presented a copy of his *Principia* to the Queen, in 1713, she begged him, and the gentlemen of the Royal Society, to *take care of the Royal Observatory*; and having received these

---

[5] Council-book, Vol. II. p. 220.

[6] Letter-book, Vol. XV. p. 26.

[7] See Professor Rigaud's remarks on Dr. Halley's Instruments at Greenwich, in the 9th volume of the *Astron. Soc. Memoirs;* wherein he says: "The same disgraceful and injurious parsimony, which had left on Flamsteed the burden of providing what was necessary for the Observatory, still continued to operate, and the claim was set up by Government to the instruments which his widow had, in full right, taken away with her."

directions from the lips of the Sovereign, the "Visitors," in endeavouring to improve the means of observing, by substituting better instruments in place of those in use (which even Flamsteed admitted to be defective), were only doing their duty. But in consequence of the unfortunate misunderstanding between the Visitors and Flamsteed, nothing was done during the lifetime of the Astronomer-Royal; and it was only when Halley succeeded him, in 1720[8], that Government furnished the Observatory with a few instruments necessary for working the establishment.

In 1711, the celebrated Committee was appointed by the Society, to report upon the respective claims of Newton and Leibnitz to the invention of *Fluxions,* now called the *Differential Calculus.*

This constitutes so interesting and important a feature in the history of the Society, and of mathematical discovery, as to render it desirable to bring forward the principal circumstances connected with the subject[9].

It is an undisputed fact, that as early as 1663, Newton had studied the works of Descartes and Wallis, and invented the celebrated Binomial Theorem. In the tract entitled *De Analysi per Equationes numero terminorum infinitas,* written by Newton in 1666, and first printed in the *Commercium Epistolicum,* it is

---

[8] It was Dr. Mead who procured for Halley the situation of Astronomer-Royal. In a letter from the former to Hearn, dated Jan. 9, 1719, preserved in the Bodleian Library, he says: "I have been so happy as to get Flamsteed's place for Dr. Halley, by means of my Lord Sunderland."

[9] Professor De Morgan has been so kind as to give me the benefit of his valuable advice and assistance, in the following pages relating to this subject.

evident that he had begun to use the notation of fluxions[10]. The tract, says Professor De Morgan, "contains a method of series, and many problems solved by application of limits to differences obtained by expansion, but no direct method of fluxions[11]."

Various letters now passed between Newton, Collins, and others, alluding to the discovery of the former. One in particular, from Newton to Collins, dated December 10th, 1672, was always stated to assert the fact, and contain one example. The Committee appointed by the Society declare that this letter was forwarded to Leibnitz.

In 1673 the latter visited England[12], and made the acquaintance of Oldenburg and other philosophers connected with the Royal Society[13]. From these he heard something of Newton's discovery, and being desirous of more precise information, he wrote to

---

[10] In allusion to the above tract, the following passage occurs in the "Life of Newton" in the *Biog. Dict*:—"His MS. was communicated to none but Mr. J. Collins and Lord Brouncker, and even this had not been done but for Dr. Barrow, who would not suffer him to indulge his modesty so much as he desired. This MS. was taken out of our author's study in 1669, entitled: *A Method which I formerly found out, &c.*; and supposing this 'formerly' to mean no more than three years, he must have discovered this admirable theory when he was not 24. But what is still more, this MS. contains both the discovery and method of Fluxions, which have occasioned so great a contest between Mr. Leibnitz and him, or rather between Germany and England."

[11] Art. *Fluxions, Pen. Cyc.*

[12] Previous to his visit, he addressed a letter to the Society, requesting to be elected into "so eminent a body:" he was unanimously elected on the 9th April, 1673. The letter, as also a second, acknowledging the honour, are preserved in the Archives.

[13] He attended the Meetings, and exhibited his Calculating Machine.

Oldenburg, on his return to Germany, requesting to have Newton's method communicated to him. Newton, at the instigation of Oldenburg and Collins, wrote the celebrated letters of June 13, and October 24, 1676, the first containing his Binomial Theorem, the second an allusion to his method of Fluxions, concealed under the anagram[14]:

*6a, cc d æ 13e ff 7i 3l 9n 4o 4q rr 4s 9t 12vx;*

meaning, that if any one could arrange six *a*'s, two *c*'s, one *d*, &c. into a certain sentence, he would arrive at the following sentence: *Data Æquatione quotcunque fluentes quantitates involvente fluxiones invenire, et vice versâ*, which literally translated is—given equation any whatsoever, flowing quantities involving fluxions to find, and *vice versâ*. "If Leibnitz," says Professor De Morgan, "could have taken a hint, either from the preceding letters in alphabetical order, or (had he known it) in their significant arrangement, he would have deserved as much credit as if he had made the invention independently." It is important to state, that Leibnitz did not receive the second letter before March 5, 1677. On the 21st June, in the same year, he wrote to Oldenburg, giving a full and clear statement of the principal use and notation of Differential Calculus[15]; and Newton immortalized Leibnitz's independent discovery ten years afterwards, by thus recognizing it in the *Principia:*

"*In literis, quæ mihi cum Geometra peritissimo G. G. Leibnitio annis abhinc decem intercedebant, cum significarem*

---

[14] This was a common practice amongst philosophers at that period, and, indeed, long antecedent to it; for Galileo, in 1610, excited the curiosity of astronomers by the publication of a logograph.

[15] Published in the *Commercium Epistolicum.*

*me compotem esse methodi determinandi Maximas et Minimas, ducendi Tangentes, et similia peragendi, quæ in terminis Surdis æque ac in Rationalibus procederet, et literis transpositis hanc sententiam involventibus (data Æquatione &c.) eandem celarem: rescripsit Vir Clarissimus se quoque in ejusmodi methodum incidisse, et methodum suam communicavit a mea vix abludentem præterquam in verborum et notarum formulus. Utriusque fundamentum continetur in hoc Lemmate*[16]."

In the course of a few years matters stood thus: that while in England the Differential Calculus had not been made clear to the scientific world, either by Newton or his friends, and was consequently but little used; it had, in the words of Professor De Morgan, "grown into a powerful system in the hands of Leibnitz and the Bernouillis[17]."

It is certain that for fifteen years Leibnitz enjoyed the honour, on the Continent, of being the inventor of his own Calculus[18]: how his claim was first disputed is now to be shown.

On the 10th April, 1695, Dr. Wallis gave the first notice to Newton that his reputation was endangered:—in a letter from Oxford of the above date, he says, "that he had heard that his notions of Fluxions pass there with great applause, by the name of Leibnitz's *Calculus Differentialis.*" "You are not so kind to

---

16 *De Motu*, Prop. VII. Scholium. I have collated the above with the original MS. and find it to be literally correct.

17 "Such was the reserve of Newton," says Professor Powell, "and so little were his methods known or followed up among his countrymen, that the first book which appeared in England, on the new Geometry (as it was called), was a treatise by Craig professedly derived from the writings of Leibnitz and his disciples." *Hist. Nat. Phil.* p. 296.

18 Printed in Raphson's *Hist. of Flux.*, p. 122.

your reputation (and that of the nation)," he adds, "as you might be, when you let things of worth lie by you so long, till others carry away the reputation that is due to you. I have endeavoured to do you justice in that." Accordingly he inserted, in the Preface to the first volume of his *Works*, that Newton's method of Fluxions had been communicated to Leibnitz in the Oldenburg letters.

Wallis's *Works* were reviewed in the *Acta Eruditorum* for June, 1696; great stress is laid on Newton's admission of Leibnitz's independent discovery, upon which Newton remarks, "I had the method of Fluxions some years before (1676);" and, in another place, "Whether Mr. Leibnitz invented it after me, or had it from me, is a question of no consequence; for second inventors have no right[19]."

In 1699 further animosity was created, by a Genevese of the name of Fatio de Duillier[20], then residing in England, publishing a mathematical treatise, in which he insinuated that Leibnitz might have borrowed his invention from Newton[21]. This was followed by Keill, stating, in 1708[22], that Leibnitz had appropriated Newton's method, changing its name and symbols. This naturally irritated Leibnitz, who complained to the Royal Society, and, at the same time, requested that the subject of dispute might be submitted to the judgment of the Members, over whom, it will be remembered, Newton then presided.

---

19 Printed in Raphson's *Hist of Flux.*, p. 122.

20 He was elected a Fellow of the Society in 1687, and contributed Papers to the *Transactions*.

21 Neither the Society nor Newton approved this charge.

22 *Phil. Trans.*, No. 317.

The Royal Society accordingly appointed a Committee on the 6th March, $17\frac{11}{12}$, "to inspect the letters and papers relating to the dispute, consisting of Dr. Arbuthnot, Mr. Hill, Dr. Halley, Mr. Jones, Mr. Machen, and Mr. Burnet." On the 20th of March, Robarts was added to the Committee; on the 27th, Bonet, the Prussian minister, and De Moivre, Aston, and Brook Taylor, on the 17th April following[23]. It should be remembered that in the preceding year, (April 5th, 1711), Newton had given from the chair of the Society "a short account of his invention, with the particular time of his first mentioning or discovering it; upon which Mr. Keill was desired to draw up an account of the matter in dispute, and set it in a just light." This account, dated May 24, 1711, was the immediate cause of Leibnitz requesting the interference of the Society.

On the 24th April, 1712, the Committee delivered in their Report:

"We have," they say, "consulted the letters and Letter-books in the custody of the R. Society, and those found among the papers of Mr. John Collins, dated between the years 1669 and 1677 exclusive, and shewed them to such as knew and avouched the hands of Mr. Barrow, Mr. Collins, Mr. Oldenburg, and Mr.

---

23 Professor De Morgan was the first to point out these additions to the original Committee. It is curious enough that all previous historical writers have overlooked the important fact. Important, because Newton declared that the "numerous Committee was composed of gentlemen of different nations," which would not hold, had the Committee consisted only of its original members. We have to thank Professor De Morgan for his weighty discovery, which he gave to the Royal Society in a short Memoir, published in the *Transactions* for 1846.

Leibnitz, and compared those of Mr. Gregory with one another, and with copies of some of them taken in the hand of Mr. Collins, and have extracted from them what relates to the matter referred to us: all which extracts, herewith delivered to you, we believe to be genuine and authentic; and by these letters and papers we find,

"1st. That Mr. Leibnitz was in London in the beginning of the year 1673, and went thence in or about March to Paris, where he kept a correspondence with Mr. Collins, by means of Mr. Oldenburg, till about September 1676, and then returned by London and Amsterdam to Hanover; and that Mr. Collins was very free in communicating to able mathematicians what he had received from Mr. Newton and Mr. Gregory.

"2dly. That when Mr. Leibnitz was the first time in London, he contended for the invention of another Differential Method, properly so called; and notwithstanding that he was shewn by Dr. Pell that it was Mouton's method, he persisted in maintaining it to be his own invention, by reason that he found it by himself, without knowing what Mouton had done before, and had much improved it. And we find no mention of his having any other Differential Method than Mouton's, before his letter of the 21st June, 1677, which was a year after a copy of Mr. Newton's letter of the 10th December, 1672, had been sent to Paris, to be communicated to him; and above 4 years after Mr. Collins began to communicate that letter to his correspondents, in which letter the method of Fluxions was sufficiently described to any intelligent person.

"3dly. That by Mr. Newton's letter of the 13th June, 1676, it appears that he had the method of Fluxions above five years before the writing of that letter; and by his analysis for *æquationes numero terminorum infinitas*, communicated by Dr. Barrow to Mr. Collins, in July 1669, we find that he had invented the method before that time.

"4thly. That the Differential Method is one and the same with the method of Fluxions, excepting the name and mode of notation; Mr. Leibnitz calling those Quantities, Differences, which Mr. Newton calls Moments or Fluxions, and marking them with the letter *d*, a mark not used by Mr. Newton. And therefore, we take the proper question to be, not who invented this or that method, but who was the first Inventor of the Method; and we believe that those who have reputed Mr. Leibnitz the first Inventor, knew little or nothing of his correspondence with Mr. Collins and Mr. Oldenburg long before, nor of Mr. Newton's having that method above 15 years before Mr. Leibnitz began to publish it in the *Acta Eruditorum* of Leipsick.

"For which reasons we reckon Mr. Newton the first Inventor, and are of opinion that Mr. Keill, in asserting the same, has been noways injurious to Mr. Leibnitz. And we submit to the judgement of the Society, whether the extracts of Letters and Papers now presented, together with what is extant to the same purpose in Dr. Wallis's 3rd Volume, may not deserve to be made publick[24]."

To which Report the Society agreed, *nemine contradicente*, and "ordered that the whole of the matter from the beginning, with the extracts of all the letters relating thereto, and Mr. Keill's and Mr. Leibnitz's letters, be published with all convenient speed that

[24] Copied from the original document in the Archives of the Royal Society.

may be, together with the Report of the said Committee[25]."

It has been truly said that the quarrel between Leibnitz and Newton, or perhaps, to speak more properly, between their respective friends, was not without advantage to mathematical science, since it produced and perpetuated the precious collection of letters known under the name of the *Commercium Epistolicum*[26]. This work appeared in 1712. It was not published for sale; very few copies were printed, and these were distributed as presents; consequently, the book was, even at that period, excessively scarce; so so much so, that Raphson says, in his *History of Fluxions*, "none are to be met with amongst the booksellers."

Leibnitz was at Vienna when he heard of the *Commercium*. In a letter to Count Bathmar, published in Des Maizeaux's *Recueil de diverses Pièces*, he says, "*J'étois à Vienne quand j'appris la publication du livre, mais assuré qu'il devoit contenir des faussettées malignes, je ne daignai point le faire venir par la poste, mais j'écrivis à M. Bernouilli, l'homme de l'Europe qui a peut-être le mieux réussi dans la connaissance et dans l'usage de ce Calcul, et qui étoit tout a fait neutre, de m'en mander son sentiment. M. Bernouilli m'écrivit une lettre datée de Bâle, le* 7 *Juin*, 1713, *où il disoit qu'il paroissoit vraisembla-*

---

[25] Journal-book, Vol. XI. p. 289. The adoption of this Report by the Society is a material point: and it is a remarkable circumstance, that all writers on this subject have omitted this very important fact.

[26] This name was frequently given to published Collections of letters about the end of the 17th century. But the title by itself is understood to apply to the above celebrated collection published by the Royal Society.

*ble que M. Newton avoit fabriqué son Calcul après avoir vu le mien*[27]."

Bernouilli's opinion led Leibnitz to sanction and circulate an anonymous letter, (believed to be by Bernouilli), which had neither the name of the author printer, or place of publication, attached to it, and in which Newton was represented as having fabricated his method of fluxions from the differential calculus.

But Leibnitz was guilty of a still greater indiscretion, which M. Biot thus details: "*Il était en correspondance avec la Princesse de Galles, belle fille du Roi George I. Cette princesse, d'un esprit très cultivé, avoit accueilli Newton avec une extrême bienviellance, elle aimait à s'entretenir avec lui, et l'honorait au point de dire souvent qu'elle s'estimait heureuse d'être née dans un temps où elle avait pu connoître un si grand génie. Leibnitz profita de sa correspondance pour attaquer Newton devant la princesse; et lui presenter sa philosophie, non-seulement comme fausse sous le rapport physique, mais comme dangereuse sous le rapport religieux: et, ce qui est plus inconcevable, il appuyait ses accusations sur des passages du traité des Principes et de l'Optique, que Newton avait évidemment composés et insérés dans les intentions les plus sincèrement religieuses, et comme de véritables professions de sa ferme croyance en une Providence divine*[28]."

The next step taken by Leibnitz was to address a letter to Mr. Chamberlayne[29] at London, which was

---

[27] Vol. II. p. 44. [28] *Life*, p. 178.

[29] John Chamberlayne. He was educated at Oxford, and was chamberlain to George, Prince of Denmark. He was elected a Fellow of the Society in 1702, and died in 1723. He had the reputation of being a learned man, and was, it is said, conversant

written on the 28th April, 1714, expressive of his entire disapproval of the Report of the Committee, and the *Commercium;* at the same time saying: "*Je n'ai pas encore vu le livre publié contre moi;*" and requesting Mr. Chamberlayne to lay his letter before the Society[30].

This request was duly brought under consideration at their Meeting on the 20th May, 1714, when the Society came to the following resolution:

"It was not judged proper (since this letter was not directed to them), for the Society to concern themselves therewith, nor were they desired so to do. But that if any person had any material objection against the *Commercium*, or the Report of the Committee, it might be reconsidered at any time[31]."

This resolution, instead of repudiating, must be regarded as adopting the Report of the Committee; and yet it is stated by Sir David Brewster, in his *Life of Newton*, (1831), that "the Society inserted a declaration in their Journals on the 20th May, 1714, that they did not pretend that the Report of their Committee should pass for a decision of the Society[32];" and Professor De Morgan, in his Life of the same philosopher, asserts, that "the Society on the 20th May, 1714, resolved that it was never intended that the Report of the Committee should pass for a decision of the Society: but others persisted in calling it so[33]."

---

with sixteen languages. He contributed three Papers to the *Transactions*.

30 An attested copy of this letter is preserved in the Archives of the Society.

31 Journal-book, Vol. XII. p. 481.

32 P. 211.

33 P. 93.

The origin of this curious and grave mistake[34], which is not confined to the above eminent individuals, may be traced without much difficulty. In a letter from Leibnitz to Chamberlayne, dated 25th August, 1714, he thanks him for laying his letter before the Society, and adds, that the extract which he sends him from their Journal, "*fait connoître qu'elle ne prétend pas que le rapport de ses Commissaires passe pour une décision de la Société*[35]." Mr. Chamberlayne evidently made such a communication to Leibnitz, as led the latter to a totally opposite conclusion to that at which the Society had arrived, and historians have followed Leibnitz's printed letter, instead of the unpublished Minutes of the Society. Nothing further of moment occurred until 1715. In that year, the Abbé Conti, a friend of Leibnitz and Newton, was in London. Leibnitz wrote to him, adverting to the treatment that he had received. This letter led Conti to interest himself in the quarrel. The following letter, from him to Brook Taylor[36], best explains what occurred. It is taken from a work entitled *Contemplatio Philosophica*, privately printed in 1793 from the manuscript of Brook Taylor, to which is appended a Life and Correspondence by his grandson, Sir Wm. Young.

"MONSIEUR, "*Paris, May* 22, 1721.

"*Je m'en vais vous expliquer en peu de mots les raisons qui m'ont engagés dans la querelle de M. Newton et de M. Leibnitz. M. Newton me pria d'assembler a la Société les ambassadeurs et les autres ministres étrangers.*

---

[34] A correction of it was inserted in the *Phil. Mag.* for 1847.

[35] This letter is published in Maizeaux's *Recueil*, Vol. II. p. 123.

[36] He was Secretary to the Society from 1714 to 1718.

*Il souhaitait qu'ils assistassent à la colation qu'on devoit faire des Papiers Originaux, qui se conservent dans les Archives de la Société avec d'autres lettres de M. Leibnitz. M. le Baron de Kirmansegger vint à la Société avec les ministres des princes; et après que la colation des Papiers fut faite, il dit tout haut que cela ne suffisoit pas; que la véritable methode pour finir la querelle, c'étoit que M. Newton luy-même écrivît une lettre à M. Leibnitz, dans laquelle il luy proposat les raisons, et en même temps luy demandat des réponses directes. Tous les ministres des princes qui étoient présent goutèrent l'idée de M. Kirmansegger, et le Roy même, à qui on la proposa le soir, l'approuva; ayant dit tout cela à M. Newton cinq ou six jours après il m'écrivit une lettre pour envoyer à M. Leibnitz à Hanover. La Comtesse de Kirmansegger la fit traduire en François par M. Costa: le Roy la lut, et l'approuva fort, en disant que les raisons étoient très simples et très claires, et qu'il étoit difficile de répondre à des faits*[37]."

The letter then goes on to say that Leibnitz was very much irritated by the communications which Conti sent him, and that he replied at great length. This closed the controversy; for Newton made no rejoinder, and Leibnitz died in November, 1716.

All the correspondence was given to Raphson, whose *History of Fluxions* was then in the press.

The letters appeared as an Appendix. In 1722 a second edition of the *Commercium* was published, with a *Recensio*, &c. prefixed, and notes by Keill and others. This edition differs in several insertions and omissions from the first[38]. In 1725 a third edition

[37] P. 121.

[38] See a Paper, entitled, *Comparison of the First and Second Editions of the Commercium Epistolicum,* by Professor De Morgan, communicated to the Royal Society in 1847.

of the *Principia* was published, in which the celebrated Scholium, upon the strength of which Leibnitz confidently appealed to Newton, was left out, and its place supplied by another, in which the name of Leibnitz is not mentioned[39]. The late Professor Rigaud says, that "Newton freely furnished Wallis with the account of his method, and of the notation which he adopted in the use of it: he did this with the express view of its being given to the public; but, although a formidable rival had already challenged the invention, he could not induce himself to communicate his own, unless under conditions which might exempt him from the danger of any personal controversy[40]."

That Newton was the earliest *inventor* of Fluxions in 1666 we cannot doubt[41], but Leibnitz has certainly the merit of having first given full publicity to his discovery of the Differential Calculus in 1673. Had Newton done this, a controversy, painful in its nature, and unsatisfactory in its results, would have been avoided. But all admit, that he laboured more for the love of truth than of fame; and this is one of the reasons why Newton is the greatest of philosophers.

During the years 1711—12 the Society were honoured by the visits of several of the foreign am-

---

39 Mr. De Morgan conceives, that as Newton allowed the Scholium to stand in the second edition when the dispute was at its height, it is possible he left the matter to Dr. Pemberton the editor, or some other person. It will be remembered that Newton was very old at this period.

40 *Essay on the first Publication of the Principia*, p. 23.

41 Dr. Young, whose opinions were not formed without deliberation, says in his *Nat. Phil.* that "Newton was unquestionably the first inventor of Fluxions." p. 191.

bassadors[42]. It was customary on these occasions to exhibit interesting experiments, the nature of which appears in the annexed extract from the Journal-book:

"Signor Grimani, the Venetian Ambassador, Signor Gerardini, Envoy from the Grand Duke of Tuscany, and the Duke D'Aumont, French Ambassador, came to the Society, and were seated in elbow-chairs near the President. Their Excellencies were then entertained with several experiments, viz.: The productiveness of light by friction; The mutual attraction of the parts of matter; The curve caused by the rising of a fluid between two glass planes; the stone called *Oculus Mundi*, which, from being put some time in fair water, from being very opaque becomes transparent. The refraction of the air, by viewing an object through a prismatical vessel exhausted of air;—filled with common air, and with compound air. Mr. Cheselden then show'd two preparations he had made of the veins and arteries of a human liver by injecting red wax into them, which was very curiously and beautifully performed. A drop of oil was placed between two glass planes *in vacuo*, so as to show the proportion of the power of gravity, and congruity or agreement of the parts, by observing at what angle the drop is observed to be stationary, and not to move towards the edge of the wedge formed by the two planes[43]."

The Minutes add, that the experiments succeeded admirably, and afforded much delight to the distinguished foreigners present.

---

[42] Many of these became Fellows of the Society.

[43] This experiment, it is stated, was proposed by Sir Isaac Newton, who occupied the chair.

In 1713, the Society received a proof of Royal favour as communicated in the subjoined letter.

"GENTLEMEN, " *Whitehall*, 7 *Feb.* 1712—13.

"HER Majesty having being graciously pleased to direct that instructions should be given henceforward to her Ministers and Governors that go abroad, to contribute all they can, in their several stations, towards promoting the design for which the Royal Society was first instituted, by corresponding, as occasion may require, with the President and Fellows of the said Society, and by procuring as satisfactory answers as possible to such enquiries as may be sent from time to time; this is to inform you thereof, and to acquaint you at the same time, that, as Her Majesty intends shortly to despatch a Minister to the Court of Mosco, if you please to prepare a draught of instructions whereby he may be usefull to you in those parts, I will lay them before the Queen for her commands. I am also directed to let you know that Her Majesty's intention is, that I should write to such of her Ministers in my department abroad, if you desire any particular to be recommended to them.

"I am, Gentlemen,

"Your humble Servant,

"BOLINGBROKE.

"*To the Council of the Royal Society.*"

The Fellows, through the medium of their President, returned their sincere thanks to the Queen for this mark of her favour. Lord Bolingbroke replied, that "Her Majesty received the compliments of the Royal Society very graciously, and was pleased to express her intention of countenancing and encouraging the studies of the Society."

In order to reap as much benefit as possible from the Queen's favour, Committees were appointed to prepare instructions for Ministers and Governors proceeding abroad[44]; and we have here a communication from Lord Cornbury, resulting from these measures, which will probably be perused with interest by the Palæontologist. It is dated from New York, and addressed to the Secretary.

"I did, by the Virginia fleet, send you a Tooth, which, on the outside of the box, was called the tooth of a Giant, and I desired it might be given to Gresham College: I now send you some of his bones, and I am able to give you this account. The tooth I sent was found near the side of Hudson's river, rolled down from a high bank by a Dutch country-fellow, about twenty miles on this side of Albany, and sold to one Van Bruggen for a gill of rum. Van Bruggen being a member of the Assembly, and coming down to New York to the Assembly, brought the tooth with him, and shew'd it to several people here. I was told of it, and sent for it to see, and ask'd if he would dispose of it; he said it was worth nothing, but if I had a mind to it, 'twas at my service. Thus I came by it. Some said 'twas the tooth of a human creature; others, of some beast or fish; but nobody could tell what beast or fish had such a tooth. I was of opinion it was the tooth of a giant, which gave me the curiosity to enquire farther. One Mr. Abeel, Recorder of Albany, was then in town, so I directed him to send some person to dig near the place where the tooth was found; which

---

[44] The "instructions," or '*inquiries*' as they are styled, are entered in the Register-books, and afford curious evidence of the meagre information existing at this period respecting foreign countries.

he did, and that you may see the account he gives me of it, I send you the original letter he sent me: you must allow for the bad English. I desire these bones may be sent to the tooth, if you think fit. When I go up to Albany next, I intend to go to the place myself, to see if I can discover any thing more concerning the monstrous creature, for so I think I may call it."

Mr. Abeel's letter runs thus:—

"According to your Excellency's order, I sent to Klaverak to make a further discovery about the bones of that creature, where the great tooth of it was found. They have dug on the top of the bank where the tooth was roll'd down from, and they found, fifteen feet underground, the bones of a corpse that was thirty feet long, but was almost all decayed; so soon as they handled them they broke in pieces; they took up some of the firm pieces, and sent them to me, and I have ordered them to be delivered to your Excellency."

In 1712 a Philosophical Society was established in the small town of Spalding, in Lincolnshire, principally at the suggestion of Newton, who was desirous that institutions of this nature should be multiplied throughout England. It was the custom of the Spalding Society to transmit their Minutes to the Royal Society; these were generally accompanied by a letter from the Secretary. In one of his communications, he gives this account of the young Society.

"Sir Isaac Newton was pleased to say, he wish'd there were a Philosophic Society in every town where there was company enough to support them. And as that great man recommended, so we have formed ourselves into a Society, and have obtained the favour of some foreign members in the most distant, as well as various parts, of the world, from whom we're favoured with answers to our enquiries, when we apply to them.

"We have a room commodiously fitted up with large presses on two sides, and in them such books, &c. as may best serve the end of such enquiries as arise in our conversation, which chiefly turns on Imbanking and Draining, Agriculture, Botany, History, Architecture, Sculpture, Musick, and such like Arts and Sciences. Adjoining, we open into a little garden of our own, where our servant (who is a curious florist) produces most flowers, fine for their sorts, in season.

"We have an Air-Pump, Microscopes, Thermometer, Barometer, &c., several large portfolios, with prints and drawings in them of various sorts; but, above all, we have raised a publick lending library of above a thousand volumes, kept clean and safe in presses, having tables of the books contained within, hanging on the doors.

"We meet together every Thursday, in the afternoon, throughout the year, and take Minutes of every thing communicated thought worth noting; and lay by, and keep carefully, all Drawings, Dissertations, Essays, &c., under their proper heads. Our Society consists at present of sixty regular members, and one hundred and fifteen honorary, who have been so good as to be our benefactors, and have encouraged our undertaking. The great Ornament of the Country and Light of the World, Sir Isaac Newton, with the Dean of Durham, Mr. Degg, Mr. Gay the poet, the Rev. S. Westley, and others, are of our number. Our fund is but small, arising only out of the contributions of twelve pence a month from each member resident here; but it is sufficient to answer the expense of such experiments as we have occasion to make, and to purchase such books as we wish for, and have more immediate concern to use; and to carry on our correspondencies of which, (as you have given us leave), we have begun with the Royal Society as the principal, trusting to their goodness in pardoning our ambition of being known to, and aim at being at all, though but in the least degree, serviceable to them, towards any of those noble purposes for which they were incorporated. A thing of some value may arise from where it is little expected, and we trust the good-will and spirit of the offerers may render our humble tribute not unacceptable to the most illustrious fraternity of learned men, and encouragers of Arts and Sciences in Great Britain, whereof we have the honour to enrol many amongst our friends and benefactors[45]."

An endeavour was made about this period to render the objects of the Society better known, and to make the scientific acquirements of the Fellows useful to the community, by the publication of a singular Pamphlet, entitled, *A List of the Royal Society of London, with the places of abode of most of its Members; as also an Advertisement, showing what subjects seem most suitable to the ends of its institution. 8vo. London*, 1718[46]. The name of Thomas Clerk appears on the title-page in manuscript, as being the Author. The Advertisement runs thus:—

"As the design of publishing this list of that illustrious body, over which the matchless Sir Isaac Newton happily presides, is to let the more inquisitive and learned part of mankind know where to find suitable correspondents; so, likewise, it is thought proper to

---

[45] Archives: Royal Society.

[46] I have only been able to meet with this Pamphlet in the British Museum.

advertise the curious of some subjects that seem, if handled with judgment and sincerity, most agreeable and useful towards the promoting of those ends for which the ROYAL SOCIETY OF LONDON was instituted by His Majesty, King Charles II., their Founder and Patron, and since, from time to time, supported by the due care and benevolence of its own Members; all along remembering, at least as a Society, not to assert any thing but what ocular demonstration would allow to be matter of fact, in spite of the hypothetical influence of Aristotelians, Cartesians, Adepts, Astrologers, and common Longitudinarians.

"It were, therefore, to be wish'd, that such as have Opportunity, Capacities, and the Advantage of good Telescopes, would be pleased to communicate all Astronomical and other Observations, whether of the Spots in the Heart of the Sun, of their situations and variation therein; of their increase and decrease; or of the Nebulæ mentioned by that universal scholar, and most acute Philosopher, Dr. Halley, in the *Philosophical Transactions*[47]; of new and strange Stars appearing, or of others disappearing; of Comets, and of all Eclipses, whether of the Sun, Moon, Stars, or Satellites.

"No less acceptable would be accurate accounts of all uncommon appearances in the Heavens; such as *Auroræ Boreales,* Thunder and Lightning; particularly noting the time between the *Flash* and the *Crack,* and the like *Phænomenas:* also Registers of the Winds and Weather, of the Thermometer and Barometer, of the quantity of Rain that falls upon any space of ground, though but a foot square; of the constant Flux and Reflux of

---

[47] See No. 347.

of Tides for some time, and how low they ebb, and the Moon's Age at the time of observing them.

"All new Discoveries in Natural History would be also very acceptable and desirable; such as good descriptions of Quadrupeds, Birds, Reptiles, Insects, Amphibious Animals, Fish, whether Testaceous, or of other kinds; of Plants, Minerals, Fossils, or the like, that are either met with but rarely, ill treated by the authors that write of them, or that have hitherto pass'd unregarded; whether they may be of any advantage to mankind as Food or Physick, and whether those, or any other uses of them, can be further improved.

"Dissections of Morbid Bodies, whether human or of other Animals, are highly wanted by the Society, with particular Relations of the Parts decay'd or affected; and all Anatomical Discoveries.

"New experiments, either in Chymistry, such as those of the learned Dr. Friend in his *Prœlectiones Chymicœ;* or in Pharmacy; such as what Medicines are easily incorporated together, and what not; and how compound Medicines may be reduc'd more simple, yet answer the same ènd.

"Improvements in Agriculture would be in like manner gratefully received, such as the best and most commodious ways to water high grounds, drain the more wet and low, to meliorate the barren, and to enrich even fertile land.

"No less valuable would be new Inventions, or Improvements, in Mechanicks, with Descriptions of Machines, Engines, Instruments, or the like; with exact Histories of all sorts of curious and beneficial Trades in any Country."

The Author then refers to Papers in the *Philosophical Transactions,* and the *Memoirs of the Aca-*

*demy of Sciences at Paris*, as models of philosophical writing, &c. for those who purposed drawing up memoirs for the Society. This is followed by the names and addresses of the most eminent Fellows of the Society, with Greek characters answering to their sciences thus symbolized:—

*Natural Philosophy and Mathematicks* ........................ α
*Astronomy* ........................................................ β
*Geometry* .......................................................... γ
*Opticks* ............................................................ δ
*Natural History* ................................................. ε
*Anatomy* ........................................................... ζ
*Chymistry* ......................................................... η
*Mechanicks* ....................................................... θ
*Husbandry, Gardening and Planting* ........................ ι
*Antiquities* ........................................................ κ

The reader will probably be interested to have a specimen of the list of names:—

Sir Isaac Newton, St. Martin Street, Leicester Fields .... α β γ δ
Edmund Halley, Prince's Street, Bridgewater Square..... α β & γ
John Machin, Astronomical Professor, Gresham College.. α β & γ
Sir Hans Sloane, Great Russell Street, Bloomsbury...... ε ι & ζ
Brook Taylor, LL.D. Secretary, Norfolk Street .......... α γ & δ
William Cheselden, Surgeon, Sadler's Hall ................ ζ
J. T. Desaguliers, Channel Row, Westminster ........... α γ & δ
Martin Folkes, Southampton Street, Covent Garden ..... α & κ
John Keill, Gray's Inn ........................................ α β & γ
Peter Le Neve, Herald's Office ............................. κ
Earl of Pembroke, St. James's Square .................... κ
Frederick Slare, Black Lion, Holborn ..................... δ ε & ζ
Abraham de Moivre, Slaughter's Coffee House, St. Martin's Lane ................................................. α & γ

Thus, in the words of the Author, "ingenious persons may at once know where to address themselves for their learned wants." The entire pamphlet is curious, and bears the impress of having been compiled and published under the sanction and patronage of the Society, for we are told, that "anything remarkable, rare, or curious, towards augmenting their repository of Rarities, may be directed to be left at the Society's House in Crane Court, for either of the Secretaries, and will be considered no unacceptable present."

In 1715 the Society acquired an accession of property, in the form of a small estate at Mablethorpe, in Lincolnshire, consisting of 55 acres, 2 roods, and 2 perches[48], bequeathed by Francis Aston, Esq., who, it will be remembered, resigned the office of Secretary in 1685. He also left a considerable number of books and some personal property to the Society, which, after paying off certain debts, amounted to 445*l.*

The Society's funds were further augmented, in 1717, by a bequest from Dr. Paget, a Fellow, of two houses in Coleman Street, which brought in a rental of about 100*l.* per annum. These houses were required in 1835 by the City authorities, in order that they might be taken down to improve the new London Bridge approaches, and a sum (according to valuation) of 3150*l.* was paid to the Society for them, which was immediately vested in the public securities.

In 1719 another bequest of 500*l.* was received from Robert Keck, Esq., who was elected a Fellow in 1713. The object of this legacy, as will be seen by the subjoined extract from the Will of the testator, was to enable the Society to carry on foreign correspondence by means of a paid officer.

"I give unto the President, Council, and Fellows, of the Royal Society of London for the encreasing

[48] This estate is still in the Society's possession.

Naturall Knowledge, five hundred pounds[49], to be by them laid out, and the profits arising, to be bestowed on some one of the Fellows, whom they shall appoint to carry on a foreign correspondence."

From this legacy originated the office of Foreign Secretary, which still forms a portion of the official establishment of the Royal Society. The Council were in some doubt as to their power of appointing under the terms of Mr. Keck's will, and consequently laid the following questions before the Attorney-General for his opinion.

"1. Whether the election of the person to be appointed for carrying on a Foreign Correspondence, be in the Society or in the Council.

"2. Whether the election of this officer is to be annual or not. To which the Attorney-General replied that:—

"1. By the Charter the Council have the making of all By-Laws, and have the government of the whole body of the Society, and to them is likewise given by the same Charter, the care, management, and disposal of the money, profits, and revenues of the Society, as well such as are established by Charter, as what are casually given by will or otherwise; it also seems as plain, from the latter part of the Charter, that the indulgence of a foreign correspondence is given to the whole Society; but the exercise and management of that power is expressly committed to the care of the President and Council; and, therefore, I am of opinion that the Council may, by virtue of the will of Mr. Keck, apply such

---

[49] The sum actually received amounted to 547*l.* 10*s.* which Sir Isaac Newton paid into the Treasurer's hands. Council-book, Vol. II. p. 272.

sums of money as they think reasonable to such person or persons as they shall from time to time appoint to manage such foreign correspondence, without the interposition of the whole body.

"2. I think the best way will be for the Council to appoint such person, from time to time, as they think proper, to manage the foreign correspondence for the Royal Society, during the pleasure of the Council[50]."

In conformity with this opinion, the Council appointed Mr. Zollman to conduct the foreign correspondence: he was styled Assistant to the Secretaries, and held the office to the period of his decease in 1748[51].

In 1724, the increasing property of the Society caused the Council to memorialize George I. for a license to purchase or hold lands, &c. in Mortmain. It is a remarkable circumstance, that no copy of this petition exists in the Journal or Council-books of the Society, although it is frequently alluded to, and is stated to have been lost or mislaid in the office of the Secretary of State; a circumstance that led the Council to draw up another, couched in similar terms.

Having heard from Mr. Lemon, of the State-Paper Office, that the above documents were preserved in that depository, I obtained permission to make copies of them. As, however, they are conceived in nearly the same words, the first in order is only transcribed:

---

[50] Council-minutes, Vol. III. p. 11.

[51] Mr. Zollman was a Fellow of the Society: his office at that period was any thing but a sinecure, as the Society received a great number of foreign communications, all of which were translated for the purpose of being read at the ordinary Meetings, and entered into the Register-book.

"To the King's Most Excellent Majesty.

"The humble Petition of the President, Council, and Fellows of the Royall Society of London for improving Natural Knowledge:

"Sheweth,

"That His Late Majesty King Charles II., by letters patent bearing date the Twenty Second day of April, in the fifteenth year of his Reign, did ordain, constitute, and appoint the Royall Society of London for improving Natural Knowledge, and did thereby grant them license to purchase in Mortmain.

"That, since the grant of the said letters patent, severall well-disposed persons have devised and granted to your Petitioners and their successors diverse lands and hereditaments, and given severall sumes of money to your Petitioners for the use of the said Corporation, and your Petitioners being desirous to invest the same money in the most desirable manner for the improvement of the said Corporation[52]:

"Your Petitioners, therefore, most humbly beseech your Majesty's Royall License to hold and enjoy the lands and hereditaments which have been devised and granted to them and their successors for ever, for the use and benefit of the said Corporation, such Manors, Lands, Tenements, and Hereditaments, as they shall think fitt to purchase, or shall receive by Will, or any Deed of Conveyance, not exceeding the yearly value of One Thousand Pounds.

---

[52] Accompanying this petition, was an affidavit by Mr. Hawksbee, clerk to the Society, setting forth the property of the Society, which is described as consisting of, two messuages in Crane Court; certain lands and hereditaments in Mablethorpe, Lincolnshire; two houses in Coleman Street, devised by the Rev. Dr. Paget in 1717; and a fee-farm in Sussex of 24*l.* per annum.

"And your Petitioners (as in duty bound) will ever pray, &c."

The Petition, which is on a large sheet of paper, bears the date of the 23rd June, 1724; it is signed by Sir Isaac Newton in a firm bold hand, and on the back are the signatures of the following members of Council, who were present when the seal of the Society was affixed:—George Parker, Hans Sloane, M. Folkes, Wm. Jones, John Browne, James Jurin, Tho. Watkins, Edmund Halley, Jo. Harwood, James Pound, and John Machin[53].

The Petition was submitted to the King on the 30th June, as appears by the memorandum written on the margin.

"*At a Court at Kensington, 30th June,* 1724;

"His Majesty having been moved upon this Petition, is graciously pleased to refer the same to Mr. Attorney, or Mr. Solicitor-General, to consider thereof, and report his opinion what His Majesty may fitly do therein."

On the 21st July, the Attorney-General delivered his opinion, which is likewise preserved in the State-Paper Office. It is as follows:—

"Having perused the Petition and Affidavit, I am humbly of opinion that your Majesty may lawfully grant to the Petitioners your Royall Licence, enabling them and their successors to hold and enjoy the lands and hereditaments which have been already devised or granted to the said Corporation, and to purchase, hold, and enjoy to them and their successors for ever, for the

[53] This interesting document is preserved amongst the domestic State Papers for the year 1724, and is numbered 37.

use and benefit of the said Corporation, such Manors, Lands, Tenements, and Hereditaments, as they shall think fit to purchase, or shall be devised, granted, or conveyed to them, not exceeding in the whole the yearly value of 1000*l*.

"All which is most humbly submitted to your Majesty's most Royall Wisdom.

"Signed, P. YORKE.

"*July*, 21, 1724."

In consequence of this favourable opinion, his Majesty granted the Society the required License, a copy of which will be found in the Appendix.

# CHAPTER XVI.

Impetus given to the Study of Meteorology—Illness of Sir I. Newton—His last Attendance at the Society—His Death—His intention of endowing the Society—Speculates in South Sea Scheme—His Order to purchase Stock—His Sun-Dial presented to the Society—Portraits of him—Original Mask of his Face—Lock of his Hair—Sir Hans Sloane elected President—Memoir of him—Proposes an Address to George II.—His Majesty becomes the Patron of the Society—Important changes made in the Statutes—Proposition to limit the number of Fellows—Opinion of Lord Chancellor Hardwicke—Certificates for Candidates first used—Their great Value—Diplomas—Foreign Members exempted from Payments—Society purchase Estate at Acton—Practice of Inoculation promoted—Prince of Wales visits the Society—Experiments made before him—Dr. Watson's Electrical Experiments—Science of Electricity originates from Royal Society—Large amount of Scientific Business—Experiments on Ether—Society at Peterborough—Donation to Museum—Botanical Specimens sent from Apothecaries' Garden at Chelsea, by order of Sir Hans Sloane—Pecuniary embarrassment of the Society—Sir Hans Sloane resigns—Thanks of Society given to him—His great attachment to the Society—Martin Folkes elected President.

---

## 1725—1745.

IN 1725 the Society gave a great impulse to the study of Meteorology, by sending barometers and thermometers to several of their correspondents abroad, who professed their willingness to make observations: as many as eighteen of these instruments were provided at the Society's expense. This subject was brought before the Council by Dr. James Jurin, who had been elected Secretary in 1721, and who undertook to transmit the instruments to their destination[1].

In the Preface to a volume of the *Philosophical Transactions*, published at this period, it is stated: "Ingenious travellers are now furnished with extraordinary accommodations, that were not known to former ages; as thermometers, barometers, hygrometers, microscopes, telescopes, micrometers, exact scales and weights promptly to weigh liquors, and with other circumstances to examine the intrinsic value of all coins and medals, or metals; pendulum-watches, instruments and indexes for magnetical variations, and inclinatory needles, and other helps to ascertain longitudes, and other mechanical contrivances for manifold uses."

The *Transactions* and Journal-books contain a great number of Meteorological Observations communicated by foreigners, who appear to have made good use of the instruments sent to them. During the year 1726, the name of Sir Isaac Newton does not occur so frequently at the Council and ordinary Meetings of the Royal Society, as in former years, but still, to within a short time of his decease, he occasionally fulfilled his duties as President. It is well known

---

[1] Dr. Jurin was a very respectable philosopher of the Newtonian school, who cultivated medicine and mathematics with equal success. He proved a very active and useful Member of the Royal Society. His communications in the *Transactions* extend from Vol. 30 to Vol. 66 inclusive Dr. Jurin was among the earliest and most able advocates for the inoculation of the small-pox, a practice at that time newly introduced into England, and which had to struggle against the prejudices and opposition, not of the vulgar only, but of a very large proportion of medical practitioners. He died on the 22nd March, 1750.

that during the last year of his life he scarcely ever went to the Mint; his office at that establishment being filled by Mr. Conduit, who lived with Newton in his house at Kensington, having married his niece, the beautiful Mrs. Catherine Barton. Sir Isaac was absent from the Anniversary Meeting in 1726, on which occasion he was, however, as usual re-elected President. On the 16th February, 1726—7, he attended an ordinary Meeting of the Society, when an unusual amount of scientific business was transacted; and on the 28th February he went into town to preside at a Meeting of the Council and an ordinary Meeting, held on the 2nd March. These were the last occasions on which he presided over Meetings of the Royal Society. According to his biographers, he suffered so much fatigue and excitement, that after returning to Kensington his former complaint (the stone) returned with redoubled violence. Mr. Conduit says, "As soon as I heard of his illness, I carried Dr. Mead and Mr. Cheselden to him, who immediately said it was stone in the bladder, and gave no hopes of his recovery. The stone was probably moved from the place where it lay quiet, by the great motion and fatigue of his last journey to London, from which time he had violents fits of pain with very short intermissions; and though the drops of sweat ran down his face with anguish, he never complained nor cried out, nor shewed the least signs of peevishness or impatience, and during the short intervals from that violent torture, would smile and talk with his usual cheerfulness. On Wednesday the 15th March, he seemed a little better, and we conceived some hopes of his recovery, but without grounds. On Saturday morning, the 18th, he read the newspapers, and held a pretty long discourse with Dr. Mead, and had all his senses perfect: but that evening at six, and all Sunday, he was insensible, and died on Monday, the 20th March, between one and two o'clock in the morning. He seemed to have *stamina vitæ* (except the accidental disorder of the stone) to have carried him to a much longer age. To the last, he had all his senses and faculties strong, vigorous, and lively, and he continued writing and studying many hours every day to the time of his last illness[2]."

Thus died Sir Isaac Newton, the most illustrious among the Presidents of the Royal Society, having filled this high office for upwards of twenty years. At the last ordinary Meeting over which he presided, he laid before the Society a letter from the then newly-established Academy of Sciences at St. Petersburg, giving an account of the founding of the Academy by the Imperial Monarch, who was desirous of following the example of the English, in encouraging and cultivating science. The letter concludes with an assurance that the Petersburg Academicians "are the more inclined to make their addresses to, and desire most to have the approbation of, the Royal Society, as being the first of the kind, and that which gave rise to all the rest[3]." To so ardent a lover of science for its own sake as Newton, this communication must have afforded sincere pleasure, testifying, as it does, to an earnest desire to imitate a Society made so illustrious

---

[2] Turnor's *Collections*, p. 166.

[3] The letter adds, that the Academy cannot fail to furnish some observations in Astronomy, in a part of the world where that science has not been much cultivated.

by his Presidency. At the Meeting of Council, held on the same day, and convened by Newton's orders, the only business recorded as having been transacted, was calling upon the Astronomer-Royal (Edmund Halley), in the name of the President, for copies of his Astronomical Observations, which he had neglected to send to the Society. According to the Minutes, Sir Isaac Newton ordered the late Queen's letter to be read, directing the Astronomer-Royal to transmit copies of his Observations to the Society, and that he (the President) "thought it proper to take this opportunity, now the Royal Astronomer was present, to put them in mind of the said precept." When it is remembered that Newton was then 85 years of age, this active zeal and energy must be admitted to exhibit a devotion to science which has never been surpassed, if, indeed, equalled[4].

Appended to the Minutes of Council, is a note recording that "Sir Isaac Newton departed his life on Monday the 20th March, 1726—7;" and the Minutes of the ordinary Meetings in the Journal-book inform us, that on the 23rd March, "the chair being vacant by the death of Sir Isaac Newton, there was no Meeting this day."

On the 28th March, the *London Gazette* an-

[4] Not long before his decease, we find him writing to the Duke D'Aumont, acknowledging a communication made to the Society:—"The little you were pleased to honour us with, came so late to the hand of the Society, that I could not sooner return you their thanks for the same. It was read in a full Meeting. And whenever any thing comes before them, which may be worth your Grace's taking notice of, they will take care to have it communicated. And in the mean time, they have desired me to signify to your Grace how exceedingly you have obliged them." MS. Letters: Royal Society.

nounced that "the corpse of Sir Isaac Newton lay in state in the Jerusalem Chamber, and was buried from thence in Westminster Abbey, near the entry into the choir. The pall was supported by the Lord High Chancellor, the Dukes of Montrose and Roxborough, and the Earls of Pembroke, Sussex, and Macclesfield, being Fellows of the Royal Society. The Hon. Sir Michael Newton, Knight of the Bath, was chief mourner, and was followed by some other relations, and several eminent persons intimately acquainted with the deceased. The office was performed by the Bishop of Rochester, attended by the prebend and choir."

The relatives of Sir Isaac Newton, who inherited his personal estate, devoted 500*l*. to the erection of a monument in Westminster Abbey, which was executed in 1731.

Hearne says, that "Sir Isaac Newton promised to become a benefactor to the Royal Society, but failed." He certainly had ample means at his disposal to effect this object, for he died worth about 32,000*l*. personal estate. Among the memorials of the great philosopher in the possession of the Royal Society, is an order, addressed by him to Dr. Ffouquier[5], directing certain sums in his possession, belonging to Newton, to be applied to the purchase of South Sea stock, at a time when that stock had nearly reached its maximum. It is written in a plain hand, and runs thus:—

[5] Dr. John Francis Ffouquier was a French Protestant, who emigrated to this country on the revocation of the edict of Nantes. He was born in 1669. From the books of the Bank of England, it appears that on the 12th August, 1699, he opened a banking account with that company to a very large amount. On the 24th Dec., 1720, he was the holder of 10,250*l*. South Sea stock. He was one of the earliest Governors of Guy's Hospital, under the will of the founder. He died in 1748.

"*Mint Office*, 27 *July*, 1720.

"I DESIRE you to subscribe for me, and in my name, the several Annuities you have in your hands belonging to me, amounting in the whole to six hundred and fifty pounds, per, and for which this shall be your warrant.

"ISAAC NEWTON.

"*To Dr. John Francis Ffouquier.*"

Dr. Wollaston[6], to whom the Society are indebted for this very interesting autograph note, observes, in his letter accompanying it, that "not knowing any such occurrence in the life of Newton had ever been made public, he was for many years unwilling to divulge the transaction; but having since found that the losses which Newton sustained by the South Sea scheme have been noticed in the biographical memoir drawn up on the authority of Mr. Conduit, he no longer hesitates to present the document, being satisfied that it will be considered by every reflecting mind an instructive instance of the soundest understanding being liable to have its judgment perverted by the appearance of enormous profit, and to forget that such profit can only be aimed at with proportionate risk of failure."

Dr. Wollaston adds that the stock originally purchased by Dr. Ffouquier in July 1720, increased in September 1721, to the enormous sum of 21,696*l.* 6*s.* 4*d.*, which was transferred in 1722, and 1723[7]. It is

---

[6] Dr. Wollaston received the above document from his father, the Rev. Francis Wollaston, F.R.S., whose mother was the daughter of Dr. Ffouquier, to whom the order of Newton is addressed.

[7] When examining some official papers in the State-Paper Office, relating to the Royal Society, I found a voluminous petition to the Crown, from the Directors of the South Sea Company, extending to 60 folio pages, and bearing the date of July 23, 1724,

thus apparent that Newton could easily have left the Society a legacy, or endowed them during his lifetime: but although he did neither the one nor the other, he bequeathed to the institution a fame which a period of 120 years has not diminished. The chair, so long filled by Newton, will always, while science numbers a votary, be regarded with feelings of honour and admiration, and to occupy it will be the proudest boast of the British philosopher[8].

---

setting forth the fearful difficulties under which they were labouring, and praying for immediate relief; and, in particular, to be exempted from the duty of ten shillings on Negroes imported into, and twenty shillings upon Negroes exported from, Jamaica; thus showing, that at the time Newton's stock was transferred, the abyss was on the point of opening in which thousands were subsequently lost.

It will be well to remember, that Newton was not the only great man implicated in this foolish adventure. Statesmen of the highest intellect participated in the speculation. Walpole, who must have been aware of the pit which was about to swallow up the public credit, profited by the public credulity until he had filled his purse. Pope is stated by his biographers to have lost a large sum of money, although, at one time, his stock was worth from 20,000*l.* to 30,000*l.* It will be recollected, that Swift in the *Voyage to Laputa*, (published in 1726,) visited with merited severity the many quack schemes which abounded at that period. The plans of their projectors were usually based upon some smattering of science, or some supposed discovery, which they had not ability to verify before setting their bubble afloat. But Swift, with the caustic energy which so strongly marked his character, shot some of his shafts where no reproof was needed. The *Voyage to Laputa* was disliked by Arbuthnot and other Fellows of the Royal Society, though the satire was probably intended by its author to be rather aimed against the abuse of philosophical science than at its reality.

[8] Voltaire visited England in the years 1726—7, when he resided principally at Wandsworth. He had the great merit of introducing Newton's theories into France. His observations on the

Another memorial of Newton in the possession of the Royal Society, is one of the solar dials made by him when a boy. This interesting relic was presented to the Society in 1844, by the Rev. Charles Turnor, F.R.S., accompanied by the following letter addressed to the Marquis of Northampton, President of the Society.

"*Cheltenham, May* 24, 1844.

"MY LORD,

"YOUR Lordship having been pleased to express a wish that I should furnish a detailed account of the Newtonian Dial which I have had the honour of presenting to the Royal Society, I beg to submit to your Lordship the following particulars. The dial was taken down in the early part of the present year from

---

the Royal Society which were made at this period, are very interesting. After giving the date of its establishment, he says: "*Elle n'a point de récompenses comme la nôtre; mais aussi elle est libre; point de ces distinctions désagréables, inventées par l'abbé Bignon, qui distribua l'Académie des Sciences en savans qu'on payait, et en honoraires qui n'étaient pas savans. La Société de Londres indépendante, et n'étant encouragée que par elle même, a été composée de sujets qui ont trouvé le calcul de l'infini, les lois de la lumière, celles de la pesanteur, l'aberration des étoiles, le télescope de réflexion, le pompe à feu, le microscope solaire, et beaucoup d'autres inventions aussi utiles qu'admirables.* Qu'auraient fait de plus ces grands hommes s'ils avaient été pensionnaires ou honoraires?" *Œuvres*, Vol. XXVI. 443. Voltaire had a far higher opinion of the Royal Society than of the French Academy. He says, that he was once asked in England respecting the memoirs of the latter body. "*Elle n'écrit point de Mémoires*," was his answer, "*mais elle a fait imprimer soixante ou quatre vingts volumes de complimens.*" It was Piron who said that a *discours de reception* at the Academy ought never to exceed three words, "*que le recepiendaire doit dire, Messieurs, grand merçi, et le directeur répondra, Il n'y a pas de quoi.*" This advice arose probably from the speeches of many being, as was whimsically said, "*long et plat comme l'épée de Charlemagne.*"

the South wall of the Manor House at Woolsthorpe, a hamlet to Colsterworth, in the county of Lincoln, the birthplace of Newton.

"The house is built of stone, and the dial, now in the possession of the Royal Society, was marked on a large stone in the south wall, at the angle of the building, and about six feet from the ground, and was reduced to its present dimensions for the convenience of carriage. The name of NEWTON, with the exception of the first two letters, which have been obliterated by the hand of time, will, on close inspection, appear to have been inscribed under the dial in rude and capital letters. There is also another dial marked on the wall, smaller than the former, and not in such good preservation. The above are the only dials about the house which I have been able to discover, nor can I find by inquiry on the spot that more ever existed, though some of Newton's biographers assert that there were several. An opinion has always prevailed, that the dials now in being were executed by Newton's own hand when a boy, which appears probable from the well-known fact, that, at a very early period of his life, he discovered a genius for mechanical contrivances, evinced, more particularly, by the construction of a windmill of his own invention, and a clock to go by water applied to its machinery. Finding, however, this latter contrivance (however ingenious) to fail in keeping accurate time, it is not improbable that, with a view to secure that object, he formed with his own hands the two dials in question; and, very probably, the dial now remaining in the wall of the house, from its inferiority in point of construction to that now in the possession of the Royal Society, was his first attempt in dial-making. The gnomons of these dials have, unfortunately, disappeared many years, but as they are described in some of the printed accounts as *clumsy* performances, it may be concluded that they were not the work of a professed mechanic, but were, probably, formed and applied by Newton himself when he constructed the dials.

"I trust your Lordship will allow me to express the high satisfaction I feel in seeing this very interesting relic in the possession of that Society of which Newton was so distinguished an ornament, and over which he presided more than twenty years.

"I must beg your Lordship's permission to add that, for the gratification which I experience on this occasion, I am greatly indebted to my nephew, Christopher Turnor, Esq., of Stoke, Rochford, to whom the manor-house and landed property of Newton now belong, and who not only permitted, but kindly encouraged me to offer this valuable relic to that Society, which he, as well as myself, consider as its fittest and most appropriate depository[9].

"I have the honour to be,
"My Lord,
"Your Lordship's humble Servant,
"CHARLES TURNOR.

"*The Marquis of Northampton,*
*&c. &c. &c.*"

A representation of this dial, or rather relic, for Time's gnawing tooth has dealt so roughly with it, as to have nearly effaced the original cutting, is inserted at the end of this chapter. It is now preserved in a well-made strong oaken box, with a plate-glass cover, and will, it is to be hoped, endure for many centuries.

---

[9] *Phil. Trans.* for 1845, p. 141.

The Society possess three portraits in oil of their former illustrious President. One, painted by C. Jervas, was presented by Newton, and is appropriately placed over the President's chair in the Meeting-room; another, in the Library, was painted by D. C. Marchand; and the third, which hangs in my Office, was presented to the Society in 1841, by Charles Vignolles, Esq., the eminent engineer, accompanied by the subjoined letter addressed to the Marquis of Northampton.

"4, *Trafalgar Square,*
*London, March* 25, 1841.

"My Lord,

"I have the honour of transmitting to your Lordship, for presentation to the Royal Society, an original portrait of *Sir Isaac Newton*, by *Vanderbank*, a Dutch painter of some note in that age. This picture has now been many years in my possession, and the tenure by which I have kept it (as the collateral descendant of so illustrious a man) was too flattering not to have been a source of great personal gratification.

"But I consider such a portrait to belong of right to the scientific world in general, and more especially to that eminently distinguished Society of which Newton was once the head, and which is now so ably presided over by your Lordship.

"I have, therefore, to request your Lordship will do me the honour to present this original portrait of Sir Isaac Newton to the Royal Society, in my humble name.

"Accident having destroyed some of the papers of my family, I am unable of myself to trace the entire history of this portrait; but, I believe, more than one Member of the Royal Society is competent to do so, and it is well known to collectors; and a small mezzotinto engraving of it was published about forty years ago. It was painted the year before Newton died, and came into the family of the celebrated Lord Stanhope, who left it by his will to my grandfather, the late Dr. Charles Hutton, a distinguished Member of the Royal Society, expressly on the well-authenticated account of that eminent mathematician having been remotely descended from Sir Isaac Newton in the following way, as I find on a family MS., viz: that the mother of the well-known James Hutton, and the mother of Dr. Charles Hutton, were sisters; and the grandmother of James Hutton, and the mother of Sir Isaac Newton, were also sisters.

"I have ever considered this very distant connexion with so great a man should not be an inducement to lead me into any but casual mention of the circumstance, that I might avoid the imputation of a vain boast; nor would it have been brought forward now, except to explain the cause by which this portrait came into the possession of an individual, who is happy in relinquishing it to grace the Hall of Meeting of the Royal Society[10].

"I have the honour to subscribe myself,

"Your Lordship's very obedient humble Servant,

(Signed) CHARLES VIGNOLLES.

"*The Marquis of Northampton.*"

Besides these portraits, which are in excellent preservation, the Society possess the original mask of Sir Isaac Newton's face, from the cast taken after death, which belonged to Roubilliac. For this truly interesting relic the Fellows are indebted to Samuel Hunter Christie, Esq., Secretary to the Society, and Professor of Mathematics at the Royal Military Aca-

[10] Archives: Royal Society.

demy at Woolwich, who presented it to them, in the year 1839. The history of this mask, as related to me by Mr. Christie, is extremely curious. Being desirous of purchasing a bust of Sir Isaac Newton, Mr. Christie entered the shop of a dealer in statues in Tichborne Street. To Mr. Christie's question, whether he had any bust of the philosopher to dispose of, the dealer replied, that though he had no bust, he had an old mask of Newton, which his father had purchased 50 or 60 years before, at the sale of Roubilliac's effects, and which he had kept on his shelves amongst various articles of his trade. It was evident that the dealer regarded the *relic* as little better than useless lumber, and this is confirmed by his having consented to dispose of it for a few shillings.

Mr. Christie, having borne off his prize, had a few casts taken from it, and subsequently enjoyed the great satisfaction of placing it in a repository, not only the most fitted for its reception, but where it will be hallowed and preserved with religious care as long as the Royal Society exists. Though much injured by rough treatment, it will be seen by those who are acquainted with the authentic portraits of Newton, that the mask, which is copied in the annexed drawing, presents the characteristic features of the Society's former illustrious President.

In addition to these memorials of the great philosopher, another of a very interesting nature has lately come into the possession of the Society, as explained in the following letter from the donor.

"Sir, "27 *Bedford Row, October 25th*, 1847.

"I am in possession of a very small lock of silver-white hair, which is wrapp'd in a paper marked 'Sir I. Newton,' and which, from the circumstances under which it came to me, I have no doubt whatever belonged to that eminent philosopher.

"As I believe the University of Cambridge possess and highly value a similar relic, I am led to presume that the Royal Society, which already possesses several memorials of that illustrious man, would feel interested in adding it to their collection.

"As I am conscious that the relic would be valueless unless its authenticity were unquestionable, I will observe that the family of Sir Isaac Newton were connected by marriage with the family of Barton, and these latter with the Burrs, who are my relations on the maternal side, and through one of whom the article came into my possession. In corroboration of which connexion I may add, a former member of the Burr family was christened Newton Barton.

"Should the Society consider this communication worthy their attention, I shall be happy to be favoured with a reply.

"I am, Sir,
"Your very obedient Servant,
"HENRY GARLING.

"*C. R. Weld, Esq.*
*Assistant Secretary to the Royal Society.*"

I immediately answered this letter, and assured Mr. Garling that the Society, I had no doubt, would highly value the interesting relic of their former President, referred to in his communication, which would be laid before the Council. Mr. Garling, in reply, invited me to his house, to examine the lock of hair, and the family documents relating to it; at the same time stating his wish to present it to the Society.

The result of my visit enabled me to report to the Council my entire belief in the genuineness of the lock of hair. It was accordingly accepted, and the thanks of the President, Council, and Fellows, specially voted to Mr. Garling for his esteemed present[11].

The death of Sir Isaac Newton was a heavy blow to the Royal Society, who were highly sensible of the great honour which his presidency had devolved upon them. They were naturally proud in the consciousness, that it was not only in England, but almost amongst themselves, that he had, in the eloquent language of Dr. Young, "advanced with one gigantic stride from the regions of twilight into the noonday of Science." Throughout all his scientific, and not always peaceful labours, they supported him with steady constancy; thus manifesting their sense of his mighty genius and brilliant discoveries.

It was fortunate for the Society, that Sir Hans Sloane had attained to so high a scientific eminence at the period of Newton's decease, as to render the choice of a President less embarrassing than it would otherwise have been. He had acted so long in the capacity of Secretary, and Vice-President, and had always manifested so lively an interest in the welfare of the Society, that there could have been little hesitation, on the part of the Council, in selecting him to fill the office of President. Nor indeed was there; for we find that on the 29th March, when a full Council assembled for the special purpose of electing a President, "Sir Hans Sloane was unanimously elected, and was accordingly declared and sworn;" and at the Anniversary Meeting in November his election was confirmed by a large majority.

---

[11] The lock is now enclosed in a small mahogany box, with a glass cover. The hair is singularly fine, and when examined under a lens, appears *irridescent*.

Sir Hans Sloane was born at Killeleagh, in the north of Ireland, on the 16th April, 1660. Though a native of the sister island, he was of Scotch extraction, his father, Alexander Sloane, having been the head of a colony of Scots, settled in Ulster by James I. From a very early period he manifested a great inclination for the study of natural history and medicine, which was strengthened by a suitable education. When about sixteen years of age, he was attacked by a spitting of blood, which threatened to be attended with considerable danger, and interrupted the regular course of his application for three years. He had already learned enough of medicine, to know that a malady of this nature was not to be suddenly removed, and he prudently abstained from wine and other stimulating liquors. By strictly observing this regimen, which he, in some measure, continued ever afterwards, he was enabled to prolong his life beyond the ordinary limits; presenting an example of the truth of his favourite maxim—that sobriety, temperance, and moderation, are the best and most powerful preservatives that nature has granted to mankind.

On his arrival in the metropolis, he commenced the study of medicine, cultivating at the same time natural history; and it was his knowledge of this science which introduced him early to the acquaintance of Boyle and Ray. After studying four years in London with unremitting severity, he visited France for further improvement. At the end of 1684, he returned to London with the intention of settling in the metropolis as a physician, and was elected a Fellow of the Royal Society on the 21st of January following. In 1685 he presented some curiosities to the Society, and in July of the same year was a candidate for the office of Assistant Secretary, but without success, as he was obliged to give way to the superior interest of Halley.

On the 12th April, 1687, he was chosen a Fellow of the College of Physicians, and on the 12th September following he embarked at Portsmouth for Jamaica, with the Duke of Albemarle, who had been appointed Governor of that island.

In the preface to his *Natural History of Jamaica*, he says:

"I had from my youth been very much pleased with the study of plants, and other parts of Nature, and had seen most of those kinds of curiosities which were to be found in the fields or the gardens or cabinets of the curious. The accounts of these strange things which I met with in collections, and was informed were common in the West Indies, were not so satisfactory as I desired. I was young, and could not be easy if I had not the pleasure to see what I had heard so much of. I thought by that means the ideas of them would be better imprinted on my mind, and that, upon occasion, the knowledge of them and their uses might be afterwards more familiar to me. These inclinations remained with me some time after I settled myself to practice physic in London, and had had the honour to be admitted a Fellow of the College of Physicians, as well as of the Royal Society.

"His Grace the Duke of Albemarle having obtain'd the supreme command of the Island of Jamaica, and other parts of English America, when he should arrive, employed Dr. Barwick, his physician, to look out for one who could take care of him and his family in case of sickness. Dr. Barwick spake to me in this matter, enquiring if any physician of my acquaintance would undertake it. This seemed to me to be such an opportunity, as I myself wanted to view the places and things I designed; wherefore, after due consideration, I resolv'd to go, provided some preliminaries and conditions were agreed to, which were all granted."

Although, in consequence of the death of the Duke shortly after his arrival in Jamaica, Sloane's sojourn in that island did not exceed fifteen months, yet he managed to collect such a prodigious number of plants, that on his return to England, Ray was astonished that one man could procure in one island, and in so short a space of time, so vast a variety.

The plants which he brought with him amounted to 800 species. Of these, he gave his friend Mr. Courten whatever he wanted to complete his collection, and the remainder, with other objects of natural history and various curiosities, formed the nucleus of his Museum, upon which he spent a large sum of money, and enriched it by every means in his power. Dr. Franklin in one of his letters says, "I had brought over a few new curiosities, among which the principal was a purse made of the *Asbestos*, which purifies by fire. Sir Hans Sloane heard of it, came to see me, and invited me to his house in Bloomsbury Square, shewed me all his curiosities, and persuaded me to add that to the number, for which he paid me handsomely[12]."

In the introduction to the second volume of the *Natural History of Jamaica*, published in 1725, nearly twenty years subsequent to the first, Sloane states, as one reason for the delay, the length of time which "putting into order his curiosities, numbering them, and entering them," occupied; and he gives the

[12] *Works*, Vol. I. p. 65.

number of his collections in natural history, &c., comprising 8226 specimens in botany alone, besides 200 volumes of "dried samples of plants."

His love for zoology is strikingly exhibited by the annexed extract from the work just quoted.

"Though I foresaw the difficulties, yet I had an intention to try to bring with me from Jamaica some uncommon creatures alive, such as a large yellow snake seven foot long, a guano or great lizard, a crocodile, &c. I had the snake tamed by an Indian, whom it would follow as a dog would his master, and after it was deliver'd to me. I kept it in a large earthen jar, such as are for keeping the best water for the commanders of ships during their voyages, covering its mouth with two boards, and laying weights upon them. I had it fed every day upon the guts and garbage of fowl, &c. put into the jar from the kitchen. Thus it liv'd for some time, when, being weary of its confinement, it shoved asunder the two boards on the mouth of the jar, and got up to the top of a large house, wherein lay footmen and other domesticks of Her Grace the Duchess of Albemarle, who, being afraid to lie down in such company, shot my snake dead. It seem'd before this disaster to be very well pleased with its situation, being in a part of the house which was well fill'd with rats, which are the most pleasing food for these sort of serpents[13]."

Sloane now applied himself sedulously to his profession. Success attended him in his practice, and he became so eminent, that he was appointed Physician to Christ's Hospital in 1694; an office which he

[13] Vol. II. p. 346.

held until his great age and infirmities compelled him to resign it in 1730. It is due to his memory to state, that during the whole of this period he never retained his salary, but always devoted it to charitable purposes. In 1716, George I. created Sloane a baronet, an hereditary title of honour to which no English physician had before attained, and, at the same time, appointed him Physician-general to the army, which office he resigned in 1727, for the appointment of Physician in Ordinary to George II. In 1719, he was elected President of the College of Physicians, which he held for sixteen years, and in 1727, he attained the highest object of his ambition, having the honour to succeed Newton in the chair of the Royal Society. His competitor on this occasion was Martin Folkes. He continued to exercise all his official duties with the greatest zeal, until he arrived at the age of fourscore, when he formed the resolution of quitting the service of the public, and of living in tranquillity. With this view, he resigned the presidency of the Royal Society, and retired to Chelsea, where he had purchased an estate. There he enjoyed in peaceful repose the remains of a well-spent life, still continuing to receive, as he had done in London, the visits of scientific men, of learned foreigners, and of the Royal family; and, what is still more to his praise, he never refused admittance nor advice to rich or poor, who came to consult him concerning their health.

Edwards, in his *Gleanings of Natural History*, has given an interesting sketch of this period of Sloane's life:—

"Sir Hans Sloane, in the decline of his life, left London and retired to his manor-house at Chelsea, where he resided about fourteen years before he died.

After his retirement to Chelsea he requested it as a favour to him (though I embraced his request as an honour done to myself), that I would visit him every week, in order to divert him for an hour or two, with the common news of the town, and with any thing particular that should happen amongst his acquaintance of the Royal Society, and other ingenious gentlemen, many of whom I was weekly conversant with; and I seldom missed drinking coffee with him on a Saturday during the whole time of his retirement at Chelsea. He was so infirm as to be wholly confined to his house, except sometimes, though rarely, taking a little air in his garden, in a wheeled chair; and this confinement made him very anxious to see any of his old acquaintance, to amuse him. He was always strictly careful that I should be at no expense in my journeys from London to Chelsea to wait on him, knowing that I did not superabound in the gifts of fortune: he would calculate what the expense of coach-hire, waterage, or any other little charge that might attend on my journeys backward and forward, would amount to, and obliged me annually to accept of it, though I would willingly have declined it."

Sir Hans Sloane enjoyed the high honour of being one of the foreign members[14] of the French Academy, by which body he was greatly esteemed. In the *éloge* pronounced at his death, it is stated that, "*il étoit de presque toutes les Académies de l'Europe, et en liaison avec toutes les personnes distinguées par leur savoir, leur naissance, ou leur génie. M. le Duc de Bourbon l'honora de sa correspondance; et pour reconnoître*

[14] The vacancy caused by his death was filled by Stephen Hales, F.R.S.

*les présens qu'il en avoit reçûs, ce Prince lui envoya son portrait dans une magnifique boîte d'or et une médaille où S.A.S. étoit représentée; le Roi même a daigné lui envoyer en présent le recueil des gravures de son cabinet, don qui ne se fait ordinairement qu'aux personnes les plus distinguées, et qui prouve à la fois et la grande reputation du* Philosophe Anglois, *et le cas que le Monarque François sait faire du mérite*[15]."

Sir Hans Sloane died on the 11th of January, 1753, after a short illness. He bequeathed his Museum to the public, on condition that 20,000*l.* should be paid to his family; a sum which is said to have scarcely exceeded the intrinsic value of the gold and silver medals, and the ores and precious stones in his collection; for, in his will, he declares that the first cost of the whole amounted at least to 50,000*l.* His library, consisting of 3566 manuscripts and 50,000 volumes, was included in this bequest. Parliament accepted the trust on the required conditions, and thus Sloane's collections formed the nucleus of the British Museum[16].

There are twenty-four Papers by Sir Hans in the *Philosophical Transactions*. He also published several works, the most important of which is his *Natural History of Jamaica*.

He presented to the Society a portrait of himself, painted by Sir G. Kneller.

One of the first acts of Sir Hans Sloane, was to propose to the Council that an address should be pre-

---

15 *Hist. de l'Acad.*, 1753, p. 319.

16 Sloane's collection was greatly increased by the bequest, in 1702, of the museum of his friend Mr. Courten.

sented to George II., praying his Majesty to become the Patron of the Society, and, at the same time, that he would inscribe his name in the Charter-book[17].

After several lengthy discussions, an Address in these terms was agreed to:—

"*To the King's Most Excellent Majesty.*

"THE humble Address of the President, Council, and Fellows of the Royal Society of London for improving Natural Knowledge.

"MOST GRACIOUS SOVEREIGN,

"WE, your Majesty's most dutiful and loyal subjects, the President, Council, and Fellows, of your Royal Society of London for improving Natural Knowledge, though deeply affected with the loss of your Royal Father, our late most gracious King and Patron, yet cannot but take part in the universal joy upon your Majesty's happy and peaceable accession to the Crown.

"We cheerfully join in the wishes, the hopes, and united prayers, of all your people, that your reign may be long and glorious, and that every blessing may be shower'd down upon your sacred person, your august Consort, and flourishing Royal Family, which the most

---

[17] At the ordinary Meeting on the 27th April, the President "acquainted the Society that there had been a design and resolution taken about ten years ago, to endeavour to obtain the favour of his Majesty, to dignify the Society, as some of his royal predecessors had done, with the grant of his name under the title of Patron; and that, accordingly, a leaf in the Charter-book had been prepared with the Arms for that purpose; yet no further steps having been taken therein, the said design proved of no effect. And whereas, he having some reason to believe that an application of this nature, at this time, might not be unsuccessful, he proposed that the said former resolution should now be resumed, which was agreed to." Journal-book, Vol. XIV. p. 76.

affectionate subjects have ever ask'd of Heaven for the best and most belov'd of Princes.

"Your Majesty's known love to science and usefull arts, gives us reason to hope that it will be one of the glories of your reign to give encouragement to learning, as has been done by the greatest Princes in all ages, and particularly emboldens us to request that you will graciously vouchsafe us the honour, granted us by your Royal Father, of declaring yourself our Protector and our Patron.

"This mark of your Royal goodness will animate and encourage us to pursue, with redoubled zeal and application, the noble design of our Royal Founder for improving those arts and sciences that tend to the general good of mankind, and the particular service of our native country."

A deputation, consisting of the Duke of Richmond, Sir Hans Sloane, and other Members of Council, presented the above Address, and it is recorded in the Journal-book, that "His Majesty was pleased to receive the gentlemen in a most gracious manner, and did the Society the honour to write his Royal name in the Statute-book as their Patron[18]; and that upon waiting on the Queen with a compliment, her Majesty had likewise been pleased to receive them very graciously."

In the early part of 1728, Sir Hans Sloane brought forward several propositions involving considerable changes in the constitution of the Society. The most important of these were:—1. To exempt the foreign Members from paying subscriptions. 2. To sue Fel-

---

[18] His Majesty's signature, in a large bold hand, is inscribed on an ornamented page with the autographs of other Sovereigns.

lows in arrear of their payments. 3. To make it compulsory for every candidate to be approved by the Council, and recommended by three Fellows, one of whom, at least, was to be a member of Council, before the candidate could be put to the ballot.

The last of these propositions was passed into a Statute, and acted on from 1728, to 1730; all the candidates being approved by the Council before being put to the ballot. In the latter year, the expediency of limiting the number of Fellows was taken into consideration; but before making any statute on this subject, a case was drawn up for the opinion of the Attorney-General, embodying these queries:— "Whether it is any infringement of the rights and privileges of the Fellows, that a candidate should be approved by the Council before being ballotted for by the Fellows generally; considering that the rejection of a candidate by the Council does not disqualify him from being put up again?" and secondly, "Whether the Council cannot, by virtue of their general power of regulating the body, limit the number of the Members thereof; or at least make such laws as may check the too great increase of the body with new Members unfit for answering the end of the Institution?"

The opinion of the Attorney-General on the first query was, that "The Charter having joined the President, Council, and Fellows together, in the elections of Fellows, as Members of the entire body, and having directed such elections to be made by a major part of them all, without giving any preference in those acts to the Council, I think the Council should not make a Statute whereby to assume a negative to themselves, which seems to me to be the effect of this

Statute. Therefore I apprehend this Statute not to be warranted by the Charter."

The opinion on the second query was: "Considering that the Charter hath left the body at large without limiting the number of Fellows, and considering also the nature of this foundation, I think the Council cannot make a Statute to limit the Fellows to a certain number. But they may make reasonable statutes, or bye-laws, to describe and ascertain proper qualifications of persons to be elected Fellows, in such manner as may best answer and promote the ends of an Institution so useful to the learned world."

This opinion, emanating from so profound a lawyer as the Attorney-General[19], had the effect of causing the Statute, requiring candidates to be approved by the Council, to be repealed, and the following to be substituted:—

"Every person to be elected Fellow of the Royal Society, shall be propounded and recommended at a Meeting of the Society, by three or more Members[20], who shall then deliver to one of the Secretaries a paper signed by themselves, signifying the name, addition, profession, occupation, and chief qualifications of the Candidate for election, as also notifying the usual place of his habitation; a fair copy of which paper, with the date of the day when delivered, shall be fixed up in the Meeting Room of the Society at Ten several Ordinary

---

19 Afterwards the celebrated Lord Chancellor Hardwicke.

20 It appears that Candidates were also expected to send in a Paper on the branch of science to which they were attached. In answer to a letter from a Candidate desiring to know the regulations, the Secretary says: "To be made a Member, you must be recommended by three Members, and send in specimens to show in what part of philosophy you are particularly conversant."

Meetings, before the said Candidate shall be put to the Ballot.

"Saving and excepting that it shall be free for every one of his Majesty's subjects who is a Peer, or the son of a Peer, of Great Britain or Ireland, and for every one of His Majesty's Privy Council of either of the said Kingdoms, and for every Foreign Prince or Ambassador, to be propounded by any single person, and to be put to the ballot for election on the same day; there being present a competent number for making elections."

This Statute was passed into a law on the 10th December, 1730, and on the 25th February, 1730—1, the first 'paper' or certificate in favour of a candidate was presented, since which period this mode of introducing candidates for election has been steadily maintained. It is important to state, that all certificates are carefully preserved in chronological order, whether the candidates have been elected or not; they form several folio volumes, and, as they generally give an account of the scientific and literary attainments of the candidates, are highly valuable records.

Previously to the introduction of this custom, diplomas were sometimes given to the Fellows on their election. The Archives of the Society are silent on this subject, but the annexed copy of one, placed in my hands by Captain Smyth, is an unquestionable testimony that the practice existed.

The document is written on parchment, headed by the Arms of the Society appropriately emblazoned, and has appended to it an impression in wax of the Society's large seal:—

"Præses, Concilium, et Sodales Regalis Societatis

Londini, *pro Scientia Naturali promovenda institutæ omnibus ad quos Præsentes Literæ pervenerint, Salutem. Sciatis virum eruditem*, Johannem Thorpe, *Medicinæ Doctorem, Oxon. in Comitiis Solennibus Regalis Societatis prædictæ trigesimo Die Novembris, Anno D*[ni] *Millesimo Septingesimo-quinto Londini, habitis in Regalem Societatem Londini prædictam electum, receptum, et admissum fuisse, omnibusque Juribus et Privilegiis, quæ ad sodalem dictæ Regalis Societatis utcunque pertinent auctum atque donatum. In cujus rei testimonium, Præsentes Literas Sigillo nostro communi muniri fecimus. Datum in Concilio Regalis Societatis Londini, nono die Novembris, Anno Millesimo Septingentesimo Decimo Tertio.*

"ISAAC NEWTON, P.R.S."

The Attorney-General having given an opinion that Fellows in arrear with their subscriptions, who had signed the obligation, might be legally proceeded against for the recovery of their arrears, immediate steps were taken to apprise defaulters of the intention of Council to enforce payment of their subscriptions, which had the effect of causing several Fellows to liquidate their debts to the Society, whilst others compounded for their past and future subscriptions by the payment of one sum. At the same time it was resolved that foreign Members, who had hitherto been classed with ordinary Members, as far as paying subscriptions, should be in future exempted from these payments[21].

The augmentation of funds, arising from these energetic measures, led the Council to seek for a more profitable investment for their capital than that afforded by Government securities; and after various

---

[21] At this period there were 79 Foreign Members upon the Books of the Society.

propositions, it was resolved to purchase a small estate at Acton, in Middlesex, a village at that period wholly in the country, although, from the rapid increase of buildings, now almost linked to Bayswater. This estate consisted originally of 48 acres, but in consequence of some portion of the land having been sold, is now reduced to 33 acres. Should our gigantic metropolis continue to increase, as it has done of late years, this small property will become, ere long, of considerable value for building ground, as the greater part of it adjoins the high road leading to London. This purchase was effected in 1732[22].

This era of the history of the Society would be imperfect, were we to omit noticing the spirited measures taken by several distinguished Members, to promote the practice of Inoculation. Sir Hans Sloane, Dr. Jurin (whose name has been mentioned in connexion with this subject), Dr. Williams, Mr. Wright, and Mr. Gale, published several Papers in the *Philosophical Transactions*, urging the importance of inoculation. The first account of this practice appears in the *Transactions* for 1714, in two Papers by E. Timoni, a physician, at that time practising at Constantinople, and J. Pylarini, a native of Cephalonia,

---

[22] The following is a copy of the Rent-roll of the Society's Estates at this period:—

| | £. | *s.* | *d.* | |
|---|---|---|---|---|
| "A Fee-farm Rent at Lewes in Sussex, producing | 24 | 0 | 0 | per an. |
| An Estate at Mablethorpe in Lincolnshire | 27 | 0 | 0 | ,, |
| A small house in Crane Court, purchased with the Society's house | 24 | 0 | 0 | ,, |
| Two houses in Coleman Street | 97 | 10 | 0 | ,, |
| An Estate at Acton | 65 | 0 | 0 | ,, |
| | 237 | 10 | 0." | |

who practised medicine in various parts of the East. "Inoculation," says Dr. Thomson, "began to be practised in London soon after the publication of these Papers[23]. But it made its way exceedingly slowly, in consequence of the violent prejudices which it had to combat[24]. In the year 1722 Dr. Nettleton began to try it at Halifax, in Yorkshire, and he published a detailof his proceedings in the *Philosophical Transactions*. The whole number inoculated in England by the year 1722, amounted, according to Dr. Jurin, to 182. Of these two died. Dr. Jurin, in an admirable Paper which he published on Inoculation, demonstrated that, in the small-pox taken the natural way, the deaths amounted to rather more than one in fourteen. Thus the superiority of inoculation became conspicuous at the very outset. For, even by the enemies of that method, the deaths were only estimated at one in ninety-one. They are, in fact, somewhat lower than that ratio[25]." Dr. Thomson then proceeds to state, that Sir Hans Sloane interested himself greatly in the subject, and was the means of causing Dr. Pylarini's Paper to be inserted in the *Transactions*. "This notice," he adds, "lay dormant till Mr. Wortley Montague, then ambassador at Constantinople, and Lady Mary his wife, inoculated their son, and brought him in safety to England. Upon this Queen Caroline, at that time Princess of Wales, begged the lives of six condemned criminals, who

---

[23] It had been practised in Wales long before the method was made known from Constantinople. The Welch called it buying the small-pox.

[24] It appears, from several letters in the Archives of the Society, that the clergy were strenuously opposed to the practice.

[25] *History Royal Society*, p. 171.

had never had the small-pox. They were inoculated, and all took the disease, except one woman. To make further trial, Queen Caroline procured half a dozen of the charity children belonging to St. James's parish, who were inoculated; and all of them, except one (who had had the disease before, but concealed it for the sake of the reward), went through it with the symptoms of a favourable kind of the distemper. Queen Caroline afterwards consulted Sir Hans Sloane about inoculating her own family. Sir Hans approved of the process, but refused to advise her Majesty to put it in practice in her own family, as not being certain of the consequences that might follow, and on account of the great importance of the persons experimented on to the public. This opinion, however, determined both Queen Caroline and King George I. The operation was performed, and the children went through the disease favourably. Sir Hans Sloane adds, that out of 200 cases of inoculation that he had seen, only one terminated fatally[26]."

The interest now attaching to this account of inoculation is simply of an historical nature[27], as the introduction of vaccination has very properly caused the former practice to be abandoned.

On the 25th November, 1731, the Society were honoured by a visit from the Prince of Wales, and the Duke of Lorraine, when the latter was admitted a Fellow of the Society, and signed his name in the Charter-book. Various experiments were made on this occasion:—

---

[26] *History Royal Society*, p. 172.

[27] There are two large MS. folios in the library of the Society, entirely filled wth the statistics of Inoculation.

"On the strength of Lord Paisley's Loadstone, formerly presented to the Society.

"On Dr. Frobenius's Phlogiston, and on the transmutation of phosphorus[28].

"Electrical experiments by Mr. Gray, which succeeded, notwithstanding the largeness of the company."

Those last mentioned consisted in showing the facility with which electricity passes through great lengths of conductors, and are remarkable as having been the first of this nature. They were repeated in 1745, when Dr., afterwards Sir William Watson, assisted by several scientific Members of the Society, made a series of experiments to ascertain how far electricity could be conveyed by means of conductors. "They caused the shock to pass across the Thames at Westminster bridge, the circuit being completed by making use of the river for one part of the chain of communication. One end of the wire communicated with the coating of a charged phial, the other being held by an observer, who in his other hand held an iron rod, which he dipped into the river. On the opposite side of the river stood a gentleman, who likewise dipped an iron rod in the river with one hand, and in the other held a wire, the extremity of which might be brought into contact with the wire of the phial. Upon making the discharge, the shock was felt instantaneously by both the observers[29]." Subsequently, the same parties made experiments

---

[28] Dr. Frobenius was accustomed to make chemical experiments before the Society, for which he was remunerated. It is worth noticing, that the phosphorus used in the above experiments, amounting to six ounces, cost 10*l.* 10*s.*

[29] Priestley's *History of Electricity.*

near Shooter's Hill, when the wires formed a circuit of four miles, and conveyed the shock with equal facility, "a distance which without trial," they observed, "was too great to be credited[30]." In the Paper detailing these experiments, printed in the 45th volume of the *Philosophical Transactions*, occurs the first mention of Dr. Franklin's name, and of his theory of positive and negative electricity. At the present day, when the astonishing properties of electricity are so prominently displayed in the electric telegraph[31], this Paper possesses peculiar interest. The science of electricity, as must be generally known, originated after the establishment of the Royal Society; almost every early electrical discovery of importance was made by its Members, and is to be found recorded in the *Philosophical Transactions*[32].

---

[30] These experiments were performed at the expense of the Society, and cost 10*l*. 5*s*. 6*d*.

[31] See a very remarkable paper in the *Spectator* upon this subject, (No. 261) taken from *Strada*.

[32] It is worthy of mention, that a Mathematical Society, which also cultivated the science of Electricity, was established in 1717, by Joseph Middleton: they met at the Monmouth's Head, in Monmouth Street, until 1725, when they removed to the White Horse Tavern, in Wheeler Street, and from thence, in 1735, to Ben Jonson's Head, in Pelham Street, Spitalfields: they afterwards occupied large apartments in Crispin Street, Spitalfields. The Members of this Society appear to have consisted principally of tradesmen and artisans, with the exception of a few persons of higher rank, amongst whom may be mentioned, Canton, Dollond, Thomas Simpson, and Crossley. The Society, according to their Minute-books, to which Mr. Williams, the Assistant-Secretary to the Astronomical Society, kindly gave me access, possessed a considerable number of philosophical instruments with which they performed various experiments. It was customary to lend these; for we find amongst the Rules, that,—"the air-pumps, reflecting tele-

During this period, the Journal-books show that an extraordinary amount of scientific business was transacted at the ordinary Meetings, which were, with few exceptions, presided over by Sir Hans Sloane[33]. Independently of papers read, experiments

---

scopes, reflecting microscopes, electrical machines, surveying instruments, &c. shall not be lent out without a book of the use of either, and borrowers shall give a note of hand for the value thereof." The first article of the Rules informs us, that "the number of Members which compose this Society shall not exceed the square of seven, except such Members as are abroad or in the country."

But it having been found that there were several candidates for admission, a new rule was subsequently enacted, increasing the number of Members to the square of eight, and more lately to the square of nine. The Members met weekly on Saturday evenings, "between eight and nine o'clock, silence being kept in the room." The Articles add, that "every Member present shall employ himself in some mathematical exercise, or forfeit one penny; and if any Member is ask'd by another a question in the mathematics, he shall instruct him in the plainest and easiest method he can, or forfeit two-pence on refusal."

These rules will be understood as applying only to the early period of the Society's existence.

This association long contributed to keep up a taste for exact science among the residents in the neighbourhood of Spitalfields, and accumulated a library of nearly 3,000 volumes. It existed until May 1845, when being on the point of dissolution, a proposition was made by the few remaining Members to present their library to the Astronomical Society, which terminated by the "books, records, and memorials of the Mathematical Society being made over to the Astronomical Society; and electing the remaining Members of the Mathematical Society Fellows of the latter body. There were nineteen in number, three of whom were already Fellows of the Astronomical Society." *Proceedings Royal Astronomical Society*, 1846. It is due to Captain Smyth (who was a Member of the Mathematical Society) to state, that he negociated this amalgamation.

[33] It should not be forgotten, that the greatest astronomical discovery of the eighteenth century was made during the Presidency of

of various kinds were made, several of which created great interest. Amongst them may be mentioned those on ether, which spirit had been noticed as early as 1540 by pharmaceutical chemists. Dr. Frobenius and Mr. Godfrey made several experiments on it before the Society, an account of which appears in two Papers published in the *Philosophical Transactions* for 1730, when the term *ether* was first adopted. At the present time, when the properties of this anæsthetic agent have been made so serviceable to suffering humanity, these Papers possess considerable historical value. Mr. Godfrey states: "That this liquor Æthereus was formerly very much esteemed and enquired into, doth clearly appear by an experiment I made formerly for my worthy Master Esquire Boyle, by the means of a metallick solution, namely, by the solution of crude mercury united with the *Phlogiston Vini*, or other vegetables, and this æther swam on the top of the solution, which I separated *per Tritorium*. This is what I have done formerly in Esquire Boyle's laboratory, and Sir Isaac Newton was very well acquainted with it too, which by reason of shortness of life was not brought to a full end, to do it so readily in quantity."

Dr. Frobenius, after describing various experiments, says: "Æther then is certainly the most noble, effica-

---

Sir Hans Sloane. I allude to the aberration of light, discovered by Bradley, and communicated to the Royal Society in an elaborate Paper published in the 35th volume of the *Transactions*. "His theory was so sound that no astronomer ever contested it, and his observations were so accurate, that the quantity which he assigned as the greatest amount of the change (one ninetieth of a degree), has hardly been corrected by more recent astronomers." Whewell's *Hist. Ind. Sci.*

cious, and useful instrument in all Chymistry and Pharmacy; *Ubi enim ignis potentialis, ibi actuali non opus est,* inasmuch as essences and essential oils are extracted by it immediately, without so much as the mediation of fire, from woods, barks, roots, herbs, flowers, berries, seeds, &c., from animals and their parts too."

In accordance with the orders of Government, new inventions were exhibited before the Society, and registered, previously to being secured to their author by Letters Patent[34]. Nor was science alone considered, for at many of the Meetings objects of archæological interest were exhibited and discussed, representations and descriptions of which frequently occur in the *Philosophical Transactions.*

A Literary and Archæological Society, established at Peterborough in 1730, appears to have contributed some articles of this nature. Le Neve, in a letter to Dr. Mortimer, dated April 1, 1735, says: "I suppose Dr. Balguy inform'd you of the nature of our Institution, for the promoting literature and friendship among us. We began six or seven years ago, are seldom less than seven or eight, or more than twelve or fourteen at a Meeting. We live in the midst of Gothick ruins, and have very many stately monuments of antiquity left entire. On one side of us we are near to one of the greatest Roman stations in Britain, where many tesselated pavements, altars, &c. are often found amongst the ruins, and where the farmer, every time he ploughs the fields, expects a plentiful harvest of medals, which

[34] It is really curious to observe how many of these inventions are similar in principle to the so-called novelties of the present day. An interesting list might easily be made from a few volumes of the Journal-book.

after a shower of rain, when the earth is lighten'd by the plough, have been found in great numbers. We have a small museum, which we have furnished with natural curiosities, as flies, insects, shells, petrefactions, minerals, &c. We have also begun a small collection of medals, of which we have betwixt three or four hundred. We amuse ourselves for two or three hours once every week in these things[35]."

Highly valuable collections in zoology, mineralogy, and botany, from America, were presented to the Society for their Museum in 1734, by John Winthrop, Hollisian Professor of Mathematics at Cambridge, in New England; and at the same time he communicated important geographical information respecting his country[36]. Considerable impulse was given to the study of botany by the receipt, in the early part of each year, of fifty specimens of dried plants from the Apothecaries' Garden at Chelsea, with a scientific description of each plant, in conformity with the directions of Sir Hans Sloane, who, on purchasing the manor of Chelsea, gave the Company of Apothecaries the entire freehold of their botanical garden there, on condition that it should be for ever pre-

---

[35] Archives: Royal Society.

[36] La Martiniere, Geographer to the King of Spain, dedicated a volume of his *Geographical Dictionary* to the Society at this period. In a letter to Sir Hans Sloane, preserved in the Archives of the Society, he says: "I intend to satisfy my own inclination by dedicating my new volume to the Royal Society. Sir, I know the rank you hold, not only among that learned Society, but also in the republic of letters. Give me leave to ask your protection for myself and for my book too. All my thoughts are bent on rendering my life and works more and more worthy of the approbation of persons of honour and truly learned."

served as a physic garden. As a proof of its being so maintained, he obliged the Company, in consideration of the said grant, to present yearly to the Royal Society, in one of their weekly Meetings, fifty specimens of plants that had been grown in the garden during the preceding year, and which were all to be specifically distinct from each other, until the number of two thousand should be presented. This number was completed in 1761[37]. The inspection and study of these plants by the Fellows is frequently alluded to in the Journal-books, and probably not a little assisted in developing that branch of botany comprising the anatomy and physiology of plants, which originated in this country, and indeed took its rise from the Royal Society.

It is not surprising that with so many attractions, the Meetings of the Society were most fully attended not only by the Fellows, but their friends, who were formally introduced, and their names inserted in the Minutes[38]. Under all these apparently favourable circumstances, we are hardly prepared to find that the financial affairs of the Society became again involved in so unhappy a condition, as to render it necessary to appoint a Committee to examine the books, and more particularly to devise some better means of recovering the arrears due from dishonourable Members, than those hitherto employed.

A Committee with these objects in view, sat during

---

[37] In 1733, the company erected a marble statue of Sir Hans, executed by Rysbrac, in the centre of the garden, with a Latin inscription, commemorating the design and advantages of his donation.

[38] The name of the Prince of Orange appears amongst the number of distinguished visitors.

the early part of the year 1740, and drew up an elaborate Report, which was presented to the Council on the 14th January, by which it appears that the total number of Fellows amounted to 293[39], composed of 152 who had compounded for their subscriptions, 2 exempted from payments[40], and 139 who had signed the obligation to pay annually, but from whom no less a sum than 1844*l*. 16*s*.[41] was due as arrears! The Committee conclude their report by saying, "that the whole revenue of the Society, exclusive of the article of contributions, amounts only to the sum of 232*l*. per annum, decreased by taxes and other expences to 140*l*. per annum, whereas it appears that the annual expenses can hardly be computed at less than 380*l*. yearly, so that the expenses must exceed the receipts by about 240*l*. annually, unless such a sum is brought in yearly by contributions, on which account they are humbly of opinion that the getting in of the contributions is, at this time, a matter of the greatest consequence to the Society, and of absolute necessity, as the business of the Society cannot possibly be carried on without."

It would be wearisome to detail the great exertions made by the Council to disengage the Society from its difficulties, which had assumed the unpleasant reality of a heavy and increasing debt. Eventually, however, they succeeded in their arduous undertaking, and restored the Society to a state of prosperity; the receipts in the following year (1741) exceeded the payments by 297*l*.

---

[39] Thirty-four of these were noblemen.

[40] Dr. James Douglas, and William Cheselden, Esq.

[41] This does not include a considerable sum due from the estates of deceased Fellows.

It is a painful task, though one of absolute duty, to record these periodical visitations of poverty, which threatened the very existence of the Royal Society; there is, however, a proportionate amount of pleasure in witnessing the triumphant manner in which the small band of philosophers, to whom the guardianship of a body already known and honoured throughout Europe was committed, extricated their institution from serious difficulties, unassisted by Royal bounty, and labouring alone on account of their love for science.

But it may be permitted us to express our surprise, that a scientific Society should ever meet with such difficulties as have been recorded, arising from want of honour amongst its Members. Happily this observation relates wholly to past years, for, in the present day, it is a most rare event for any Fellow of the Royal Society to be behind with his subscription[42]. But it is a well-known fact, that other scientific bodies suffer sadly from the defalcations of their members[43], who allow their names to remain on the books, and in many cases avail themselves of the privileges of the institution, without at all contributing to its support. Such a course entails formidable difficulties on the executive of the Society. Naturally trusting to the honour of the Members, who undertake to comply with the statutory regulations, certain measures are taken and expenses incurred which

---

42 And this, be it remembered, without the intervention of a collector, for such an officer does not exist at the Royal Society.

43 The Society of Antiquaries presents a strong instance, 2,700*l.* having been lately lost to the funds of the institution, from Members of that body.

can only be defrayed by the current annual subscriptions, hence it follows, as a matter of course, that should these fail, the legitimate objects of the Society cannot be carried out.

There are indispensable obligations on all who associate themselves with any scientific Society. Those who do not comply with them incur disgrace instead of honour, for a title can only be regarded as a reproach by those who fail to deserve it; nor can they claim a share in the reputation of a Society, who never in any manner contribute to its advancement.

It is indeed startling, to hear periodically of the large amounts due in the form of arrears from members of various Societies, the non-payment of which must necessarily occasion the most crippling and disastrous effects.

At the latter end of 1741, the declining health of Sir Hans Sloane compelled him unwillingly to resign the office of President. On the 16th November, 1741, Martin Folkes, Esq., V. P., stated at a Meeting of Council, that "he was charged with a message from the President, who desired him to bear his respects to them, and to acquaint them that the weakness in his limbs, which has now so long continued, and the precarious state of his health, he finds will render it impracticable for him to give that attendance on the Society which his office requires, and therefore he desires them to think of some other proper person for that office in the ensuing election."

The Council were extremely desirous to prevail upon Sir Hans Sloane to retain office, and appointed a deputation to wait on him, in order to convey their sentiments, and to "devise some measures to reconcile, if possible, his holding the office, without injury to his health;" but he was so fully impressed with the propriety of connecting the performance of the Presidential duties with that high office, that being really unable to undertake these, he firmly requested to be allowed to retire, and sent the annexed communication to Martin Folkes, which was read to the Society.

"Sir Hans Sloane is very sensible of the many benefits he has received by being for a great many years present at the Meetings of so many knowing and learned persons of the Society, and of the honours conferred upon him by the several offices with which he has been intrusted by them.

"He is very sorry that the bad state of his health will not longer permit him to enjoy the advantage and satisfaction of so constantly attending their Meetings: but he will endeavour to do the Society all the services in his power, by communicating, from time to time, any curious notices which he shall receive, either at home or from abroad, concerning natural knowledge, during the small remainder of his life."

The Society "Resolved, that their thanks should be given to Sir Hans Sloane for his many and great favours done to the Society during his continuance in the chair, and for his constant and diligent attendance at their Meetings, notwithstanding the great business he was otherwise engaged in;" and they ordered "Martin Folkes, Esq., Vice-President, Dr. Mead, James West, Esq., Treasurer, and Mr. Machin and Dr. Mortimer, the Secretaries, to wait on Sir

Hans Sloane with the foregoing resolution, and to assure him the Society still hope he will attend their Meetings when his health shall permit, though he declines being again elected President; as also that they promise themselves the advantage of his friendship and assistance on all occasions, with the benefit of the advices he daily receives in his extensive and learned correspondence." It is almost needless to add, that although Sir Hans ceased to hold office, his interest in the Royal Society continued unabated to the end of his life. The Journal-books contain a great number of communications made by him to the Society, which were read at the ordinary Meetings, and which are remarkable for their value and originality. Indeed, he appears to have regarded the Society with an interest extinguished only by death, and on the other hand, the Society were fully impressed with the advantages which they derived from their connexion with so eminent a man[44].

The Society had now to select a President, and their choice fell on Martin Folkes, who was elected

---

[44] "The sense entertained by the Society of his services and virtues, was evinced by the manner in which they resented an insult offered to him by Dr. Woodward, who (as the reader is aware) was expelled the Council. Sir Hans was reading a paper of his own composition, when Woodward made some grossly insulting remarks. Dr. Sloane complained, and moreover stated, that Dr. Woodward had often affronted him by making grimaces at him; upon which Dr. Arbuthnot rose, and begged to be "informed what distortion of a man's face constituted a grimace?" Sir Isaac Newton was in the chair when the question of expulsion was agitated; and when it was pleaded in Woodward's favour, that "he was a good natural philosopher, Sir Isaac remarked, that in order to belong to that Society a man ought to be a good moral philosopher, as well as a natural one." Wadd's *Memoirs*, p. 232.

at the Anniversary, held, as usual, on St. Andrew's day. He had long been a Vice-President, and when Sir Hans Sloane was unable, on account of his infirmities to attend, had presided in his place.

SOLAR DIAL MADE BY NEWTON.

# CHAPTER XVII.

Memoir of Martin Folkes—His Acquirements more Literary than Scientific—Sir John Hill's Review of the Royal Society—Death of Halley—Formation of Royal Society Club—Originally entitled 'Club of Royal Philosophers'—Their Rules—Receive presents of Venison, &c.—Cost of Dining—Present Rules—List of Members—Philosophical Club—Their Rules—Original Members—Fairchild Lecture instituted—Dr. Knight receives Copley Medal—Discovery of Nutation by Bradley—Harrison's Chronometers—The Copley Medal awarded—Authorities request the assistance of the Society to ventilate Newgate—Sanitary Measures taken by Sir John Pringle—Canton's Method of making Artificial Magnets—He receives the Copley Medal—Dr. Gowan Knight's method—Controversy between Canton and Michell—Letter from Dr. Priestley—Change of Style—Tables prepared by Mr. Daval, Secretary to the Society—Assistance afforded by Father Walmesley—Alterations proposed by Lord Macclesfield in the mode of publishing the *Transactions*—Committee of Papers appointed—Advertisement in the Volume for 1753—Cost of Printing the *Transactions*—Translation of the *Transactions* published in Italy—Resignation of Martin Folkes—Earl of Macclesfield chosen President—Mr. Folkes leaves the Society in a prosperous condition—Large number of Visitors to the Meetings—Stukely's notice of the Meetings—His description of a Geological Soirée—Conceives Corals to be Vegetables.

---

## 1745—55.

MARTIN FOLKES was the eldest son of Martin Folkes, Esq., Barrister-at-Law, and was born in the parish of St. Giles in the Fields, on the 29th October, 1690[1]. He was sent when a boy to the University of Saumur; where, under the superintend-

---

[1] *Life*, by Dr. Birch.

ence of Mr. Cappel, son of the celebrated Lewis Cappel, he acquired considerable knowledge of the Hebrew, Greek, and Latin languages. On the suppression of that university, in January, 1694—5, he returned to England, and entered at Clare Hall, Cambridge, where he added the study of philosophy and mathematics to that of the ancient languages. "The progress," says Dr. Birch, "which he made at the university, and after he left it, in all parts of learning, and particularly mathematical and philosophical, distinguished him at so early an age, that when he was but three and twenty years old he was esteemed worthy of a seat in the Royal Society, into which, having been proposed as a candidate on the 13th December, 1713, he was, on the 29th July following, elected[2]."

In 1723 he was appointed a Vice-President of the Society, by Sir Isaac Newton, and when declining health no longer permitted that illustrious philosopher to preside, the chair was often occupied by Mr. Folkes. At the first anniversary election after the death of Newton, he was competitor with Sir Hans Sloane for the office of President; his interest was supported by a great number of Fellows, but the choice, as we have already seen, fell upon Sir Hans.

In 1733 he set out with his family for Italy, for the purpose of improving himself in classical antiquity. He remained abroad two years and a half, and lost no opportunity of acquiring information upon archæological and classical subjects. On his return to England, he presented the Royal Society with his *Remarks on the Standard Measure preserved in the Capitol*

---

[2] *Life*, by Dr. Birch.

*of Rome*, and the model of an ancient globe in the Farnese Palace, which model was made at Rome, under his direction: the original sphere, in stone, supported by an atlas, was supposed to have been made in the year of the Christian era 112, towards the end of the Emperor Trajan's reign.

In 1742 he was elected a Member of the French Academy. His election is thus alluded to in the *éloge:*—

"*La mort du célèbre M. Halley ayant fait vaquer parmi nous en* 1742 *une place d'Associé Etranger, l'Académie crut ne pouvoir mieux réparer la perte qu'elle venoit de faire qu'en nommant M. Folkes à cette place, et il fut élu le* 5 *Septembre de la même année.*

"*A peine en avoit-il reçu la nouvelle, que voulant apparemment faire voir qu'il se croyait attaché désormais à la France sans cesser cependant de l'être à sa patrie, il lut une mémoire également interessant pour les deux nations; ce fut la comparaison des mesures et des poids de l'une et de l'autre, qu'il donna à la Société Royale avec tout le détail de ce qu'il avait fait pour s'en assurer*[3]."

On the death of Algernon, Duke of Somerset, President of the Society of Antiquaries, in February, 1750, Mr. Folkes, then one of the Vice-Presidents, was elected to succeed his grace in that office, in which he was continued by the charter of incorporation of that Society, November 2, 1751[4].

On the 26th September, 1753, according to a manuscript memoir of Folkes in the British Museum[5], he was seized with a palsy which deprived him of the use of his left side. In this unhappy situation he languished until June 28, 1754, when a second stroke

[3] *Hist. de l'Acad.*, 1754, p. 174. [4] *Archæologia*, Vol. I. p. 38.

[5] Additional MSS. No. 4222.

put an end to his life. He was buried at Hillington, near Lynn, in Norfolk, under a black marble slab, with no other inscription than the date of his death and his name, in conformity with the express direction of his last will. By his wife, Lucretia Bradshaw, who had been an actress, he left two daughters. He contributed ten Papers to the *Transactions.*

Martin Folkes was a man of extensive knowledge, which has however rendered more service to archæology than to science; the latter being chiefly enriched by his work on the intricate subject of coins, weights, and measures[6].

The sale of his library, engravings, coins, and medals, lasted fifty-six days, and produced 3090*l.* 5*s.* His numerous manuscripts not being in a fit state for publication, were destroyed by his orders a short time previous to his death.

Immediately after he was elected President, he presented the Society with 100*l.*; and at his death bequeathed them 200*l.*, a magnificent portrait of Lord Chancellor Bacon, and a large cornelian seal-ring bearing the Arms of the Society, for the perpetual use of the President. Dassier struck a medal in his honour in 1740; and two years afterwards another was struck at Rome, bearing the motto *Sua sidera morunt,* with a pyramid and sphinx. In 1792 a handsome monument was erected to his memory, in Westminster Abbey, on the south side of the choir. Dr. Jurin, Secretary to the Society, dedicated the 34th volume of the *Trans-*

[6] Mr. Folkes purposed illustrating this work, and had engraved 42 plates, which were, however, in an incomplete state at the time of his death. These, with the copyright of the book and tables, were purchased by the Society of Antiquaries, and the whole published under the care of Dr. Gifford in 1763.

*actions* to him, and concludes the dedication with the compliment, "It is sufficient to say of Mr. Folkes, that he was Sir Isaac Newton's *friend*, and was often singled out by that great man to fill his chair, and to preside in the assemblies of the Royal Society, when the frequent returns of his indisposition would no longer permit him to attend them *with his usual assiduity*."

Mr. Folkes presented the Society with a fine portrait of himself, painted by Hogarth.

It is an undeniable incident, that in selecting Mr. Folkes to fill the chair so lately occupied by Newton, the Society ran some risk of their Meetings assuming more of a literary than a scientific character, for at that period the business of the Society, as well as the publication of Papers in the *Transactions*, were considerably influenced by the President. Dr. Thomson, in his short account of the Society, says, "If any person will take the trouble to examine the volumes of *Transactions* published during the presidency of Martin Folkes, and to compare them with the rest of the work, he will find a much greater proportion of trifling and puerile papers than are any where else to be found." This remark, it must be admitted, is true; and the charge of puerility may also be applied to the proceedings of some of the ordinary Meetings recorded with great minuteness in the Journal-books[7].

[7] A remarkable instance occurs in the 19th volume of the Journal-book, in the case of an old woman described as having been found burnt to death. She was regarded by many of her neighbours as a witch, and it is stated, that the gentlemen who furnished an account of the event, also mentioned "some strange magical experiments pretended to be performed at the same time by a farmer, who thought himself injured by her, and who imagined that what he did in the burning of some of his sheep, must have been the cause of what happened to her; which articles not seeming

"It was during this period," says Dr. Thomson, "that Sir John Hill published his *Review of the Works of the Royal Society of London*, in which he endeavours, with all the humour and all the knowledge he was master of, to throw ridicule upon the labours of that illustrious body. He was induced, it is said, to take this step, in consequence of being disappointed in an attempt which he had made to be elected a Fellow. The story is by no means improbable, though he himself formally disavows its truth[8]. It cannot be denied that he has selected and exposed a variety of trifling and absurd papers. But his own humour is coarse and poor, and in more instances than one the statements contained in the papers which he attempts to ridicule, are much more accurate than his own. He affirms that at the time he wrote, the Society was entirely under the management of Mr. Henry Baker, a man of acknowledged abilities, but whose knowledge and pursuits were too circumscribed to qualify him for superintending such a Society with advantage[9]."

---

to come within the verge of natural knowledge, created an unusual demur about returning thanks, but which when considered was overruled, as being, in truth, a relation of fact, that there is, however, such an opinion about her; whereupon it was ordered that thanks be returned to these gentlemen." p. 300.

[8] These are his words: "The elections into the Royal Society are in great form; a recommendation is drawn up in writing, signed by several of the Members, who declare the person worthy of that great honour; this is hung up in the room of their Meetings a quarter of a year, and at the end of that time it is put to the ballot whether the Candidate shall be received. If it were true that the author of these animadversions was ever so recommended, or so ballotted for, the paper must remain, and those at least who gave the negative balls would remember that they did so." *Preface*, p. 6.

[9] It does not appear that the gentleman here alluded to, took

Sir John Hill's book, which is more curious in a literary than any other point of view, is directed principally against Mr. Folkes, to whom, indeed, it is dedicated. In the dedication the author says, "you have so natural a right to the patronage of these animadversions, that it were at once unjust and ungrateful to rob you of the honour. It is to you alone that the world owes their having been written; the purport of the more considerable of them has been long since delivered to you in conversation; and if you had thought the Society deserved to escape the censure that must attend this method of laying them before the world, you might have prevented it by making the necessary use of them in private." The whole book, as Dr. Thomson justly observes, is a poor attempt at humour, and glaringly exhibits the feelings of a disappointed man. It is probable, however, that the points told with some effect on the Society; for shortly after its publication the *Transactions* possess a much higher scientific value.

---

any active part in the official business of the Society: at least, his name does not occur in the Journal-books as so doing. His certificate states him to be a person "well versed in mathematicks and natural knowledge, particularly eminent for his great skill and happy success in teaching persons born deaf, and consequently dumb, to speak, (having improved upon that great invention of the late famous Dr. Wallis), author of a very beautiful poem called the *Universe*, with many curious notes regarding natural history, and one who hath communicated some useful papers to the Royal Society." He was recommended by Sir Hans Sloane, Dr. Mortimer, and Mr. Folkes, and elected on the 12th March, 1740—1; and in 1744 he received the Copley Medal, by adjudication of Sir Hans Sloane, for "curious experiments relating to the crystallization or configuration of the minute particles of saline bodies dissolved in a menstruum."

Sir John Hill was not alone in his attempts to cast ridicule on the scientific labours of the Fellows. The author of a pamphlet entitled, *A Dissertation on Royal Societies, in three letters from a Nobleman on his travels*, written about the same period, endeavours to malign the Society, and writes of the Fellows as "personages acting the importants, and solemnly met to trifle away time in empty forms and grave grimace." As a contradiction to this statement, the following extract from an article on the Society, published in the *Monthly Review* of the same date, will be perused with interest:—

"We ourselves were present," says the writer, "some few nights ago at a Meeting of the Society, when a paper model of a cell in an honey-comb was produced, which had been sent by that great ornament to mathematical knowledge, Professor M^c^Laurin. Several strangers, introduced by some of the Fellows, (who are allowed to bring their friends occasionally), began to discover in their faces a mixture of mirth and contempt, at seeing an object so trivial, which had been transmitted as far as from Scotland. But when the Professor's treatise, which accompanied the model, had demonstrated that it was beyond all mathematical power to assign another figure that would compose an equal number of cells in the same given space, their tittering gave place to silent confusion and astonishment; and the Great Creator, from this little piece of modelled paper, received the honour due to his immense wisdom, which had infused into the little architects of the honey-comb a kind of knowledge more than human.

"But, to do further justice to this respectable body, it is impossible, in the nature of things, that the importance of several of their communications should appear at once. The hints of one year may the next be carried on to experiments; and those experiments gradually open either a new, or an improved field of natural knowledge. The design of the Society is to incite the learned in all parts of the world to improve upon their labours, to correct them when necessary; in short, to make what use of them they please; so that natural and mathematical knowledge may be promoted: and he that will take upon him to aver that the Royal Society of London have not made the noblest contributions to the advancement of these most useful sciences, must have more hardiness than either modesty or learning. He must utterly have forgot that there ever existed among them a Boyle, a Ray, or (*ille! O! Newton! Quot Aristoteles!*) the greatest philosopher the world ever did, or, it is to be feared, ever will see."

Pursuing the chronological order of events, from which the notice of Sir John Hill's book has led us away, we must revert to the year 1742, which was marked by the death of Halley, in whom the Society lost one of their most distinguished Members. For some years before his death, he had suffered from an attack of paralysis, which manifested itself in the first instance in 1737; he still continued, however, to attend the weekly Meetings of the Society, until his disorder increasing, he expired on the 14th January, 1741—2, in the 86th year of his age. He was buried at the church of Lee, in Kent. The inscription records that, with his dearest wife, there reposes by far the chief astronomer of his age, *Astronomorum sui sæculi facilè princeps:* and adds, "That you may know, reader, what kind of, and how great a man he was, read the multifarious writings with which he has illustrated, adorned, and amplified nearly all the arts and sciences." This tomb, says Captain Smyth, "was recently opened to receive the corpse of Mr. Pond, the late Astronomer-Royal[10]."

During the time that Halley was at the head of the Greenwich Observatory, he observed the heavens with the closest attention, hardly ever permitting a day to pass without making observations, and performing, unassisted, the entire business of the Observatory. His writings are very numerous, comprehending no fewer than twelve distinct works, and eighty-one Papers in the *Philosophical Transactions.* His portrait is amongst the Society's collection.

Halley was succeeded as Astronomer-Royal by Dr. James Bradley, who was principally indebted to George, Earl of Macclesfield, for his appointment. This nobleman interceded strongly in Bradley's favour, and addressed the subjoined earnest letter to Lord Chancellor Hardwicke.

"MY LORD, "*Shirburn, Jan.* 14, 1741—2.

"YESTERDAY I received notice that Dr. Halley could not hold out longer than a day or two, and I hope your Lordship will pardon my troubling you with this in behalf of my friend Mr. Bradley, whom you formerly seemed inclined to serve whenever Dr. Halley's death should make a vacancy at Greenwich.

"It is not the salary annexed to that professorship which makes me so desirous that Mr. Bradley should succeed Dr. Halley in it, but there is a credit attending such a professorship when possessed by a man of real merit; and it is a disappointment to, and a sort of slight put upon, such a person, when, upon a vacancy, he is neglected, and a person much inferior to him is

---

[10] *Cycle of Celestial Objects*, Vol. I. p. 55.

preferred before him: and give me leave to say, that must be Mr. Bradley's case whosoever, except himself, succeed Dr. Halley; and, besides Mr. Bradley's abilities, he has so very great a liking to the practical part of astronomy, the making observations, that, on that score, it would be extremely agreeable to him, and the science would have the greatest reason to expect to receive very considerable improvements from his observations.

"But it is not only my friendship for Mr. Bradley that makes me so ardently wish to see him possessed of the professorship, it is my real concern for the honour of the nation with regard to science. For, as our credit and reputation have hitherto not been inconsiderable amongst the astronomical part of the world, I should be extremely sorry we should forfeit it all at once by bestowing upon a man of inferior skill and abilities the most honourable, though not the most lucrative, post in the profession, (a post which has been so well filled by Dr. Halley, and his predecessor), when, at the same time, we have amongst us a man known by all the foreign, as well as our own astronomers, not to be inferior to either of them, and one whom Sir Isaac Newton was pleased to call the best astronomer in Europe. This will, I flatter myself, plead my excuse, if I should appear a little importunate in pressing your Lordship to intercede early and earnestly in favour of Mr. Bradley. Nor can I apply on this occasion to a more proper person than your Lordship. For as this place has no relation to any department of the administration, but its sole business and view is the advancement and improvement of the science that is of use to mankind in general, but more particularly so to us, as a trading nation, and the chief of the maritime powers: this, I say, being the nature of the place, to whom can the recommendation to it more properly belong than to your Lordship, who, not only in private character, but by your public office likewise, are the patron of learning and learned men in general? It was upon this foot that my father, when in the post which you now enjoy, took upon him to recommend Dr. Halley to the Royal professorship at Greenwich, and Mr. Bradley to the Savilian at Oxford, and succeeded in both his recommendations; and he always thought it for his honour to have recommended two such able men. And I dare assure your Lordship, that if you shall be pleased to espouse Mr. Bradley's interest, you will have the satisfaction to find your recommendation of him approved and applauded universally by those who are versed in those studies, both at home and abroad.........But, my Lord, we live in an age when most men, how little soever their merit may be, seem to think themselves fit for whatever they can get, and often meet with some people, who by their recommendations appear to entertain the same opinion of them; and it is for this reason that I am so pressing with your Lordship not to lose any time, as I am confident you would be sorry the professorship should be given to a person unqualified for it, and the finest instrument perhaps in the universe put into the hands of a person unable to make a proper use of it, and this to the prejudice of the best qualified and most able astronomer, that not only this nation, but, probably, all the world can at present shew[11]."

Lord Macclesfield succeeded in his desire, and although not supporting Government, the weight

---

[11] The original of this letter is preserved at Shirburn Castle, and was first printed in Bradley's *Works*, edited by the late Professor Rigaud.

which was justly attributed to his opinion on scientific subjects, very properly caused the minister to appoint Bradley, to whom, it will be remembered, we are indebted for two of the most beautiful discoveries of which the science of Astronomy can boast,—the aberration of light, and the nutation of the earth's axis. In 1748, the Society prevailed on Government to expend 1000*l.* on new astronomical instruments for the Observatory, which were constructed by the celebrated artists, Graham and Bird. Bradley made a most assiduous use of these instruments.

In 1743 the Royal Society Club was founded under the designation of the "Club of the Royal Philosophers." Through the kindness of Joseph Smith, Esq., Treasurer to the Club, who has, with the permission of the Members, placed the Minutes in my hands, I am enabled to quote the original regulations. They are dated October 27, 1743:—

"RULES AND ORDERS TO BE OBSERVED BY THE THURSDAY'S CLUB, CALLED THE ROYAL PHILOSOPHERS.

"A Dinner to be ordered every Thursday for six, at one shilling and sixpence a head for eating. As many more as come to pay one shilling and sixpence per head each. If fewer than six come, the deficiency to be paid out of the fund subscribed.

"Each Subscriber to pay down six shillings, viz. for four dinners, to make a fund.

"A pint of wine to be paid for by every one that comes, be the number what it will, and no more, unless more wine is brought in than that amounts to."

The names appended to these rules are, Mr. Postlethwaite, Rev. Thomas Birch, Mr. Colebrooke, Mr. Dixon, Mr. Watson, Captain Middleton, Mr. R. Graham, and Mr. Burrow. Mr. Colebrooke was appointed Treasurer.

From this period to the present time, the Club have continued to meet with great regularity. The original Members were soon increased by various Fellows of the Society; amongst whom was Martin Folkes, who was elected President of the Club, Lord Macclesfield, Lord Charles Cavendish, Sir John Pringle, &c. In 1748 the following additional rules were made:—

"Ordered, that the under-mentioned notice be hung up in the Club room, viz.

"It is thought proper to inform the Gentlemen who dine here, that all those who are not Subscribers themselves must be introduced by a Subscriber present each time they dine here.

"Resolved, that the number of Members to this Club shall not exceed forty for the future; that the election of Members, to supply any vacancy that shall happen by death or otherwise, be made annually, on the last Thursday in July, by ballot, by the Members then present, and that no person be deemed chosen who hath five negatives.

"That those Gentlemen who have not attended for twelve months, nor sent an excuse, be deemed no longer Members, and that their places be filled up out of those Gentlemen that are Candidates, whose names are to be put to ballot, according to their precedence on the list which is ordered to be kept for that purpose."

In 1749 it was "Resolved, *nemine contradicente*, that no strangers be admitted to dine here for the future, except introduced by the President."

In 1760 it was "Resolved by ballot, that no person be deemed chosen a Member of this Society who shall have three negatives." Some of the entries in the oldest Minute-book are very curious. Under the date of May 3, 1750, it is recorded: "Resolved *nem. con.* That any nobleman or gentleman complimenting this company annually with venison, not less than a haunch, shall, during the continuance of such annuity, be deemed an Honorary Member, and admitted as often as he comes without paying the fine, which those Members do who are elected by ballot." At another Meeting in the same year a resolution was passed, "That any gentleman complimenting this Society annually with a turtle, should be considered as an Honorary Member;" and that "the Treasurer do pay keeper's fees and carriage for all venison sent to the Society, and charge it in his account." Such a resolution seems to have been desirable, for I find very frequent entries of gifts of venison, which are thus recorded: "Paid keeper's fee and carriage of half a buck, from Hon. Philip York, 14*s.*; ditto from Earl of Hardwick, 1*l.* 5*s.*" The more general entry is simply fees for venison; but the Club were not regaled by venison alone. Presents of salmon and turtle are also duly chronicled, and the gift of good old English roast beef was not despised, as appears by the subjoined minute, under the date of June 27, 1751, when Martin Folkes presided.

"William Hanbury, Esq. having this day entertained the company with a Chine of Beef which was 34 inches in length, and weighed upwards of 140 pounds, it was agreed, *nem. con.*, that two such chines were equal to halfe a Bucke, or a Turtle, and entituled the Donor to be an Honorary Member of this Society."

The Minutes record that the Club met at the Mitre Tavern in Fleet Street, from the date of their institution until December, 1780, on the 21st of which month, the members dined for the first time at the Crown and Anchor, in the Strand, where they continued to meet until that Tavern was converted in 1847 into a club-house[12]. It is interesting to observe the periodical increase in the charges for dinner, &c. From 1743 to 1756, the cost was 1*s*. 6*d*. per head. In the latter year, it was resolved to give 3*s*. per head for dinner and wine, the commons for absentees to remain at 1*s*. 6*d*., as before. In 1775, the price was increased to 4*s*. a-head, including wine, and 2*d*. to the waiter; in 1801, to 5*s*. a-head exclusive of wine, the increased duties upon which made it necessary for the members to contribute an annual sum for the expense of wine[13], over and above the charge of the Tavern-bills; and in successive years the sum increased to 10*s*., which is the amount now paid per head, any deficiency being made up by the annual subscriptions of members.

The following are the Rules now in operation:—

"RULES FOR THE ROYAL SOCIETY CLUB.

"I. THE Club shall consist of Forty ordinary Members, who must be Fellows of the Royal Society, exclusive of the following, who shall be Members *ex officio;* viz. the President; the Treasurer; the two Secretaries; the Foreign Secretary; and the Astronomer Royal;—

[12] They now dine at the Freemason's Tavern, in Great Queen Street.

[13] The price of wine was ordered to be limited to one shilling and sixpence a bottle.

not only those for the time being, but also those who may have filled any of these offices. Provided, however, That upon any Gentleman becoming entitled, from his official station in the Royal Society, to be a Member of this Club, the Treasurer of the Club be instructed to ask him, whether it be his intention to take advantage of such privilege, and to become permanently a *Subscribing* Member.

"II. Every Candidate must be proposed by one Member of the Club, and seconded by another.

"III. The Annual Election of Members shall be held on the Thursday in the week following that on which the Royal Society Meetings close for the Vacation.

"IV. The Candidates shall be ballotted for in the order in which they have been proposed.

"V. No person shall be deemed elected as a Member, unless he shall have three-fourths, or more, of the Votes in his favour.

"VI. Any Member who has not attended the Club at least *once* between, and exclusive of, the two Anniversary Meetings, shall no longer be considered as a Member.

"VII. Provided, however, That any Member declaring his intention of going abroad, shall be considered as a supernumerary Member during his absence, without paying his annual contribution; but his vacancy shall be filled up. And on his return he shall be admitted to the Meetings of the Club on the usual terms, and be admitted as a regular Member on the first Vacancy, on his signifying a wish to that effect.

"VIII. Any Member who may resign his seat on account of leaving the kingdom, shall, on being regularly proposed and seconded for re-admission, have a preference to other Candidates in the order of ballot.

"IX. The Meetings of the Club shall be continued every THURSDAY throughout the year, unless a special Resolution be made to the contrary[14].

"X. The Treasurer shall lay his accounts before the Club at the Anniversary Meeting, when the amount of the next succeeding year's Subscription shall be fixed; and it is expected that every Member will pay his Subscription on that day, or on the day when he next attends the Club, in order to prevent arrears.

"XI. Every newly elected Member of the Club, whether by ballot, or *ex-officio*, shall pay an admission fee of Two Guineas, in addition to the annual contribution, to defray the expenses of the Club.

"XII. Every Member of the Club shall have the privilege of introducing one Visitor; but the President, or, in his absence, the Chairman, shall not be so limited.

"XIII. Every Member bringing a Visitor, shall write his name under his own, to be laid on the table; and no Visitor can be admitted into the room till this regulation shall have been complied with.

"XIV. No Visitor shall, on any account, be admitted on the Anniversary of the Club[15].

"XV. It is expected that those Members who may bring their servants, will order them to assist generally in waiting at table."

---

[14] During the vacation the Club meet only on the first Thursday in each month.

[15] This is held on the Thursday succeeding the last Meeting of the Society for the Session. All the business of the Club is transacted at this Meeting. Mr. R. Brown informs me, that when Solander was Treasurer, the Members had the privilege of voting by proxy.

*The subjoined are the present Members of the Royal Society Club.*

*EX-OFFICIO* MEMBERS.

The Marquis of Northampton. *President.*

P. M. Roget, M.D.
W. T. Brande.
S. H. Christie.
George Rennie.
J. G. Children.
Charles König.
Col. Sabine.
Capt. Smyth, R.N.

ORDINARY MEMBERS.

Sir John Barrow, Bart.
John Barrow.
Col. Batty.
Admiral Beaufort.
Nicholas, Lord Bexley.
R. E. Broughton.
Robert Brown.
Rev. C. P. Burney, D.D.
C. G. B. Daubeny.
Hart Davis.
John Dickinson.
George Dollond.
Charles Elliott.
Thomas Galloway.
Edward Hawkins.
Dean of Hereford.
Sir John F. W. Herschel, Bart.
Sir Robert H. Inglis, Bart.
Rev. Philip Jennings, D.D.
Sir Alexander Johnston.
Lieut.-Col. Leake.
John G. S. Lefevre.
Lord Lyttelton.
Thomas Mayo.
Sir Roderick I. Murchison
Richard Penn.
William H. Pepys.
Rev. Baden Powell.
Sir John Rennie.
Sir Martin A. Shee, P.R.A.
Joseph Smith.
Sir George T. Staunton, Bart.
Charles John, Lord Teignmouth.
Travers Twiss, D.C.L.
James Walker.
Dean of Westminster.
Charles Wheatstone.

SUPPLEMENTAL LIST.

General Colby.
Herbert Mayo.
Sir John Franklin, R.N.

The President of the Royal Society is elected President of the Club. In his absence, the chair is taken by the senior Member present. There are always more candidates for admission than vacancies, a circumstance that had some influence in leading to the formation of a new Club in 1847, composed of eminent Fellows of the Society. The designation of this new Association is the 'Philosophical Club,' and although anticipating the order of events, it has been thought advisable to introduce their constitution in this place:—

"OBJECTS AND RULES OF THE PHILOSOPHICAL CLUB.

"I. THE purpose of the Club is to promote as much as possible the scientific objects of the Royal Society, to facilitate intercourse between those Fellows who are actively engaged in cultivating the various branches of Natural Science, and who have contributed to its progress; to increase the attendance at the Evening Meetings, and to encourage the contribution and the discussion of papers.

"II. The number of Members shall be limited to forty-seven, of whom thirty-five, at least, shall be resident within ten miles of the General Post Office. With the exception of scientific Foreigners, temporarily visiting this country, no strangers are to be present at any of the Meetings.

"III. With the exception of the President of the Royal Society for the time being, those only shall be eligible as Members of the Club who are Fellows of the Royal Society, and authors, either of a paper published in the Transactions of one of the chartered Societies, established for the promotion of natural science, or of some work of original research in natural science.

"IV. The Meetings of the Club shall take place once

a month, on Thursdays, from October to June, both inclusive, except the Anniversary Meeting, which shall take place on the last Monday in April. The chair shall be taken at half-past five o'clock precisely, and quitted at one quarter past eight; each Member presiding in turn in alphabetical order. It shall be the duty of the Chairman to regulate and control all ballots and discussions in the Club; to announce to the Meeting, previously to seven o'clock, the subject of the paper to be read at the Royal Society that evening, and to bring forward, or to invite Members of the Club to bring forward, any correspondence or scientific subjects worthy of consideration. Members will be expected afterwards to attend the Meeting of the Royal Society, unless unavoidably prevented. The times of Meeting shall be notified each year by a circular from the Treasurer, and also by a note to each Member one week before every Meeting.

"V. At the first Meeting of the Club, a Treasurer, and a Committee of four, shall be appointed, who shall together advise on the general management of the Club.

"VI. The Committee shall subsequently be elected by ballot at each Anniversary Meeting. Two Members of the Committee shall retire by seniority each year, being however re-eligible after the space of one year.

"VII. The Treasurer shall be elected annually, at the same time and in the same manner with the Committee[16]; he shall not remain in office longer than three years, though re-eligible after the space of one year. The Treasurer shall, with the concurrence of the Committee, order and arrange the time and place of meeting, and summon Special Meetings when deemed neces-

---

[16] Mr. Grove, who has kindly placed a copy of these rules in my hands, is the first Treasurer.

sary. The Treasurer shall receive the subscription of Members, and keep the accounts of the Club; he shall prepare and issue all notices, regulate the dinners, and act as Vice-Chairman whenever he is present. It shall also be the duty of the Treasurer to keep a register of all Meetings of the Club, to make a minute of all resolutions which may be adopted, and, whenever practicable, to furnish the Chairman with the title of the Paper to be read at the Royal Society on the evening of the Meeting of the Club.

"VIII. The subscription shall not exceed twenty shillings per annum, to be paid to the Treasurer at the first Meeting of the session; the price of each dinner shall not exceed ten shillings.

"IX. Candidates for election shall be proposed in writing by three Members of the Club, not being Members of the Committee. The certificates, stating the grounds of eligibility from personal knowledge, shall be transmitted to the Treasurer, read by him at the next Club Meeting, and retained by him until the Anniversary. If the number of Candidates exceed the number of vacancies, the Committee shall at this Meeting report to the Club those of the Candidates whom they consider the most eligible to fill the vacancies, and the Candidates so recommended shall be entitled to priority of ballot. In case of no excess in the number of Candidates, or of the non-election of those recommended by the Committee, the ballot shall take place according to priority of the date of proposal. The ballot shall take place at the Anniversary, provided fifteen Members be present; one black ball in five to exclude. If less than fifteen be present, the Meeting shall adjourn, and the ballot shall take place at the first adjourned Meeting at which fifteen shall be present.

"X. All new Rules shall be proposed by at least three Members, notified in writing, to the Committee, and read to the Club at least two Meetings previously to the Anniversary. They shall afterwards be taken into consideration, and put to the vote at the Anniversary Meeting. If four-fifths of those present (the quorum being fifteen) be in favour of the proposed new Rule, it shall be adopted, otherwise not."

ORIGINAL MEMBERS OF THE PHILOSOPHICAL CLUB.

David Thomas Ansted.
Francis Beaufort.
Thomas Bell.
William Bowman.
William John Broderip.
Robert Brown.
Proby Cautley.
Samuel Hunter Christie.
H. T. De la Beche.
P. de M. G. Egerton.
Hugh Falconer.
Michael Faraday.
Edward Forbes.
J. P. Gassiot.
John Goodsir.
Thomas Graham.
John Thomas Graves.
J. H. Green.
William Robert Grove.
William Snow Harris.
J. F. W. Herschel.
J. D. Hooker.
William Hopkins.
Leonard Horner.
Charles Lyell.
James MacCullagh.
William Allen Miller.
William Hallows Miller.
Roderick Impey Murchison.
George Newport.
Richard Owen.
Richard Partridge.
Jonathan Pereira.
John Phillips.
George Rennie.
John Richardson.
J. F. Royle.
Edward Sabine.
Adam Sedgwick.
William Sharpey.
William Henry Smyth.
Edward Solly.
William Spence.
William Henry Sykes.
W. H. F. Talbot.
Nathaniel Wallich.
Charles Wheatstone.

In 1746 a lecture or sermon on the "wonderful works of God in the Creation," was instituted, which had its origin in a bequest of 25*l*., left by Mr. Thomas Fairchild, of Hoxton, in the parish of St. Leonard, Shoreditch. The will was proved in 1729, and, at the same time, it was resolved by the Council, that a subscription should be opened to increase the above sum to 100*l*. This amount was made up in 1746[17], and laid out in the purchase of South Sea stock, the accruing dividends being given annually to a lecturer appointed by the President and Council, in accordance with the terms of the bequest, which were: "To preach a sermon in the parish-church of St. Leonard, Shoreditch, in the afternoon of the Tuesday in every Whitsun week in each year, on The wonderful Works of God in the Creation; or on the Certainty of the Resurrection of the Dead, proved by certain changes of the animal and vegetable parts of the Creation."

It was the custom for the President, accompanied by several Fellows of the Society, to hear this sermon preached. Stukeley records: "Whitsunday, June 4, 1750, I went with Mr. Folkes, and other Fellows, to Shoreditch, to hear Dr. Denne preach Fairchild's sermon, On the Beautys of the Vegetable World. We were entertained by Mr. Whetman, the vinegar-mer-

---

[17] Among the subscribers are Sir Hans Sloane, Lord Charles Cavendish, Dr. Alexander Stuart, and Dr. James Douglas; but the sum collected being insufficient, Archdeacon Denne afterwards added 29*l*. out of the money he had received for preaching the sermon. He was the first lecturer, and one of the Trustees of Mr. Fairchild's bequest. Some of his lectures are published. The present lecturer is the Rev. John Joseph Ellis, M.A.

chant, at his elegant house by Moorfields; a pleasant place encompass'd with gardens well stored with all sorts of curious flowers and shrubs, where we spent the day very agreeably, enjoying all the pleasures of the country in town, with the addition of philosophical company[18]."

Though but of slight importance in an historical point of view, it is worthy of notice, as indicative of the times, that in 1746 the hour of the Anniversary Meeting was changed from 9 a.m. to 10 a.m.; at the same time, the place of dining was altered from Pontock's Tavern in Abchurch Lane (which it is stated was inconveniently situated for the majority of the Fellows), to the Devil Tavern, near Temple Bar[19]. It appears by the Minutes of Council, that it was customary to make a collection after dinner, and if this fell short of the expenses incurred, the deficit was supplied from the funds of the Society. It is almost unnecessary to say, that this custom has long since

---

[18] MS. Journal.

[19] This Devil Tavern, on the site now occupied by Child's Place, was the resort of several of the wits and literati of the day. At Dulwich College are preserved some of Ben Jonson's Memoranda, which prove that he owed much of his inspiration to good wine, and the convivial hours he passed at the Devil Tavern. "Mem. I laid the plot of my *Volpone*, and wrote most of it, after a present of ten dozen of palm-sack from my very good Lord T——; that play I am positive will live to posterity and be acted, when I and Envy be friends, with applause." "Mem. The first speech in my *Catalina* spoken by Sylla's Ghost, was writ after I parted with my friend at the Devil Tavern: I had drank well that night, and had brave notions. There is one scene in that play which I think is flat. I resolve to drink no more water with my wine." "Mem. Upon the 20th May, the King (Heaven reward him!) sent me 100*l*. At that time I often went to the Devil, and before I had spent forty of it wrote my *Alchymist*."

been discontinued. At the Anniversary in 1747, Dr. Gowan Knight received the Copley Medal for his experiments in Magnetism[20], performed before the Society on various occasions[21]. Mr. Machin, who had filled the office of Secretary for the long period of twenty-nine years, retired on account of declining health, and Peter Daval was elected in his place. The other officers were continued.

In 1748 Bradley made a communication, announcing his discovery of Nutation, which was published in the *Transactions*, and honoured by the award of the Copley Medal. On this occasion, the President, Martin Folkes, delivered an Address of considerable length at the Anniversary Meeting, when the Medal was given, explanatory of Bradley's discoveries.

The bestowal of the Copley Medal still rested with Sir Hans Sloane, but it appears that he had no hesitation in conferring it upon the illustrious astronomer, as Mr. Folkes says: "Sir Hans Sloane, who, in the autumn of a long life, laboriously spent in the service of mankind, has at last dedicated the remainder of his days to contemplation, in a studious and philosophical retirement, still attentive to the progress of the sciences, and to the welfare of this Society, over which he so long and worthily presided, and of which he is now the affectionate father: this venerable person, I say, when I gave him some account of these important discoveries, immediately judged it incumbent upon him, as the surviving Trustee of the late

---

[20] These related to the making of artificial magnets, of which we shall have more to say presently.

[21] The President delivered an Address at the time of the award. See Journal-book, Vol. xx. p. 358.

Sir Godfrey Copley, to recommend it to your Council, to present to Dr. Bradley the next annual Medal." Then addressing the Astronomer-Royal, the President said, "I do not doubt but that you will set the justest value upon such a testimonial of the respects of a Society, to whose service you have devoted your studies, and whose honour and reputation you have so eminently promoted by your constant pursuit of, and application to, that knowledge, for the advancement of which this Society was instituted by their Royal Founder.

"I am indeed well assured that every one of my brethren here present sincerely joins with me in the true satisfaction which I have to see so able an astronomer worthily seated in the place where the great Dr. Halley and Mr. Flamsteed sat before him: in the most honourable post to which merit can here raise a professor of this science, that of his Majesty's Royal Astronomer; and in which I have still the greater pleasure, from the near prospect I have of soon seeing him possessed, through his Majesty's royal bounty, of an Observatory every way well fitted for the purposes of astronomy, and of a set of instruments capable of affording the most accurate observations, wherewith he will be further enabled to pursue the great work that lies before him, and to add new lustre to that science which he has already so greatly improved[22]."

Bradley's discoveries throw great lustre on the Royal Society, and strongly redeem the body from the satirical but superficial strictures of Sir John Hill, and others.

---

[22] Journal-book, Vol. xx. p. 599.

"*C'est à ses deux découvertes de Bradley,*" says Delambre, "*que nous devons l'exactitude de l'astronomie moderne. Sans elle, il était impossible à l'astronome le plus soigneux de faire accorder ensemble les ascensions droites observées d'une même étoile à* 50 *ou* 60″ *près, et les déclinaisons à une demi minute. Ce double service assure à son auteur la place la plus distinguée après celle d'Hipparque et de Kepler, et au-dessus des plus grands astronomes de tous les ages et de tous les pays; ces travaux auraient suffi à sa gloire*[23]." The discovery of Nutation was made when Bradley was at Greenwich; where "he continued, with great assiduity, observations of the same kind as those by which he had detected Aberration[24]."

It may be mentioned here, that astronomers were greatly assisted in their researches about this period, by the improvement made in clocks and watches. This fact is worthy of record, because the Royal Society had some share in the merit of encouraging artists in their endeavours to improve these important auxiliaries to astronomical observation. In 1749 their attention was directed to the improvements effected in the construction of watches, or, as we now call them, chronometers, by John Harrison, who submitted his inventions on several occasions to the President and Fellows. These appeared of so remarkable a nature, that the Society awarded the Copley Medal to Harrison at the Anniversary Meeting in 1749, on which occasion the President gave an account in his Address of Harrison's inven-

---

23 *Astronomie au Dix-huitième Siècle*, p. 420.

24 Whewell, *Hist. Inductive Sciences*, Vol. II. p. 266.

tions, and delivered the Medal to him with these encouraging and laudatory words:

"I do here, by the authority and in the name of the Royal Society of London for improving natural knowedge, present you with this small but faithful token of their regard and esteem. I do, in their name, congratulate you upon the successes you have already had, and I most sincerely wish that all your future trials may every way prove answerable to those beginnings; and that the full accomplishment of your great undertaking may at last be crowned with all that reputation and advantage to yourself, that your warmest wishes can suggest, and to which so many years, so laudably and so diligently spent in the improvement of those talents which God Almighty has bestow'd upon you, will so justly entitle your constant and unwearied perseverance[25]."

Thus encouraged by the Royal Society, and doubtless also animated by the hope of sharing the reward of 20,000*l.* offered by Parliament for the discovery of the longitude, Harrison continued his labours with unwearied diligence, and produced, in 1758, a timekeeper, which was sent for trial on a voyage to Jamaica. After 161 days, the error of the instrument was only one minute five seconds, and the maker received from the nation 5000*l.*[26] The Com-

---

[25] Journal-book, Vol. XXI. The Discourse, of which the above is an extract, gives a very interesting account of Harrison and his inventions. An abstract of it is published in the *Connaissance des Temps* for 1765.

[26] Stukeley writes of Harrison and his clock: "I pass'd by Mr. Harrison's house at Barrow, that excellent genius at clock-making, who bids fair for the golden prize due to the discovery of the longitude.

missioners of the Board of Longitude subsequently required Harrison to construct under their inspection chronometers of a similar nature, which were subjected to a trial in a voyage to Barbados, and performed with such accuracy, that, after having fully explained the principle of their construction to the Commissioners, they awarded him 10,000*l.* more[27]; at the same time Euler of Berlin, and the heirs of Mayer of Göttingen, received each 3000*l.* for their lunar tables.

The great attention now paid to the very important subject of ventilation, renders it interesting to notice that in 1750, a Committee was appointed by the Society to investigate the wretched state of ventilation in jails, which produced the well-known distemper called 'Jail-Fever.' In the above year, the Lord Mayor of London, two of the judges, and an alderman on the bench, in consequence of the state of the felons tried at the Old Bailey Sessions, were seized with this fatal distemper, and died. This roused the magistrates to resolve on adopting some measures for rendering Newgate more healthy, and the assistance of the Royal Society was requested. Sir John Pringle and Dr. Hales recommended the

---

longitude. I saw his famous clock last winter at Mr. George Graham's; the sweetness of its motion, the contrivances to take off friction, to defeat the lengthening and shortening of the pendulum through heat and cold, and to prevent the disturbance of motion by that of the ship, cannot be sufficiently admired." MS. Journal.

[27] For a detailed account of Harrison's Chronometer, see Montucla's *Histoire des Mathématiques*, Vol. IV. pp. 554—560. It is proper to mention, that the Royal Society were consulted by the Admiralty respecting their instruments, and that the latter acted entirely in conformity with their recommendations.

use of a ventilator, invented by the latter. The machine was erected, and the salutary effects became speedily apparent. The deaths in Newgate were reduced from seven and eight in a week, to about two in a month. Some idea may be formed of the fearful state of the prison, from the fact that out of the eleven men employed to erect the ventilator, seven were attacked by the malignant fever, and one of these died[28].

It is difficult at the present time to imagine the state of prisons a century ago. They were exceedingly small, ill-contrived, and consequently most fatal to the health of the miserable wretches doomed to languish in them[29]. "The enlightened exertions of Dr. Hales and Sir John Pringle," says Dr. Thomson, "first turned the attention of mankind in a forcible manner to the importance of ventilation, which led to the subsequent improvements introduced into our ships and our prisons by Cook and Howard."

On the 17th January, 1750—1, a Paper was read before the Society by Mr. Canton[30], entitled, "Method

---

[28] See a very interesting Paper on this subject, by Sir John Pringle, *Phil. Trans.*, Vol. XLVIII.

[29] It is not a little gratifying to read the following paragraph taken from an Edinburgh Paper, of July 13, 1847: "A great sanitary fact is at present being exemplified in Glasgow. While fever rages around, the prisons containing about six hundred inmates have *not one* fever patient." This is at once a reward to the friends of prison-discipline, and an encouragement to those of sanitary advancement.

[30] Mr. Canton was born at Stroud, Gloucestershire, in 1718. He was apprenticed to a broad-cloth weaver, but early shewed a distaste for the occupation, which he relinquished for scientific pursuits. These he followed with great success, and eventually kept an Academy in Spital Square, London, which he continued during his

of making Artificial Magnets without the use of natural ones[31]." "This Paper, which had been written some time before, would," says his son, Mr. William Canton, "have been communicated earlier to the Society, had not the author apprehended that the publication of it might be injurious to Dr. Gowan Knight, who procured considerable pecuniary advantages by touching needles for the mariner's compass, and kept his method a secret. But Mr. Canton having shewn his experiments to Mr. Folkes, that gentleman was of opinion, that a discovery of such general utility to mankind ought not to be withheld from the public on any private consideration[32]." Accordingly, Mr. Canton exhibited his experiments before the Society on the occasion of reading his Paper. They gave great satisfaction to the Fellows, who "Ordered their thanks to Mr. Canton for his very curious experiments, and for his free and very candid communication[33]."

---

his life. Electricity engaged his principal attention, and he greatly assisted in advancing this science by his discoveries. He was elected a Fellow of the Society on the 22nd March, 1749—50, and in 1751 was chosen a member of the Council. Dr. Thomson calls Canton "one of the most successful experimenters in the golden age of electricity."

[31] Published in the *Phil. Trans.*, Vol. XLVI. p. 578. He used a poker and tongs to communicate magnetism to steel bars. "He derived his first hint from observing them one evening, as he was sitting by the fire, to be nearly in the same direction with respect to the earth as the dipping needle. He thence concluded that they must from that position, and the frequent blows they receive, have acquired some magnetic virtue, which on trial he found to be the case, and therefore he employed them to impregnate his bars, instead of having recourse to the natural loadstone." *Life*, by Kippis.

[32] *Biog. Brit.*, Vol. III. p. 216.

[33] MS. Journal-book, Vol. XXI. p. 499.

For these experiments, and the Paper which accompanied them, the Copley Medal was awarded to Canton at the Anniversary in 1751. This award proves that the Council have not always considered priority of publication as an indispensable condition in bestowing this honorary distinction. It appears, from original unpublished documents now lying before us, that as early as 1747 Canton had turned his attention with complete success to the production of powerful artificial magnets, principally in consequence of the expense of procuring those made by Dr. Knight, who, as has been stated, kept his process secret[34]. Though Canton on many occasions exhibited his results, he nevertheless for several years abstained from communicating his method even to his most intimate

[34] Dr. Gowan Knight's method of making artificial magnets was first communicated to the world by Mr. Wilson, in a Paper published in the 69th Vol. of the *Philosophical Transactions.* He provided himself with a large quantity of clean iron filings, which he put into a capacious tub about half full of clear water; he then agitated the tub to and fro for several hours, until the filings were reduced by attrition to an almost impalpable powder. This powder was then dried and formed into paste by admixture with linseed-oil. The paste was then moulded into convenient shapes, and exposed to a moderate heat until they attained a sufficient degree of hardness. "After allowing them to remain for some time in this state, he gave them their magnetic virtue in any direction he pleased, by placing them between the extreme ends of his large magazine of artificial magnets for a second, or more, as he saw occasion. By this method, the virtue they acquired was such, that when any one of those pieces was held between two of his best ten-guinea bars, with its poles purposely inverted, it immediately of itself turned about to recover its natural direction, which the force of those very powerful bars was not sufficient to counteract." We may mention, that Dr. Knight's powerful Battery of Magnets is in the possession of the Society, having been presented by Dr. John Fothergill, in 1776.

friends. In the beginning of 1750, the Rev. J. Michell published a *Treatise on Artificial Magnets*, in which he described several new processes; and within twelve months afterwards, Canton, strongly urged thereto by his friends, communicated, as above stated, his methods to the Royal Society, some of which are very analogous to those of Mr. Michell. In consequence of this resemblance, Mr. Michell and his friends did not hesitate to assert that Canton had borrowed the most important of his processes from Michell's *Treatise*. This charge of plagiarism, though unpublished, gave Canton considerable uneasiness, and on his death-bed in 1772, he made his friend Dr. Priestley promise that he would not allow his memory to be unjustly reflected on in this affair. In the *Biographia Britannica*, published in 1784, it is stated that Canton's Paper was read before the Society on the 17th January, 1750, and that it procured him on the 22nd March, 1750, the honour of being elected a Fellow. These dates, from an ambiguity in dating the year according to old style, appear from the Registers of the Society to be erroneous, for Canton was elected in March, 1749—50, nearly ten months prior to the communication of his experiments to the Society. Mr. Michell hereupon addressed a letter to the Editors of the *Monthly Review* (Nov. 1784), complaining of "misrepresentations tending to mislead the public," and stating that from the looseness of the dates, several persons had been erroneously led to apprehend that Canton's experiments were prior to the publication of his treatise, whereas the contrary was the case: he then adds, "Mr. Canton's experiments are so nearly the same with mine, that no one who will take the trouble of comparing them together, can well doubt from whence they were borrowed;" and mentions some circumstances calculated to confirm his suspicions. This accusation naturally occasioned considerable uneasiness to the surviving members of Mr. Canton's family, and his son immediately occupied himself in procuring documentary evidence of his father's claims to originality. This evidence is conclusive as to the point that Canton possessed the means of making very powerful artificial magnets several years before the publication of Mr. Michell's treatise. Dr. Priestley, who was a personal friend and neighbour of Mr. Michell's, endeavoured to prevail upon him to do justice to Canton's memory: the result will be best shown by the subjoined unpublished letter of the Doctor's to Mr. William Canton.

"DEAR SIR, "*Birmingham, August* 20, 1785.

"I HAVE been endeavouring to engage Mr. Michell to retract what he has advanced in the *Monthly Review;* and he says, he has no objection to say, what he never meant to deny, that your father had a method of making artificial magnets before the publication of his treatise. What he insists upon is, that this was not the method which he afterwards published. This, I tell him, I think highly improbable; that I knew Mr. Canton well, and believe him to be incapable of any such thing; and besides, as his method, whatever it was, perfectly succeeded, he could have no interest to publish any other, and especially one that would subject him to the charge of plagiarism. We shall see what he says in the next *Review,* in which I shall expect a letter from him. In the mean time, it cannot be improper to do what you propose in the *Gentleman's Magazine.*

"It is impossible, I fear, to produce any *positive proof* that your father's original method was the same that he published. This is one of the inconveniences attending *secrets*, of which your father sincerely repented.

"I am, dear Sir,

"Yours, &c.

"*Mr. William Canton.*" "J. PRIESTLEY.

In consequence of Dr. Priestley's advice, supported by that of other friends, the matter did not go any further.

It will be remembered that the alteration in our Calendar, or, as it was called, Change of Style, took place in 1752. The Earl of Chesterfield is said to have been the original promoter of the assimilation of the British Calendar to that of other countries. The Duke of Newcastle "was too deeply impressed with the favourite maxim of Sir Robert Walpole and his royal master, *tranquilla non movere*, to relish a proposal calculated to shock the civil and religious prejudices of the people;" but the question being strongly supported by Lord Chancellor Hardwicke and the Earl of Macclesfield, was eventually carried[35].

The authority for the alteration emanated from Parliament, but the Royal Society had considerable share in effecting the change[36].

---

[35] Coxe's *Memoir of Pelham's Administration*, Vol. II. p. 178.

[36] It had been long perceived that the day of the Equinox was slowly receding, and that the Moveable Feasts were gradually losing their connexion with it. In the middle of the thirteenth century, Roger Bacon pointed out to Clement IV. the change, its cause and remedy. From this time, reformers of the Calendar were never wanting, until, in 1582, in the Popedom of Gregory XIII., the change was actually made. The accumulated errors of the old cycle were corrected by the omission of eleven nominal days, the day following the fourth of October being called, not the fifth, but the

Lord Macclesfield, in his speech in the House of Peers, on the second reading of the Bill, "for regulating the commencement of the year," stated that "the Bill was drawn, and most of the Tables prepared, by Mr. Daval, a barrister of the Middle Temple[37], whose skill in astronomy, as well as in his profession, rendered him extremely capable of accurately performing that work; which was likewise carefully examined and approved of by two gentlemen, whose learning and abilities are so well known, that nothing which I can say can add to their characters; I mean *Mr. Folkes,* President of the *Royal Society,* and *Dr. Bradley,* his Majesty's Astronomer at Greenwich. Upon this authority do the new Tables and Rules stand; and as to the Bill itself, no endeavours have been wanting to make it as complete and as free from objections of all kinds as possible[38]."

---

fifteenth. At the same time, a new Lunar cycle was introduced, in which, with considerable skill, the defects of the old one were avoided. A beautiful and emblematic Papal medal was struck on this occasion, designed by Parmegiano. For an account of the avowed departures from astronomical accuracy, made on theological grounds, which exist in this cycle, the reader is referred to the article in the *Companion to the Almanac* for 1845.

[37] He was at the time Secretary to the Society, to which office he had been elected on the 30th Nov. 1747. There are several Astronomical Papers by him in the *Transactions.* A note in the *Gentleman's Magazine* for 1751, mentions: "Peter Daval, Esq. of the Middle Temple, Secretary to the Royal Society, who drew the Bill (and prepared most of the tables), under direction of the Earl of Chesterfield, the first former of the design. And the whole was carefully examined and approved by M. Folkes, Esq., President of the Royal Society, and Dr. Bradley, his Majesty's Astronomer at Greenwich, who composed the three tables at the end of this bill."

[38] This speech was published in 1751.

In a Paper by Lord Macclesfield, "On the expediency of altering the Calendar," published in the 46th volume of the *Transactions*, particular allusion is made to Mr. Daval's labours in compiling the necessary Tables, which, it appears, had considerable influence in leading to the desired change.

Professor De Morgan, in a very lucid and interesting Paper in the *Companion to the Almanac for* 1846, entitled, "On the earliest printed Almanacs," says, "It was casually brought to our notice that Charles Walmesley[39], who was well known as a mathematician, and had just been brought into the Royal Society[40], was said to have been one of those who were consulted by the framers of the bill. Combining the characters of a priest and an astronomer, he had probably made himself acquainted with the details of

---

39 "Father Charles Walmesley was an English Benedictine Monk and Roman Catholic Bishop of Rome. He was also senior Bishop and Vicar Apostolic of the Western District, as well as Doctor of Theology in the Sorbonne. He died at Bath in 1797, in the 76th year of his age, and the 41st of his episcopacy. He was the last survivor of those eminent mathematicians who were concerned in regulating the chronological style in England, which produced the change of style in this country in 1752. He is the author of five Papers on astronomical subjects in the *Philosophical Transactions*, besides which he published several separate works both on mathematics and theology. By the fire at Bath, at the time of the riots, several valuable manuscripts which he had been compiling during a well-spent life of labour, and travelling through many countries before his return to England, were irretrievably lost." *Abridg. Phil. Trans.*, Vol. II. p. 17.

40 He was elected Nov. 1, 1750. His certificate describes him as "a gentleman of very distinguished merit and learning," and is signed by the Duke of Richmond, Mr. Folkes, Dr. Mortimer, and Mr. Askew. He was also strongly recommended by Buffon, d'Alembert, Le Monnier, and de Jussieu.

the reformed calendar. In the short obituary which is given in the *Gentleman's Magazine*, he is mentioned as the last survivor of the mathematicians who were consulted on the change of style. But not a trace of mention of his name can we find at the time, for which it is not difficult to conjecture the reason."

By this Professor De Morgan means, that the religion of Father Walmesley would have made the change of style (if known to have been in any way brought about by him) much more disliked by some parties than it was.

"There is much reason to suppose," says the Professor in another Paper, 'On the Ecclesiastical Calendar,' "that this violent change placed as great a difficulty in the way of Protestant governments acceding to the new Calendar, as religious feeling. When in England, in the eighteenth century, it was at last introduced, the mob pursued the minister in his carriage, clamouring for the days by which, as they supposed, their lives had been shortened: and the illness and death of the astronomer Bradley, who had assisted the government with his advice, were attributed to a judgment from heaven[41]."

The change occurred in England on the 2nd September, 1752, "eleven nominal days being then struck out, so that the last day of old style being the 2nd, the first of new style (the next day) was called the 14th instead of the 3rd. The same legislative enactment which established the Gregorian year in England in 1752, shortened the preceding year 1751 by a full quarter. Previously, the ecclesiastical and legal year was held to begin with the 25th March, and the

---

41 *Companion to the Almanac*, 1845.

year A.D. 1751 did so accordingly; that year, however, was not suffered to run out, but was supplanted on the 1st January by the year 1752, which it was enacted should commence on that day, as well as every subsequent year[42]."

The expediency of changing the style had frequently been agitated by distinguished Members of the Society prior to this period, and Canton is stated, by Dr. Kippis, to have furnished Lord Macclesfield with several memorial canons for finding leap year, the dominical letter, &c.

In 1752 an important amelioration was made with respect to the publication of the *Transactions*. It originated from Lord Macclesfield, who was probably sensible that the *Philosophical Transactions* had not for some time been equal to those formerly published, and he was consequently led to consider whether any steps could be taken to improve their character.

On the 15th February, 1752, he brought forward the following propositions, of which he had given the usual notice; and after they had been seriously considered at two Meetings of Council, they were laid before the Fellows at an ordinary Meeting, held on the 27th February, for their approbation[43].

"That it is the opinion of the Council, that it would tend to the credit and honour of the Society, if, for the future, they should so far take under their care and inspection the publication of such papers as shall have been read before, or communicated to them at their weekly Meetings, as to appoint a Committee who should

---

[42] Herschel's *Astronomy*, p. 413.

[43] Lord Macclesfield brought his motion forward at an ordinary Meeting, in the first instance.

from time to time, as occasion should require, assemble together and select from the said papers (which should be referred to the said Committee for that purpose) such of them as they should think proper to be printed, and to order that no other papers should be published in the *Philosophical Transactions* than such as shall have been so selected by the said Committee.

"That it is the opinion of the Council that the President, Vice-President, and Secretaries, should be constantly Members of the said Committee; the several Meetings whereof should be appointed by the President, or, in case of his sickness or absence, by one of the Vice-Presidents; and that due and a sufficient notice of such Meeting should be sent previously thereto to every Member of the said Committee."

It was also proposed that not less than five Members of the Committee should be a quorum.

These resolutions, with others explanatory of the manner in which the Council proposed to carry out the new measures, met with the approbation of the Fellows, and passed into laws.

The next volume of the *Transactions*[44] which appeared in 1753, was in consequence published under the superintendence of a Committee of the Council, by whose orders the following Advertisement was inserted[45].

---

[44] This volume contains papers communicated to the Society from 1750 (when the previous volume was published under the superintendence of Dr. Cromwell Mortimer) to 1753.

[45] The names of the Members of Council existing when this important change was made, were: Martin Folkes, President; James West, Treasurer; the Rev. Thomas Birch and Peter Daval, Secretaries; Francis Blake; John Canton; John Ellicott; Dr. William Heberden; Gowan Knight; Earl of Macclesfield; John Ward; William Watson; Lord Willoughby, of Parham; Lord Charles

"The Committee, appointed by the Royal Society to direct the publication of the *Philosophical Transactions*, take this opportunity to acquaint the public, that it fully appears, as well from the Council-books and journals of the Society, as from the repeated declarations which have been made in several former *Transactions*, that the printing of them was always, from time to time, the single act of the respective Secretaries till this present 47th volume. And this information was thought the more necessary, not only as it has been the common opinion that they were published by the authority, and under the direction, of the Society itself, but also because several authors, both at home and abroad, have in their writings called them the *Transactions of the Royal Society*. Whereas, in truth, the Society, as a body, never did interest themselves any further in their publication, than by occasionally recommending the revival of them to some of their Secretaries, when, from the particular circumstances of their affairs, the *Transactions* had happened for any length of time to be intermitted. And this seems principally to have been done with a view to satisfy the public that their usual Meetings were then continued for the improvement of knowledge and benefit of mankind—the great ends of their first institution by the Royal Charters, and which they have ever

---

Charles Cavendish; Nicholas Mann; Richard Mead; Dr. Cromwell Mortimer; Sir Hans Sloane, Bart.; Charles Stanhope; Daniel Wray, Esq., and James Burrow. In a notice upon the Society, in the *Gentleman's Magazine* for 1752, it is stated, that "the Council resolved upon taking the publication of the *Transactions* into their hands for their honour and reputation, which has been much injured by an enemy to that illustrious body, of which he attempted, *but in vain*, to be a member; and when convinced of mistakes, refused to correct them."

since pursued. But the Society being of late years greatly enlarged, and their communications more numerous, it was thought advisable that a Committee of their Members should be appointed to reconsider the papers read before them, and select out of them such as they should judge most proper for publication in the future *Transactions;* which was accordingly done upon the 26th of March, 1752[46]. And the grounds of their choice are, and will continue to be, the importance or singularity of their subjects, or, the advantageous manner of treating them; without pretending to answer for the certainty of the facts, or propriety of the reasonings contained in the several papers so published, which must still rest on the credit or judgment of their several authors. It is likewise necessary on this occasion to remark, that it is an established rule of the Society, to which they will always adhere, never to give their opinion, as a body, upon any subject, either of nature or art, that comes before them. And therefore the thanks, which are frequently proposed from the chair to be given to the authors of such papers as are read at their accustomed Meetings, or to the persons through whose hands they receive them, are to be considered in no other light than as a matter of civility, in return for the respect shewn to the Society by those communications. The like also is to be said with

---

[46] It is a highly creditable feature in the history of the Society, that the *Transactions* have been most punctually published every year, and that in no case have papers, read before the Society and ordered for publication, been delayed more than a year (generally only a few months) between their being read and appearance in the *Transactions*. In this respect the Society is far in advance of the French Institute, where papers frequently remain unpublished for two, three, and even four years.

regard to the several projects, inventions, and curiosities of various kinds, which are often exhibited to the Society; the authors whereof, or those who exhibit them, frequently take the liberty to report, and even to certify in the public newspapers, that they have met with the highest applause and approbation. And therefore it is hoped that no regard will hereafter be paid to such reports and public notices; which, in some instances, have been too lightly credited, to the dishonour of the Society."

It appears from the Council-book, that orders were given to print 750 copies of the *Transactions*, for the benefit of the Fellows, and that all copies exceeding the numbers of Fellows were sold to the Society's bookseller at 25 per cent. under the selling price to the public. To meet the increased expenditure incurred by printing the *Transactions*, this statute was passed:—

"Whereas the great charge of printing so large a number of copies as may happen to be demanded by the present and future Fellows, must be defrayed out of the Stock or Fund of the Society, and it is but reasonable that all persons who for the future shall be propounded to be, and shall be, admitted Fellows of the Society, and, consequently, will become entitled to receive gratis a copy of the *Transactions* printed after their respective admissions, should contribute in some measure towards defraying the said extraordinary expense; it is hereby ordered and enacted, that no person whatsoever, who shall be propounded after the eighth day of April next to be a Fellow of the Society, shall be admitted as a Fellow thereof, until he shall have paid into the hands of the Treasurer of the

Society, or his Deputy, the sum of 3*l.* 3*s.*, over and above the sum of 2*l.* 2*s.* which have hitherto been paid as an admission fee by every Fellow of the Society, previously to his admission as a Fellow thereof[47]."

The Council-minutes subsequently show that the Society had no reason to regret the course they had taken regarding the *Transactions* in a pecuniary point of view; for, in 1754, it is stated, that "an account was read of the gain or loss to the Society by printing the *Transactions*, by which it appears that they are gainers by the new established method of printing them[48]."

We may mention here, that about this period several volumes of the *Philosophical Transactions* were translated into the Italian language, and published at Naples. The fourth volume of this translation is dedicated to the *Conte di Arconate*[49], and it would appear, that the former volumes were favourably received by the savans in Italy, for we are told that, "*Quella parte delle varie Materie Filosofiche, trattate nella celebrë Società di Londra, le quali, trasportate dalla Lingua Inglese nella nostra Italiana furono pubblicate col nome di* Transazioni, *è stata universal-*

[47] Council-minutes, Vol. IV. p. 82.

[48] It may not be uninteresting to give the cost of printing the 47th Vol. of the *Transactions*, the first published by the Society. It contains 571 pages, small 4to.

| | £. | *s.* | *d.* |
|---|---|---|---|
| Mr. Richardson for Printing .................... | 76 | 11 | 0 |
| Mr. Mynde for Engraving ........................ | 34 | 18 | 6 |
| Mr. Johnson for Paper .......................... | 39 | 18 | 3 |
| | 151 | 7 | 9 |

[49] The dedication affirms that the Count's ancestors presented several of Leonardo da Vinci's original MS. works to the Ambrosian Library.

*mente riceruta con piacere sì grande per le molte naturali osservazioni quasi a gara dagli ingegni più felici del secolo da ogni parte dell' Europa ivi radunate, che io volendo dare a V.S. Illustrissima un' attestato della mia divozione, con presentarle alcuno de' Libri che vengono alla luce da queste Stampe, ho creduto niun' altra Opera più di questa poterle venire gradita.*" Scientific men may be interested to learn, that these volumes contain, under the form of an Appendix, a great number of letters from English and other philosophers, giving an account of the state and progress of science. A French translation of the *Transactions* had been undertaken by M. de Bremond. Montucla says, alluding to the *Transactions;* "*On ne sauroit trop regretter que cette précieuse collection soit encore si rare parmi nous, soit en original, soit dans une langue plus commune aux savans que la langue angloise. Ces raisons avoient engagé, vers* 1744, *M. de Bremond, de l'Académie Royale des Sciences, à en donner une traduction Françoise, et il en publia les années* 1744—7, *avec un volume de tables indiquant de diverses manières le contenu des volumes antérieurs à* 1744. *On ne sauroit trop louer la disposition de ces tables. La mort de M. de Bremond ayant interrompu ce travail, M. Demours, de la même Académie, s'est proposé long-temps de le continuer, mais ses occupations, et peut-être la difficulté de faire imprimer un si volumineux recueil, sont cause que ce projet n'a point eu d'exécution, et grace à la tournure actuelle de l'esprit françois, il n'y a pas d'apparence qu'il en ait jamais*[50]."

At the Anniversary in 1753, Martin Folkes resigned

---

[50] *Histoire des Mathématiques,* Tom. II. p. 556.

the Presidency[51]; he was succeeded by the Earl of Macclesfield, who, as we have seen, evinced from the time of his election into the Council a warm interest in the Society. It is but just to Mr. Folkes to state, that he left the Society in a much more flourishing condition than when he was elected President; for, at the time of his resignation, their funded capital amounted to 3,000*l.* A careful examination of the voluminous Minutes of the ordinary Meetings, extending over the eleven years that he was in office, enables me to state, that he was scarcely ever absent from the chair, and that the Meetings were honoured by a greater number of visitors than usual, numbering frequently as many as thirty or forty. Indeed, so much inconvenience was occasionally experienced by the crowds desiring to be admitted, that the President was obliged to request the Fellows to exercise a little discretion in bringing visitors, and to enforce the standing order, precluding their admission, until leave had been obtained from the Society in the usual manner.

---

[51] At the Anniversary Meeting Lord Cavendish stated, "That having attended the President, he had declared to him, that the weak condition to which his present indisposition had reduced him, having rendered him incapable of attending to the business of the Society, he was therefore desirous of declining the office of President."

The Society "Resolved, That their thanks be returned to Martin Folkes, Esq., their worthy President, for the many and great services which they have received from him, both as Member and as President, of which they shall retain the highest sense. And that he be assured of the just concern which they feel that his ill state of health will not permit him any longer to discharge the office of President, which he has so many years filled with so much credit to himself, and advantage to the Society." Journal-book, Vol. XXII. p. 195.

It was during the Presidency of Martin Folkes, that Stukeley thus alludes to the Meetings of the Society: "They are a most elegant and agreeable entertainment for a contemplative person;—here we meet, either personally or in their works, all the genius's of England, or rather of the whole world, whatever the globe produces that is curious, or whatever the heavens present. My custom is, when I return home and take a contemplative pipe, to set down the memoirs of what entertainments we have had there[52]." He frequently alludes to the "splendid company" at the Meetings.

Stukeley's account of a Geological Soirée which he gave at this period to several Fellows of the Society, is so curious as to merit insertion here. It is contained in his Manuscript Journal:

"28 *Aug.* 1751.—Celebrated the dedication of my library[53]; present, the President, Mr. Folkes, Mr. Fleetwood, Dr. Parsons, Mr. Pard, M. De la Costa, Mr. Baker, Mr. Sherwood, &c. &c. At the little window, which I called the sideboard, we began the entertainment with three sorts of plumb-pudding stone, with other natural and antique curiosities. The great window was spread over entirely with fossils of all kinds, which were extremely admired. The great lump of *Coralliam Tubulatum,* found in the river Ribble; another lesser lump, white; another, filled full with juice of black flint, which I picked up from the pavement of pebbles before my neighbour Curtis's door at Stamford. Two black flints I pick'd up the other day in a bank in our fields: one has a white shell in it, the other a

---

[52] MS. Journal.

[53] His house was in Great Ormond Street.

piece of bone. I shew'd the bone I took out of the stratum of brick-earth in digging at Bloomsbury; many periwinkles, and all kinds of shells, fluors, petrifactions, incrustations, &c. &c.

"I also show'd many sorts of *cornu ammonis;* a model of Stonehenge, some of the stone, the common sort polish'd; a Roman cup and saucer, entire, of fine red earth, dug up at Trumpington; Bishop Cumberland's clock, with the first long pendulum. After this dry entertainment we broach'd a barrel of fossils from the isle of Portland.

"Lastly, to render it a complete rout, I produc'd a pack of cards made in Richard II.'s time; and shew'd the British bridle dug up in Silbury Hill, probably the greatest antiquity now in the world."

It is to be feared that Stukeley's love for Geology did little to advance the science: for it appears that shortly after the above period he communicated some Geological Papers to the Society, containing so many absurd hypotheses, that even at that period, when geology was so little understood, the Council determined they should not be printed. He also made several communications, in which he asserted, in the most positive manner, that corals were vegetables. These papers were likewise rejected, which made the sturdy antiquary very angry. He gives vent to his feelings in forcible language; and concludes: "Whoever has eyes must see that they are vegetables."

END OF THE FIRST VOLUME.

For EU product safety concerns, contact us at Calle de José Abascal, 56–1°,
28003 Madrid, Spain or eugpsr@cambridge.org.

www.ingramcontent.com/pod-product-compliance
Ingram Content Group UK Ltd.
Pitfield, Milton Keynes, MK11 3LW, UK
UKHW010353140625
459647UK00010B/1031